ROUTLEDGE HANDBOOK OF LAW AND SOCIETY

I0894587

This innovative handbook provides a comprehensive, and truly global, overview of the main approaches and themes within law and society scholarship or social-legal studies.

A one-volume introduction to academic resources and ideas that are relevant for today's debates on issues from reproductive justice to climate justice, food security, water conflicts, artificial intelligence, and global financial transactions, this handbook is divided into two sections. The first, 'Perspectives and Approaches', accessibly explains a variety of frameworks through which the relationship between law and society is addressed and understood, with emphasis on contemporary perspectives that are relatively new to many socio-legal scholars. Following the book's overall interest in social justice, the entries in this section of the book show how conceptual tools originate in, and help to illuminate, real-world issues. The second and largest section of the book (42 short well-written pieces) presents reflections on topics or areas concerning law, justice, and society that are inherently interdisciplinary and that are relevance to current – but also classical – struggles around justice. Informing readers about the lineage of ideas that are used or could be used today for research and activism, the book attends to the full range of local, national and transnational issues in law and society. The authors were carefully chosen to achieve a diverse and non-Eurocentric view of socio-legal studies.

This volume will be invaluable for law students, those in inter-disciplinary programs such as law and society, justice studies and legal studies, and those with interests in law, but based in other social sciences. It will also appeal to general readers interested in questions of justice and rights, including activists and advocates around the world.

Mariana Valverde is a Sociolegal Scholar, who has taught at the University of Toronto's Centre for Criminology, and Sociolegal Studies Canada for 25 years.

Kamari Maxine Clarke is a Professor at the University of Toronto in Criminology and Legal Studies, Canada, with a cross-appointment in Diaspora and Transnational Studies.

Eve Darian-Smith is a Professor and the Chair of Global & International Studies Department at the University of California, Irvine, USA.

Prabha Kotiswaran is a Professor of Law & Social Justice at King's College London, UK.

ROUTLEDGE HANDBOOK OF LAW AND SOCIETY

Edited by

Mariana Valverde, Kamari Clarke, Eve Darian-Smith, and Prabha Kotiswaran

Routledge
Taylor & Francis Group
LONDON AND NEW YORK

First published 2021
by Routledge
2 Park Square, Milton Park, Abingdon, Oxon OX14 4RN

and by Routledge
52 Vanderbilt Avenue, New York, NY 10017

Routledge is an imprint of the Taylor & Francis Group, an informa business

© 2021 selection and editorial matter, Mariana Valverde, Kamari Clarke, Eve Darian-Smith, and Prabha Kotiswaran; individual chapters, the contributors

The right of Mariana Valverde, Kamari Clarke, Eve Darian-Smith, and Prabha Kotiswaran to be identified as the authors of the editorial material, and of the authors for their individual chapters, has been asserted in accordance with sections 77 and 78 of the Copyright, Designs and Patents Act 1988.

All rights reserved. No part of this book may be reprinted or reproduced or utilised in any form or by any electronic, mechanical, or other means, now known or hereafter invented, including photocopying and recording, or in any information storage or retrieval system, without permission in writing from the publishers.

Trademark notice: Product or corporate names may be trademarks or registered trademarks, and are used only for identification and explanation without intent to infringe.

British Library Cataloguing-in-Publication Data
A catalogue record for this book is available from the British Library

Library of Congress Cataloging-in-Publication Data
Names: Valverde, Mariana, 1955- editor. | Clarke, Kamari Maxine, 1966- editor. | Darian-Smith, Eve, 1963- editor. | Kotiswaran, Prabha, editor.
Title: The Routledge handbook of law and society / edited by Mariana Valverde, Kamari Maxine Clarke, Eve Darian-Smith, and Prabha Kotiswaran.
Description: Milton Park, Abingdon, Oxon ; New York, NY : Routledge, 2021. | Includes bibliographical references and index.
Identifiers: LCCN 2020040882 (print) | LCCN 2020040883 (ebook) | ISBN 9780367234249 (hardback) | ISBN 9780429293306 (ebook)
Subjects: LCSH: Sociological jurisprudence. | Law--Social aspects.
Classification: LCC K376 .R675 2021 (print) | LCC K376 (ebook) | DDC 340/.115—dc23
LC record available at https://lccn.loc.gov/2020040882
LC ebook record available at https://lccn.loc.gov/2020040883

ISBN: [978-0-367-23424-9] (hbk)
ISBN: [978-0-367-69468-5] (pbk)
ISBN: [978-0-429-29330-6] (ebk)

Typeset in Bembo Std
by KnowledgeWorks Global Ltd.

Dedicated to Sally Engle Merry for
her lifelong quest for greater understanding of law's social life.

CONTENTS

EDITORS

Mariana Valverde earned a PhD in Social and Political Thought from York University (Toronto) in 1982, but spent much of her doctoral and post-doc years in grassroots feminist, socialist, and gay/lesbian rights endeavours. Her first and only commercially successful book, *Sex, Power and Pleasure* (1985) was originally published by a local feminist collective press. From the late 1980s onward she has mainly worked in the academy, first on issues of sexuality and gender and later on urban governance and urban law, historically and in the present. She has taught at the University of Toronto's Centre for Criminology and Sociolegal Studies for 25 years and has published very widely, recent books being *Chronotopes of Law: Scale, Jurisdiction and Governance* (2015) and *Michel Foucault* (2017). 'Smart city' projects are an important current research interest.

Kamari Maxine Clarke is a Professor at the University of Toronto in Criminology and Legal Studies, with a cross appointment in Diaspora and Transnational Studies. Her work spans the emergence of various transnational legal domains, especially international criminal tribunals and the export and spread of international legal norms. She is the author of over fifty articles and eight books, including her 2009 publication of *Fictions of Justice: The International Criminal Court and the Challenge of Legal Pluralism in Sub-Saharan Africa* (Cambridge University Press, 2009) and to her most recent, *Affective Justice: The International Criminal Court and the Pan-Africanist Push-Back* (Duke University Press, 2019). Over the years, Clarke has been involved in a range of politically engaged projects related to the building of African institutional capacity, where she serves as a technical advisor to the African Union. One such project is related to the building of African judicial capacity through the operationalization of the African Court of Justice and Human and Peoples' Rights.

Eve Darian-Smith is a Professor and the Chair of Global & International Studies Department at the University of California, Irvine, USA. Trained as a lawyer, historian, and anthropologist, she is interested in postcolonialism, legal pluralism, global governance, authoritarianism and socio-legal theory. Her first book *Bridging Divides: The Channel Tunnel and English Legal Identity in the New Europe* won the Law & Society Association Herbert Jacob Book Prize. A more recent book, *Laws and Societies in Global Contexts*, won the Kevin Boyle Book Award. With the global rise of authoritarianism, she has been an actively involved with Scholars at Risk, an international network that supports academic freedom and helps scholars find institutional homes outside their home countries.

Prabha Kotiswaran is a Professor of Law & Social Justice at King's College London. She has authored *Dangerous Sex, Invisible Labor: Sex Work and the Law in India* (Princeton, 2011) and edited books on sex work, sociology of law, anti-trafficking law, and governance feminism. She has long worked with the sex workers' movement in India. She has written extensively on anti-trafficking law and policy in India and internationally and has co-founded with like-minded academic-activists the Open Democracy blog *Beyond Trafficking and Slavery.* She advises a DFID funded, ILO managed project Work in Freedom which seeks to prevent trafficking of women and girls in South Asia and the Middle East. She is currently engaged in a materialist feminist study of women's reproductive labour in sex work, erotic dancing, commercial surrogacy, paid domestic work and unpaid domestic work in India.

CONTRIBUTORS

Ifrah Abdillahi. University of Toronto. Toronto, Canada.

Donatella Alessandrini. University of Kent, UK.

Kirsten Anker. Faculty of Law, McGill University, Canada.

Diamond Ashiagbor. University of Kent, UK.

Antonio Azuela. Instituto de Investigaciones Sociales, Universidad Nacional Autónoma de México, Mexico.

Pallavi Banerjee. Department of Sociology, University of Calgary, Alberta, Canada.

Leticia Barrera. CONICET, Argentina.

Adelle Blackett. Faculty of Law, McGill University, Montreal, Canada.

Nicholas Blomley. Geography Department Simon Fraser University, Canada.

Irus Braverman. University at Buffalo, School of Law, The State University of New York.

Matthew Canfield. Van Vollenhoven Institute, Leiden Law School, The Netherlands.

Lynnette J. Chua. National University of Singapore, Singapore.

Kamari Maxine Clarke. University of Toronto, Canada.

Amy J. Cohen. The Ohio State University Moritz College of Law, Ohio.

Rosemary J. Coombe. Tier One Canada Research Chair in Law, Communication and Culture. Departments of Anthropology and Social Science, York University, Canada.

Brenda Cossman. Faculty of law, University of Toronto, Canada.

Susan Bibler Coutin. University of California, Irvine.

Eve Darian-Smith. Global & International Studies Department at the University of California, Irvine USA.

Margaret Davies. Flinders University, Adelaide South Australia, Australia.

Anver Emon. Faculty of Law, Department of History, Institute of Islamic Studies, University of Toronto, Canada.

David M. Engel. School of Law, University of Buffalo.

Michael Fakhri. The Ohio State University, Moritz College of Law, Ohio.

Bryant G. Garth. Center for Empirical Research on the Legal Profession, University of California-Irvine.

Ben Golder. Faculty of Law, UNSW, Australia.

Carmen G. Gonzalez. Loyola University School of Law, Chicago.

Sarah Hunt. School of Environmental Studies, University of Victoria, Canada.

Engin Isin. Queen Mary University of London and University of London in Paris.

Aisha M. Beliso-De Jesús. Department of Spanish and Portuguese, Princeton University.

Miranda Johnson. University of Otago, New Zealand.

Sarah Keenan. Centre for Research on Race and Law, Birkbeck, School of Law University of London.

Prabha Kotiswaran. King's College London.

Sarah Lamble. School of Law, Birkbeck, University of London.

Sergio Latorre. Universidad del Norte, Colombia.

Sida Liu. Department of Sociology, University of Toronto, Canada.

S. Ali Malik. Socio-legal Studies, York University, Canada.

Renisa Mawani. Law and Society Program, University of British Columbia, which is located on the unceded territory of the Musqueam peoples.

Shaun McVeigh. Melbourne Law School, The University of Melbourne, Australia.

Sally Engle Merry. Department of Anthropology, New York University, New York.

Tamar Meshel. University of Alberta, Faculty of Law, Canada.

Pedrom Nasiri. Department of Sociology, University of Calgary, Alberta Canada.

Usha Natarajan. American University in Cairo, Egypt.

Vasuki Nesiah. NYU Gallatin, New York.

Pat O'Malley. School of Sociology. Research School of the Social Sciences. Australian National University. Canberra, Australia.

Nicola Palmer. King's College London.

Pooja Parmar. Faculty of Law, University of Victoria, Canada.

Shiri Pasternak. Ryerson University, Canada.

Jemima Pierre. Department of African American Studies and Department of Anthropology. UCLA Los Angeles, California.

Jothie Rajah. American Bar Foundation, Chicago.

Bhavani Raman. Department of History, University of Toronto, Canada.

Jennifer Raso. Faculty of Law, University of Alberta, Canada.

Rachel Rebouché. Temple University Beasley School of Law. Philadelphia, Pennsylvannia.

Galit A. Safarty. Canada Research Chair in Global Economic Governance. Allard School of Law, University of British Columbia, Canada.

Dayna Nadine Scott. Osgoode Hall Law School and Environmental Studies, York University.

Nofar Sheffi. UNSW Law School, Sydney, Australia.

Jacob Livingston Slosser. University of Copenhagen. Faculty of Law, Danish National Research Foundation's Centre of Excellence for International Courts (iCourts), Denmark.

Maximo Sozzo. Universidad Nacional del Litoral, Santa Fe, Argentina.

Gavin Sullivan. University of Kent, UK.

Mariana Valverde. Centre for Criminology and Sociolegal Studies. University of Toronto.

Di Wang. University of Wisconsin-Madison, Wisconsin.

Brenda S.A. Yeoh. Department of Geography, National University of Singapore, Singapore.

CONTESTED LAWS, CONTESTED SOCIETIES

Introductory remarks

Mariana Valverde, Kamari Maxine Clarke, Eve Darian-Smith, and Prabha Kotiswaran

In February 2020, as we were editing many of the contributions to this book, a story about state law's contested legitimacy dominated the headlines in Canada. The story had started in British Columbia, where a proposed expansion to an existing natural gas pipeline had been met with a road blockade led by the hereditary chiefs of the Wet'suwet'en nation, who refused to let the bulldozers and other equipment pass through their territory. The hereditary or traditional chiefs insisted that their land was 'unceded', that is, unlike much of the land in Eastern Canada, it had never been bought or otherwise acquired by Canada. The company obtained a judicial injunction, and once this was formally read out, the police with jurisdiction in the area – the Royal Canadian Mounted Police (RCMP or known as 'the Mounties'), a force with strong colonial roots – arrested a few of the leaders and ensured that the pipeline work could proceed. In the meantime, Canadians of non-Indigenous background got quite the education, via news articles, about the fact that the system of elected 'band chiefs' imposed on Indigenous communities in the late 19th century had not managed to ever completely eliminate the existence of a competing authority, the traditional or hereditary chiefs, at least in some parts of the country.

In previous years, the story would have ended with the enforcement by police of the injunction – unlike in many previous occasions, the protesters were not charged or detained, and a limited amount of violence was used by the police. (Previous Indigenous blockades had resulted in deaths). But for reasons that the Canadian political and journalistic classes tried desperately to understand and explain, in early 2020, the story did not end there.

Thousands of kilometres across the country, Mohawks in communities near Belleville, Ontario, blockaded the train line linking Toronto (and mid-Western US) with Montreal. In this case, the freight-rail company, CN, also got an injunction from the court to demand the end of the blockade. But the Mohawks refused and the police force that has jurisdiction around the Mohawk Tyendinaga territory, the Ontario Provincial Police (OPP), did not want to be seen extending the history of colonial violence against First Nations people. Instead, they sent some patrol cars and making their presence felt, essentially stood back, and waited for various politicians to solve the problem. The OPP had been highly criticised about 20 years earlier, when an Indigenous effort elsewhere in Ontario to defend sacred land resulted in one police killing and lots of criticism and aggravation for the OPP. So, at the train tracks – tracks which carry the most passenger as well as freight traffic in the country – the OPP read out the court-approved injunction, but did not push or remove anyone, and did not seize the trucks that had been used to create the blockade. It's highly unlikely that the OPP today has somehow shed its historic racism; probably, they merely wanted to protect their force against the political risks of direct physical action against Indigenous activists.

To up the ante, CN cancelled all freight trains in Eastern Canada – which gave rise to great handwringing about economic consequences – while VIA rail, the passenger service, cancelled almost all trains around the country, including in places that had seen no protests at all. Clearly, the private sector was turning up the heat on both police and politicians, while on-the-ground police deferred decision-making to the politicians.

At the time that the initial blockades went up in BC and then Ontario Prime minister, Justin Trudeau, was invested in maintaining a progressive reputation. The initial British Columbia blockade having been removed, he assumed the problem was over. Instead, he was involved in currying favour on an African tour with African leaders in an attempt to solidify Canada's quest for a seat on the United Nations Security Council. And since being on the heels of the historic Truth and Reconciliation Commission strengthened his quest to be seen addressing histories of harm against First Nations people, he did not mobilise the federal authority against the Mohawk blockade. The opposite happened. In the following days, powerful solidarity protests expanded and shook the cities of Victoria and Vancouver, and new blockades appeared – not only in the usual 'hotspots', such as Mohawk reservations, but also in the suburban and conservative-voting municipality of Vaughan, Ontario, and even in normally touristy-peaceful Niagara Falls, Ontario. Trudeau cancelled another planned UN Security Council-related trip to Caribbean states and returned to Ottawa. While on his way home from his African tour, he declared, 'We are a country of the rule of law', adding that in such a country, politicians don't tell police how to do their job. The message was clearly directed against right-wing opposition leader Andrew Scheer, who had called on the federal government to tell police to take down all blockades right away.

What is the vague principle of 'the rule of law' doing here? The events on the ground prompted questions not usually asked in peaceful democracies: Which 'rule'? And whose law? Oil and gas companies, supported by right-wing politicians, opined that the 'rule of law' meant that a court injunction had to be followed by quick police action. For them, the injunction was the law – even though in an injunction hearing only the party seeking the injunction is represented and heard. Prime Minister Trudeau had said that the 'rule of law' means that politicians don't interfere with operational police decisions – meaning that in his view the police would themselves decide if and when and how to enforce the court-approved injunction. But what about Indigenous people in affected regions, all around the world, who are struggling to defend their land from extractive industries? What about their 'rule of law'?

Oil and gas companies and right-wing politicians such as Alberta Premier Jason Kenney had repeated ad nauseam that the pipeline was legitimate because 20 elected 'band council' chiefs had signed agreements allowing for the construction work on the gas pipeline in dispute. (An 'Indian band' is Canadian legalese for what elsewhere might be called a tribe). But elected 'band council' chiefs are leaders of their people only by virtue of Canada's Indian Act – the key piece of settler colonial legislation, dating back to 1875, which despite many patches and amendments over the years remains the legal backbone of Canadian settler colonialism.

In some Indigenous communities, the elected 'band council' chiefs do enjoy legitimacy. In some communities, though, such as in Mohawk territories and in large parts of British Columbia and in the North, the settler colonial apparatus of indirect rule through elected 'band council' chiefs is under serious challenge from those who reject the legitimacy of the Indian Act. Sometimes, the traditional-law chiefs live side-by-side with the 'official' elected chiefs. But at other times, there is direct conflict between the 'elected' and 'hereditary' chiefs as to who has authority to represent

Indigenous communities[1]. The point here is that the Wet'suwet'en hereditary chiefs – among whom there is at least one woman, contrary to the colonial myth that the hereditary system is always patriarchal – are not protesters. They are standing up on their land for their land, which in their legal system is sacred, and standing up for their law. They further say that their territory has never been part of Canada, legally. Like Indigenous leaders all over the world, they say that the land does not belong to them but rather the opposite: they belong to the land and must therefore protect it as one would protect one's family or one's house or one's mother. They are the keepers of Wet'suwet'en law, and that law has not ceased to exist, even if it did not always have nationally visible spokespeople. Their rule of law does not recognise whatever the elected, official 'band council' chiefs agreed to with the extractive industries and/or the government.

Train service eventually resumed; but the initial bone of contention was not removed. How pipelines, mines, and other infrastructures located on Indigenous traditional territory can be legitimately built remains a huge unsettled issue in Canada as well as in other countries. And as the entry on 'extractive industries' in this volume explains, we can see that the blockades and demonstrations in Canada in the early 2020 cannot be dismissed as a fleeting one-time dispute. It is something deeper; it is a clash of incompatible legal systems, each of which has its own rules of legitimacy, its own view of what is 'the rule of law'.

Serious questions about the very meaning and scope of the term 'law' – the questions that this collection addresses – are also visible in situations quite removed from Indigenous activism. When the government of France bans hijab-wearing Muslim women from working in the public sector, and bans niqab-wearing women from most public spaces, for example, that raises profound issues, not just about the conflicts between different versions of Islamic law and French law, but also about the very meaning of the 'equality' that the 1789 Revolution invented and that remains inscribed on French government buildings to this day (equality, fraternity, liberty). Similarly, when, in the U.S., President Trump declared, in relation to the measures designed to prevent the spread of the COVID virus, that the president of the United States has 'absolute power', this was quite an insult to the early modern political thinkers who created the 'rule of law' doctrine precisely through struggles against absolutist monarchies.

This volume offers readers tools with which to think about what 'law' is in the first place. When we say 'law' without any qualifier, whose law are we actually referring to? Who exercises 'juris-diction', that is, the power to say the law, the power to say that something is law? What happens when different legal systems are found in the same space at the same time? Is law whatever duly appointed judges say it is? Is law whatever the president of the republic says it is? What are the effects of the power relations that are embedded in legal tools large and small, tools ranging from constitutions all the way down to property titles and site-specific injunctions? These are the questions, the fundamental questions, addressed by the contributors to this collection. They all address the same deep questions about what constitutes and is recognised as law in the first place, even though the substantive areas they cover are highly diverse.

The law and society tradition, today

To locate our collection in historical context we can say that 20 to 30 years ago, a 'law and society' collection would have emphasised the political and social character of state law, pointing out, for example, that a seemingly neutral law regulating housing mortgages had a clear racist effect – or

1 One example from the academic world is that one doctoral student at the University of Toronto had her thesis work delayed by several months because it was not clear which leadership should be consenting to her research, for university research ethics purposes.

showing that as women obtained legal equality in various countries, they did not achieve either economic equality or in-practice reproductive freedom. The law and society movement was very closely linked to the international effort to promote not just formal equality rights but deeper social and economic rights. But the standard categories of race, class, and gender that came to dominate critical legal and social studies from the 1970s onward, while remaining central, do not completely suffice to delve into the fundamental issues about what is law and whose law counts that the initial story about the Wet'suwet'en people and their land raises. Informed by Indigenous legal thought and post-colonial studies, as well as critical inter-disciplinary scholarship from across the social sciences and humanities, leading-edge socio-legal scholarship today must take seriously the complex and overlapping assemblages of law and authority that inform our world, a world that holds multiple incompatible legal systems and a variety of incompatible views about what law is for and who has the authority to proclaim 'the' law. This plurality of law and legal meaning has always been present, even within supposedly legally homogenous nation-states. In the past, modern national governments carefully marginalised and often silenced competing legal systems, especially within settler colonial contexts. The theory that the world is meant to be made up of coherent, autonomous states, each of which has an effective and legitimate monopoly over law-making and law enforcement in its territory was always a fiction. But in recent years, it has become blindingly obvious that this central fiction of mainstream 'law' is more out of keeping with reality than ever, as it is apparent to all that a range of different transnational challenges and conflicts are brought to bear on local communities. Phenomena such as sanctuary cities, climate change activism, global Indigenous networks, permanent refugee encampments, tech companies exercising surveillance capitalism across borders, and the mass migrations of over 70 million people around the world are all destabilizing a world order that was thought of as based on singular and stable national law systems, through which individuals were supposed to be all equally governed as citizen-subjects.

It is readily apparent from the day's headlines that what one could call the sovereign fiction of mainstream views of law is fraying under the stress of often unanticipated forces. What does this mean for the field of law and society? How does one study law when the what, which, why, and who of law are profoundly in question? This is the current challenge of our times and these questions frame our collection. They are intended to provide a domain of rethinking and a resource for students, scholars, and practitioners. The volume explores legal terms and subjects, some of which are standard and predictable in a volume on law and society and others less so.

Topics and authors

The process that led to this book started at the 2017 annual meeting of the US-based Law and Society Association, in a conversation between Mariana Valverde and Routledge commissioning editor Colin Perrin. Colin wanted to expand the offerings of the Routledge Handbook series in the socio-legal field. Mariana agreed to edit the 'law and society' collection on condition that a diverse group of socio-legal scholars representing different experiences and standpoints be assembled to jointly make editorial decisions.

One thing that our editorial team has in common is that all of us have a variety of transnational experiences, both academic and personal. Three of the four editors do not live and work in the country in which they were born, and all of us spend periods of time and do at least some of our research not just across borders but across oceans. All of us have extensive experience with academic work in different countries, as well as, importantly, with social and political movements in different parts of the world. And while none of us are primarily activists, we all support social justice movements either through personal participation or through our writing or both.

In keeping with our shared post-colonial, anti-racist, and feminist perspectives, and our common desire to have socio-legal teaching and scholarship catch up with what contemporary political movements are doing, we have worked hard to come up with a table of contents and a list of contributors that would reflect recent critical approaches to questions of law and justice. Along these lines, the collection includes several entries on Indigenous perspectives and issues – not just the token chapter on Indigenous legal traditions that older books often have – and also features a number of innovative contributions on climate justice, food security, and struggles over water, issues that particularly affect people in the global South. Importantly, these perspectives are not limited to the contributions specifically devoted to these issues. For example, the entry on feminist socio-legal scholarship spends some time on the issue of hunger, one that women in the global South are particularly affected by, and not one that the pioneering legal feminists of the 1970s included under their purview. To that extent, we believe this collection is fairly 'global', if one can use that overused word. Existing collective works tend to reflect the interests of scholars in the global North, especially the American law and society tradition.

However, we did not want to produce a work devoted exclusively to global South perspectives or to newer socio-legal topics, such as big data, climate justice, and food and water security. We set out to produce a diverse and timely collection that could be used as a teaching tool in classes taught in the global South, but also in the global North. And we certainly value the important contributions made to law and society scholarship by the largely white and largely American pioneers of the field – many of whom, such as Susan Coutin, Bryant Garth, David Engel, and Sally Merry are represented here. But we simultaneously solicited and received contributions from people in the global South (and global South scholars working in the North), both for newer topics, such as 'Indigeneity' or 'post-colonial legal studies' but also to provide a fresh take on old topics. For example, the entry on 'Actor Network Theory' is co-authored by two Latin American scholars; the entry on 'censorship' is co-authored by two Chinese scholars; the entry on 'cities and urbanisation' is authored by an influential Mexican legal sociologist whose work has rarely appeared in English; and the entry on 'gender' departs from mainstream feminist legal studies by reflecting a global South sensibility as well as recent developments in queer theory. These are just some examples showing that careful attention to the complex relationships between particular authors, their background and location, and the state of the relevant literature can provide our readers with material that is informative, original, and timely.

We hope the collection will be used to teach courses in social science programs, whether specific 'law and society' courses or other courses that include socio-legal scholarship. But we also feel it can be very useful in law school teaching. To law professors, however, we offer a preliminary caution: throughout, the four of us agreed that we would avoid replicating law-school categories. Thus, there are no entries that are specifically about 'contract', 'family law', or 'criminal law'. Many of the topics traditionally covered in law school curricula are indeed covered so that one could use this collection in a 'regular' law school course, but instructors will have to choose a selection of entries that were chosen from a perspective that does not take law's categories for granted. For example: there are two entries that focus on property. One, by Rosemary Coombe and graduate student co-author Ali Malik, covers current issues in intangible property with particular attention to struggles in the global South over patenting plant products; the other one, a less subject-specific one, does not seek to discern the essence of legal property, but instead undertakes a relational study of how property works in law in relation to two other key terms, 'person' and 'community'.

In keeping with our aim to better reflect issues of our time that were not part of the original, 1960s law and society movement, we have entries on data and on artificial intelligence – both of these, not surprisingly, by younger scholars. Similarly, we have an entry on 'infrastructure', not a traditional topic in the socio-legal canon. And discussions about climate and climate justice are by no means limited to specifically named entries, just as Indigenous critiques of settler colonialism and economic imperialism are found in numerous entries.

Finally, we encouraged authors to take not just legal disputes and legal arenas as their source of material, but to also pay attention to justice-seeking social movements, whether or not they specifically seek law reform or otherwise directly engage with formal law. There is no specific entry on 'resistance' or on 'social movements' – because the concerns and the innovations made by social justice movements inform the whole work.

What constitutes social justice today is a central question as countries pull apart at the seams under the strains of partisan politics and often right-wing populist movements. For some people, supporting authoritarian strongmen offers a better sense of 'justice' than the conventional markers of a democratic legal system with fair elections, uncensored media and communications, and an impartial judiciary. In short, the very goals and objectives of the law and society movement, which have been largely taken for granted for decades, are no longer clear. This has become abundantly apparent as the horrifying impact of the COVID-19 pandemic plays out as we write these lines, in the (northern hemisphere) spring of 2020. As social fears mount and death figures increase, the fragility of democratic legal systems is revealed, as government leaders in the global North and South seek to legitimise their increasingly authoritarian powers. What 'law' is and whose interests it serves has long been a central question for sociolegal scholars, but it is particularly apparent now that this concern is not relevant only to marginalised communities. As walls go up, worldviews close, and certain groups of people are abandoned by the state, the very meaning and scope of the term 'law' takes on new urgency. It is with this reality that we offer this volume as a contribution to contemporary issues in socio-legal studies.

PART I

Contemporary perspectives and approaches

1

ACTOR-NETWORK THEORY AND SOCIO-LEGAL ANALYSIS

Leticia Barrera and Sergio Latorre

As the extensive scholarly production in socio-legal research has demonstrated, legality is not a field that can be studied independently from the social. Different theoretical approaches have emerged attempting to explain the relationship between the law and its contexts of production, realization, and reception. Geertz's 'legal sensibility', Feeley's 'legal culture', Ewick and Silbey's 'legal consciousness', and Brunneger and Faulk's 'legal subjectivity' are examples of the many analytical categories that socio-legal scholars have elaborated to address the experiences and meanings of legality to different actors in a plurality of settings.

Recently, in the effort to understand how society and law hold together, and in particular, the place of law and legal knowledge in shaping the world and being shaped by it, some researchers have borrowed from the field of Science and Technology Studies (STS) the insights of Actor-Network Theory (ANT), which focuses on the rich web of associations between persons or humans on the one hand and things or non-humans on the other, and on the crucial role these relationships play in the production of knowledge. In ANT's view, knowledge is made through assemblages or networks of human and non-human agents (*actants*), that is, by tying together material objects or technological devices (such as microscopes or file folders), scientific concepts, human experts (scientists, judges); and references made to natural phenomena (such as the hole in the ozone layer or global warming).

Certainly, ANT's understanding of knowledge production and circulation challenges those theories that endorse modernist epistemologies that draw sharp divisions between objects and subjects, nature and culture, and separate the natural sciences – which are assumed to deal with objects – from the social sciences or humanities, which allegedly care about subjects/social agents. In contrast to this traditional set of binary oppositions, as John Law and Bruno Latour explain, ANT considers knowledge, the product of science, as 'a process of heterogeneous engineering' in which bits and pieces from the social, technical, conceptual and textual are fitted together, and so converted (or 'translated') into a set of equally heterogeneous scientific products. From this perspective, legal knowledge, just like scientific knowledge, can be understood as one of the ways in which the world is assembled. The word 'legal' does not refer to an inherent quality of certain people or objects but is rather an attribute that is attached to all sorts of events, people, documents, and other objects when they become part of the decision-making processes involved in the law.

Through the lens of ANT, and in particular its key notion of the network, legal instruments, such as documents, forms, files, leases, deeds, patents, and other legal materials that are an important part of the processes of the making and operation of the law but are usually taken for granted by both legal actors and legal scholars, are turned into distinct objects of analysis. Methodologically, in following the networks of human and non-human actors that arise out of and constitute knowledge-making processes, scholars discover the dynamics of specific practices and problems. The purpose of ANT analysis is generally to describe the dynamics of a specific network, not to generate a static model of how 'law' or any other system works, in general.

For ANT, the materials that one may encounter in the field are not fixed units; rather, they move, create new social relations, and even may become something else. ANT literature provides different examples of this kind: water pumps, documents, microbes, doors – all may be different entities depending on the set of relations they create in different moments and places.

Therefore, according to ANT, facts and objects are conceived as inherently relational, complex, fluid, and ontologically multiple. In this vein, the analysis remains mobile and engaged in detailed and close descriptions of technical and materials aspects, rather than being a universal theory based on general statements or diagnostics. This, however, does not mean that ANT advocates for the absence of a critical analysis. Rather, it points towards a shared sense that the fields of moral, ethical, or political valuation and activity are shifting and should be themselves scrutinized.

In the remainder of this essay, each of us will provide, in the first person, an example of crucial research findings that challenged our previous assumptions about legal knowledge production and that encouraged us to borrow ANT methods and insights. These findings – in the context of the Argentine Supreme Court of Justice, in Leticia Barrera's work, and at *Incoder* (the Central Land Office) and two rural field sites in Colombia, in the case of Sergio Latorre – brought to our research on law-making practices an unexpected dimension of the law: its materiality. In our fieldwork, we saw law manifested through the assembling of various papering practices, forms, agency, theory, and even infrastructure. ANT worked as the critical lens to bring law's materiality into the centre of our analysis and allowed us to ask about the quotidian and mundane operation of law. It also allowed us to take this question beyond legal thinking to inquire into its sociological effects.

The file (Leticia Barrera)

On December 19, 2005, I was conducting archival research in the Argentine Supreme Court's Library for my doctoral dissertation when I received a phone call from a law clerk, whom I had met a couple of weeks before, telling me that he had been informed that there was a file with my name at the Court's General Administration office. I then decided to pay a visit to this office and check the accuracy of the information I was given. At the front desk, an officer confirmed that there was indeed a dossier named 'Dossier 3737/05 Barrera, Leticia (Doctorate at Cornell University Law School) on internship to the Nation's Supreme Court of Justice'. I asked the officer if I could read the dossier, but he replied that I could not have access to it at that moment because it was ready to be circulated to the then seven Supreme Court Justices for consideration. I was shocked about this news because I had never applied for an internship in the Court. Then I asked him who had made such a petition on my behalf. He looked at his computer records and answered that a letter had been sent to his office from one of the Court Justice's offices (*vocalías*) requesting a legal opinion; that this office's legal counselling division (*Asesoría Jurídica*) had concluded that all the Court Justices should review the letter before granting consent for my 'internship' (*pasantía*). He also made it clear that, as in any other case that the Court decides, my petition would need a majority of five affirmative opinions out of the seven Justices to be approved.

The letter that the Administration Office employee referred to was one of the introductory letters addressed to each of the seven Justices that I had brought to the field in fulfilment of Cornell University's policy on the ethical use of human subjects in research. These letters were signed by my thesis supervisor and the Dean for Graduate Studies, and provided a very brief description of the ethnographic research that I would intend to conduct in the Court, including the methods I would apply, my course of action within the institution, and above all, my commitment to confidentiality. The letters also included informed consent documentation.

To skip procedural formalities that could delay my research inside the Court, I had delivered these letters in person to each of the seven Supreme Court judges' offices rather than submitting them to the Court's front desk. Nonetheless, the fact that I had chosen to do it in writing meant to judicial agents that I had filed a petition, which like legal requests generally should be reviewed and decided upon according to the rules and procedural mechanisms that regulate the Court's workings. To put it differently, what I had assumed as a matter of just a research-related formality to get access to the 'field' was translated by Court officials into a matter of law.

Anthropologist Jennifer Shannon recounts a similar experience in encountering bureaucratic proceedings in her field research and describes a type of relationality that informed consent documents may enact in this context. Drawing on Annelise Riles's insight that 'documents anticipate and enable certain actions by others', Shannon describes the consent form as an 'actant that sets people into action, as well as an institutional symbol in an existing cultural context'. She suggests that the more bureaucratically regulated the informed consent becomes, the more it takes on the form of established legal practices, and hence the more it leads to a 'legal relationality' between researchers and research participants. Like Shannon, my initial concern regarding my file was about the potential effect that the introduction of 'legally oriented' documents endorsed by the university's representatives might have on some relationships that I had been making in the field so far. Before I was informed about the existence of the dossier, I had met many Court agents and conducted several interviews in the tribunal, mostly with law clerks. Nonetheless, what I foresaw at the outset as disruptive and negative towards the development of my fieldwork relationships was the ordinary effect of my own engagement with the judicial apparatus, whether this interaction is articulated through informed consent letters, or through any kind of written document submitted to the Court. In other words, when I believed that Court officials had interpreted and translated into a legal matter a formal introduction for research ethics purposes and related consent forms, I did not realize that, in fact, I was dealing with a bureaucracy's everyday response to any written request, that is, the making of a file or a dossier. In my case, such a response was even more commonsensical due to the aesthetics and content of the letters delivered to the judges: they were written in a formal style, on the university's letterhead, with the institutional logo and references, and were signed by school authorities who backed my position and research. Consequently, it should not be surprising that the state's response to a petition is the framing of such a request by its own terms and proceedings, that is, according to the legal bureaucracy's definitions (an 'internship', in my case), and following 'carefully scripted routines', to quote Riles.

As Latour's ground-breaking work on the French *Conseil d'État* has demonstrated, there are instances of judicial practice, other than the final court decision, through which legal knowledge can be accessed and apprehended. In my research, it was the quintessential bureaucratic practice of file-making which allowed me to access the field and engage in a fine-grained study of the practices of knowledge production and circulation in the Argentine Supreme Court from a different perspective. In the process of following up on the status of my personal Court dossier, I also became myself a Court petitioner with expectations to get a favourable decision to my request. When my petition for an 'internship' – a petition I had never actually made – was refused by the Court, I requested, unsuccessfully, an explanation for this ruling. However, regardless of the outcome, the experience of being both a researcher and petitioner pushed me to reflect on aspects of legal knowledge that remain a blind spot for socio-legal studies because their mundane and instrumental character does not make them attractive to critical and socio-legal analysis. According to Cornelia Vismann, legal studies lack any reflection on their tools. Files, as she states, are the basis for legal work but 'remain below the perception threshold of the law … and certainly do not turn into objects of scientific investigation'. Indeed, files and documents are ordinarily regarded as nothing but mediators between the law and the decision. However, it is this intermediary position that files are said to occupy in the legal process which actually turns them into artefacts to access and apprehend the practices of legal knowledge-making. Files speak of events; they record processes and are constituents of a number of relations. They even impact on the actors' subjectivities and set the physical and epistemological boundaries of reality, of legal knowledge's reality. Files are instances of knowledge creation, expansion, and even concealment.

The land title and the seal (Sergio Latorre)

My encounter with the land title document came during the early stages of writing, after having a chapter of my dissertation rejected. In the search for theoretical avenues to push further my analysis, I was encouraged to start again from scratch. Rather than taking for granted the problems of access to land, which in my case reaffirmed law's inherent hegemonic power and understood law's procedures and institutional bureaucracy as another instantiation of such power, my adviser's suggestion was simple: to leave aside the question of who wins and who has power, and to start first describing what I had in front of me.

During the discussion of the failed chapter, the land title document had appeared several times in many of the stories of land struggle I had collected in my three field sites: Marialabaja, in Colombia's Caribbean coast; El Socorro, in the country's Andes region; and the offices of *Incoder*, in Bogotá. Until then I had not managed to tell a compelling story about this document, even though I intuitively knew that it had something hidden that allowed this piece of paper to connect the struggles of *campesinos* (peasants) to one another and also with the bureaucrats in Bogotá whose daily practices also involved working with land titles. The energies of *campesinos* and bureaucrats alike were spent in caring for, fixing, crafting, and fighting over land titles.

I started to approach the land title through an aesthetic analysis of the document in front of me. For a moment, I put aside the background information I had about the document I was looking at. It belonged to a *campesino* family involved in a land acquisition dispute that had lasted over a decade, in which two *campesinos* were killed. Instead of focusing on the content or on the background story, however, I turned to describe the material aspects of the document. The kind of paper: a notarized stamped paper. Its textural quality: thicker and heavier than most documents. I also paid close attention to its design, a framed rectangular box from which perpendicular lines came out forming a lined sheet of paper. Atop the lines was the careful display of the text in a typewritten letter with commas, accents, capital letter, hyphens, and bold lettering. Each box formed by the lined format was used in full. Any space left blank by the text was completed with additional hyphens until they met the margin. At the end of the document, the signatures of the parties involved in the transaction and their fingerprints. Immediately below, the signature of the notary public accompanied by the notarial stamp.

This stamp was one among a number of official stamps you can see on the title, each with a different meaning. However, the most important stamp in the document, placed on its upper left corner, was not a stamp produced with a rubber stamp but a seal, difficult to spot by the naked eye as it was now part of the printed notarial paper.

The seal is an antique technology that was used either over lumps of clay to secure the content of bags or it stamped over melted wax to close folded documents, thus guaranteeing the secrecy and authenticity of something regarded as having great value. In this case, the seal found in the land title opened up a network of relations that could help me explain not only the effects caused by the formalization of property over land but also important features of property and legal knowledge more broadly.

Attention to the seal in the title, apart from revealing the colonial underpinnings of the old system of property (since it was the Spanish royal seal, not the seal of the modern republic of Colombia), would allow me to understand and connect the materiality of legal documents, such as land titles, to present-day situations in rural Colombia. In some villages, *campesinos* most often had titles that needed to be updated or had no titles at all, but in other places, they had documents with official stamps that they thought were land titles but were considered by legal expert to be worthless. These documents were known as '*carta de ventas*'. These documents were also crafted at the notary but wrongly regarded by *campesinos* as land titles. Unlike property titles, they did not bear the seal of the land title but only other stamps. In fact, many stories regarding formalization of land tenure could be understood as a struggle of documents themselves to achieve the status that made them legal land titles.

The relationships between the seal and the land title allowed me to understand the property title, with its particular features, as an important legal technology for property. The stamps, the signatures, the lined boxes filled out in full (with dashes used to fill any empty space at the end of a line), the lettering, and the serial numbers were all technical devices that spoke of the importance of this particular paper and the power it held within it. All these technical devices constituted a design aimed at ensuring the documentary authenticity and uniqueness necessary to guarantee property. The land title document contained a particular relationship between people (in this case, the buyer and seller) in which each party was individualized using devices such as signatures, fingerprints, and national identification numbers – devices that identified them by their external features. The plot of land was also made individual by its external features, that is, in relation to other plots of lands, by establishing its boundaries in the document. The authority of the State – which conferred to the notary the power to bear witness to the transaction – was also ratified by parties. Finally, the document of land title itself, as a technology incorporating technical devices, became individualized. Therefore, this document came to embody the piece of land, creating the pathway for new relationships (i.e. collateralization of land as an economic asset) to take place. The land title had finally revealed its great value.

In short, the description and analysis of the materiality of the seal and the land title inspired by ANT methods allowed me to see that behind this document there is a careful configuration of social and non-human elements. The latter too exert agency and play a key role in the configuration of property and formalization of land ownership.

Conclusion

ANT's versatility attracts scholars engaged in rethinking the complexity of law's immanent operations and its links to other fields such as sciences, politics, literature, the environment, and the economy. Scholars working with ANT's toolbox have made significant contributions to discussions on the role of the quotidian and mundane operation of the state apparatus and how these operations are instantiated and reproduced in disperse institutional and social networks of power. Their interventions have influenced current Law and Society discussion on topics such as the making of governance, everyday legality and legal consciousness; the role of experts and expertise; and the workings of bureaucracies and bureaucratic practices. Certainly, ANT opens up many possibilities for critical and socio-legal studies.

Further readings

Barrera, Leticia. 2018. 'Gatekeeping: Documents, Legal Knowledge and Judicial Authority in Contemporary Argentina'. *PoLAR: The Political and Legal Anthropology Review* 41(1): 90–107. https://doi.org/10.1111/plar.12242

Law, John. 1992. 'Notes on the Theory of the Actor-Network: Ordering, Strategy, and Heterogeneity'. *Systems Practice and Action Research* 5 (4): 379–393. https://doi.org/10.1007/BF01059830

Latour, Bruno. 1993. *We Have Never Been Modern*. Cambridge, MA: Harvard University Press.

Latour, Bruno. 2004. *La Fabrique du droit. Une ethnographie du Conseil d'État*. Paris: La Decouverte/Poche.

Latorre, Sergio. 2015. 'The Making of Land Ownership: Land Titling In Rural Colombia, a Reply to Hernando De Soto'. *Third World Quarterly* 36(8): 1546–1569. https://doi.org/10.1080/01436597.2015.1046984

Riles, Annelise. 2000. *The Network Inside Out*. Ann Arbor, MI: University of Michigan Press.

2

CRITICAL LEGAL STUDIES

A curious case of hegemony without dominance

Vasuki Nesiah

On February 27, 2020, a blustery winter day, the founders of Critical Legal Studies (CLS) gathered in Princeton University for a two-day conference on CLS history. The gathering was billed as a CLS retrospective marking the inauguration of a CLS archive in Princeton's Mudd Library. The last formal CLS gathering had taken place over 25 years before that, in 1992.

The original CLS conference was convened in 1977 at the University of Wisconsin, Madison. After that, the pioneers of CLS met regularly in conferences, summer workshops and reading groups, until 1992. The silence since 1992 was thus noticeable, indeed deafening – perhaps most of all to the Crits (as CLS scholars came to be known) who had been standing closest to the original CLS 'explosion'.

By 1992, unceremoniously, CLS was declared dead – 'they crushed us like bugs' were the words used by CLS notable Duncan Kennedy. In the domestic US conversation, right wing political figures like Ronald Reagan and Jeff Sessions mobilized opposition to CLS. In legal academia, key tenure cases were lost. Law and economics had become ascendant and within the legal profession, and legal professions organizations like the Federalist Society were established to steer the legal world decisively to the right.

The obituaries and epitaphs to CLS that its founders had been delivering since the 1990s were rehearsed and restated in Princeton in 2020 by its principal protagonists. Yet this time, there was concerted push-back from those in the audience – a younger generation of legal academics who had come to this event because CLS had and continued to be so significant to their intellectual projects and their lives in the legal academy. Participants cited CLS's 'wide and deep influence', CLS's impact on their own scholarship directly and also through the mentorship of senior CLS faculty, 'sleeper cells' of critical thinking in law schools all over North America, and discussed intellectual traditions of critical legal theory that had influenced even mainstream legal education and served as inspiration for a heterodox style of critical engagement with the law. This included Critical Race Theory scholars and 'femcrits' (feminist legal scholars) who spoke of the constitutive debates regarding race, gender, and sexuality that were central to these distinct yet allied critical traditions that emerged in conversation with, and in some cases, in juxtaposition to, CLS. Then there are all the transnational reverberations of CLS, from Colombia to Kenya, including the movement known as TWAIL – a transnational scholarly enterprise with the ironically old fashioned moniker 'Third World Approaches to International Law', but which has resonated in important ways in the global South and amongst the global South/Third world diaspora in the global North.

The subtitle of this chapter is 'hegemony but not dominance' because even though CLS as a movement may no longer exist, it is worth pondering on its rise and decline: the insights and analytical techniques of CLS have had extraordinary influence within the legal academy, nationally and internationally. To borrow from Mark Twain, reports of the 1992 death of CLS turned out to have been widely exaggerated.

The following account of CLS insights and preoccupations is neither a comprehensive account of the intellectual legacies of CLS, nor a history of the movement. Instead, I present CLS in terms of a series of constitutive debates and tensions that have travelled a plurality of paths in different areas of law on the basis of a powerful and internally diverse repertoire of political and legal critique of the dominant approaches and legacies of legal liberalism.

The central insight most associated with CLS is the *indeterminacy thesis* – better described as the dialectic between structure and indeterminacy. Received liberal legalist conceptions of legal development and interpretations draw on formalist conceptions of the rule of law to present an idealized model of objective legal decision-making through deliberations that take into account all the relevant facts, laws, and policy considerations. The formalist view understands litigation and adjudication as processes that expel special interest in favour of logical deduction from those relevant sources, and is guided by neutral rules of procedure in ways that are more or less complex depending on whether they are easy cases where the authoritative legal conclusion is 'obvious' or so-called hard cases.

In contrast to formalism, CLS treats law as a terrain of politics. Legal proceedings mobilize particular vocabularies and traditions defined through doctrine, precedent and the ideological commitments embedded into the legal framework, as well as the background institutional arrangements and the consciousness of legal actors, including lawyers and judges. Rather than yielding one correct answer, CLS holds that legal doctrine and precedents are riven with 'gaps, conflicts, and ambiguities', so that the process of legal reasoning can yield multiple, even competing, interpretations. These ambiguities can be due to the footprint of past struggles over resources and meanings, the complexity and contradictions of society, our own irreconcilable phobias and desires, and also our intersecting and diverging identities, interests and normative commitments. The upshot is that a legal system is not machine-like and internally coherent and complete; rather, there are many possible answers, and internal tensions can have radical yield if a critically minded legal actor creatively explores new possibilities. In sum, the structure of dominant institutional arrangements and the indeterminacies that characterize the law and legal framework are in a dialectical relationship. Understanding this dynamic is central to CLS' analysis of the situation in any particular context.

For CLS, the recognition of this indeterminacy does not mean that the decision is wide open and anything goes. CLS also understands the legal framework as having a *structural tilt* in favour of powerful actors along a range of pivotal social axes, including class, race, and gender. The structural tilt lends a gravitational pull to legal decisions, advancing and reproducing those structures. However, this usually takes place not through explicit discrimination but rather through the unfolding of liberal legalism. For instance, in looking at the validity of a contract, if there is no violence or illegal coercion, the contract is assumed to be the valid expression of the will of free and equal parties exercising their freedom of contract. In contrast, CLS deconstructs the illusion of fair and free contract by tracing the work of 'formal freedom' in obscuring structural coercion and legitimizing maldistribution.

This does not mean that in each and every legal dispute those who carry class, race, and gender privilege are always victorious; structure works in more complex ways than allowed by a functionalist account of law as exclusively servicing the powerful. Rather, CLS is attentive to how there may not be direct lines between 'the structure' or a dominant set of institutional arrangements and the law and policy that will reproduce or develop those structures. Moreover, there may not be any direct lines between any particular group's interests and the law or policy that will advance those interests. There may not be direct lines between any particular legal dilemma and the interpretation of the relevant law; and in addition, there are always unexpected exigencies that may result in unexpected consequences.

Influenced by Marxism on the one hand, and post-structuralism on the other, the structure-indeterminacy dialectic emerged as central to CLS analytics not only in reading legal history and past inheritances, but also in plotting legal strategy for future interventions. Mobilizing law for interventions aimed at social change requires careful political analysis and nimble legal reasoning that can mine the indeterminacy of law in these different registers for subversive and counterhegemonic legal meanings that forms the heteredox counterpart to the dominant doctrinal regime. For instance, Joseph Singer has argued that a wide variety of existing legal rules about property give expression to what he calls 'the reliance interest in property', which foregrounds social relationships over possessive individualist interests. Many of these rules may be marginalized within dominant property law ideology, but they constitute an important jurisprudential tradition that provide an important doctrinal source for counter-hegemonic legal interventions.

If attention to the structural tilt of the socio-legal field is a marked feature of CLS analysis, its corollary has been attention to *distributional analysis*. Such analysis is central to a legal project that often has an activist dimension. For instance, in a context where many gay and lesbian rights advocates were arguing for legal recognition of sexual identity as a fixed identity similar to race (which in the US triggers strict constitutional scrutiny) in order to empower new grievances brought to anti-discrimination legal venues, Janet Halley takes a step back to perform a distributional analysis of the work of coherentist assumptions of identity. She argues that such strategic moves stabilize identity with 'possible coercive effects' on queer identities and activities that did not fall within the purview of the US legal requirement that a minority seeking protection from discrimination be able to prove it is 'discrete and insular' (a greater challenge for those challenging gender binaries, or who claim gender fluidity for instance). Essentialist identity politics benefits from the early efforts to legitimize LGBT rights by claiming that sexual orientation is inborn, while those perceived as deviant in relation to those essentialist assumptions lose out, including bisexual and trans people. Distributional analysis begins with analyzing transactions and their legal architecture to understand who profits and who is dispossessed, in the short term and the long term.

The material and intangible benefits that flow to those who benefit from the existing legal structures can be hidden by a range of mechanisms that naturalize the received social structure and the extant distribution of wealth and status. One of the mechanisms through which CLS unpacks 'stealth' distribution is by tracing the braided relationship between different domains of law that are seen as distinct. For example, in a domestic (national) labour negotiation, the bargaining power of the contractual parties may be shaped not only by the national economic structure but also by international trade rules governing transnational capital mobility, as well as immigration law rules and their governance of labour mobility. Thus, a contract is negotiated between a union and the employer in the shadow of a multi-faceted, multi-scalar legal regime, which impacts the leverage of both parties, tilting the parameters, paths, and outcomes of these negotiations.

There is a parallel argument that can be made in relation to women's unpaid labour in the home, drawing attention to the fact that domestic (household) negotiations about resources and power take place under shadow of the law. Noting the institutional underpinnings of specific family/household negotiations helps to reveal contradictions internal to the public/private distinction, while also enabling strategic intervention in areas that range from remuneration for domestic labour to accountability for domestic violence. For instance, Fran Olsen, the noted Marxist-feminist CLS scholar, authored a series of influential interventions in the 1980s that analyzed the basic dynamics of what is often referred to as 'the public-private split' and deconstructed the 'coherence' of this distinction and how it played out in defining the legal and ideological contours of family and market.

The work of the public/private distinction can also help to illuminate the ways in which CLS analysis has focused on the role of *legal consciousness*, that is, the ways in which ordinary people as well as legal actors come to have certain, often culturally specific, expectations about law and certain attitudes about what law can and can't do. Certain background assumptions are legitimized as a taken for granted dimension of the routine unfolding of law, and legalism itself becomes a way of masking these background power relations.

Within liberal legalism, rights discourse is central. One of the most familiar ways in which law's neutrality is promised and its illusion achieved is through the sedimentation of the notion of individual rights. Liberal legalism attributed a normative status to rights that claims to be prior to law, while also being the conduit, as it were, of law. CLS points to how rights translate political claims into almost 'natural' phenomenon, so that liberal legalism proceeds as if rights 'exist' prior to any interpretation of a legal conflict, awaiting recognition and upholding through correct legal reasoning. The work of rights in producing legitimacy is central to the CLS critique of liberal legalism. Rights present the particular interests as universal ones – as when a corporate actor's interests in de-regulation is produced as a universal right to the free market, or an abuser's interest in not having the state interfere with domestic abuse travels as a universal right to privacy in one's home. Even when deployed in left-liberal advocacy projects on behalf of vulnerable communities (e.g. the right to collective bargaining), CLS attends to how the political and legal machinery of rights interpellates people as liberal subjects legible to individualist assumptions of liberal legalism. In sum, the CLS critique of rights is not against rights discourse; rather, attentive to its many enchantments, CLS grounds the assessment of any particular invocation and deployment of rights talk in strategic and politically anchored analysis of the particular legal terrain on which a struggle will be fought, including a distributional analysis of what is gained and lost and the opportunity costs for alternative vocabularies of justice.

While critique is what CLS is most associated with, it needs to also be understood as a ground clearing operation that creates the condition for a forward-looking *experimental ethos*. On the one hand, the CLS critique of dominant interpretations of doctrine and precedent steer it away from received policy solutions. On the other hand, its attention to indeterminacy makes CLS particularly attentive to the possibilities of alternative interpretations that are immanent in the existing legal framework. For instance, in a case seeking judicial recognition of the 'reliance interest in property' that we referenced earlier, critically oriented lawyers collaborated with unions and other civil society groups in creatively deploying an argument making a claim about a reliance interest in supporting workers trying to force their employer (US Steel) to sell the factory to workers rather than demolishing the plant, laying off workers, and moving overseas. While this particular fight was ultimately unsuccessful, it is a window into how CLS has mined the critical potential of legal indeterminacy to advance unorthodox and radical reconfigurations of the legal issues that are seen as pertinent to a case. Protection of the interests of tenants over owners, which can and does happen in many judicial or quasi-judicial settings, may provide another example of how social justice lawyers have strategically exploited the contradictions internal to the legal regime of property to foreground a heterodox repertoire of marginalized doctrinal sources that support tenant rights. There are similar arguments regarding other areas of law. For example, Karl Klare has argued that American labour law history has both a dominant tradition that reinforces managerial power and a minority tradition that has sought to advance industrial democracy. Labour lawyers have drawn on the latter in advancing union rights and a radically redistributive view of the background legal regime of labour markets.

CLS's emphasis on the possibilities for radical change partially explains how these possibilities have been thwarted through an analysis of the role of mainstream legal education in stunting political imagination. Duncan Kennedy argued that 'legal education' naturalizes hierarchy, and that the law school was organized 'to train them [law students] in detail to look and think and act like all the other lawyers in the system'. Against this indictment of traditional legal education, CLS scholars invested heavily in teaching and mentorship to train students to think against the grain, develop scholarship that challenged hegemonic jurisprudential conceits, resist dominant legal consciousness, and form political solidarities.

We began this chapter with an account of the February 2020 Princeton conversation that catalyzed many refutations of the death of CLS, drawing on the claims of many younger legal scholars of the vibrancy of critical legal thinking in their own work. This is partly a testament to the enormous investment of CLS scholars in mentorship. Many of those younger scholars, some of whom were present at Princeton, traced a scholarly lineage through law school classes with CLS-oriented professors, summer workshops, reading groups, conference encounters, feedback on papers, and a range of other avenues through which 'crits' of many generations invested in their work and opened academic doors.

We have already noted the important examples of CRT and feminist legal scholarship in this context, but perhaps the third most significant legacy in this regard is critical approaches as they have travelled in international law. Here, there is no figure more significant than David Kennedy and the work that unfolded in the 1990s under the moniker of New Approaches to International Law (NAIL), and then since 2010, under the institutional sponsorship of the Institute for Global Law and Policy (IGLP). Indeed, this work has generated its own alternative tributaries – in particular Third World Approaches to International Law (TWAIL), in solidarity with a range of radical traditions from the global south, global feminisms, and critical race theory. Addressing the injustices of empire while also renewing traditions of resistance have been pivotal drivers of TWAIL scholarship.

CLS's influences thrived partly through debate and these other traditions we have cited also generated important critiques of CLS that have been central to its trajectory. Feminist scholars such as Claire Dalton, Fran Olsen, Elizabeth Schneider, and Mary Joe Frug made early interventions problematizing the fact that early CLS reflected White, male elite dominance. CRT scholars, such as Patricia Williams and Kimberly Crenshaw, advanced important critiques of the racial blinders in CLS scholarship; For instance, Williams made an influential critique of the CLS critique of rights as itself built on assumptions about the rule of law that were predicated on White privilege. The scholarship of TWAIL scholars – such as Bhupinder Chimni and Anthony Anghie – have challenged the American-centric preoccupations of CLS and foregrounded intellectual traditions, political perspectives and legal problems anchored in the global south, and the legacies of colonialism and slavery. There are also criticisms of the institutional location of CLS, the outsize influence of Harvard Law School, and the underappreciation of other important centres of CLS scholarship even within the US academy, including SUNY Buffalo, Northeastern Law School, University of Wisconsin (Madison), and UCLA.

These other critical traditions stand in a complex relation to CLS, marking the influence of CLS but also with critical and dialectical distance from it. Indeed, the lines of influence have been bi-directional and CRT, TWAIL, and Feminist and Queer legal theory have profoundly influenced the work of CLS scholars. Perhaps one of the most important characteristics of CLS is that ideas about law, legal institutions, and the scholarly vocation have important stakes and are inextricably intertwined with the most central political questions regarding class, race, gender, sexuality, and empire. The founders of CLS were themselves shaped by the civil rights struggle, second wave feminism, the fight against the Vietnam War, and the ongoing struggle for economic justice. The movement catalyzed by the 1977 Madison conference, mentioned at the outset as one possible 'birth' of CLS, exploded onto the US legal academy, but its reverberations were felt far and wide. As Roberto Unger reflected, when CLS landed in the legal academy, 'they [the law professors] were like a priesthood that had lost their faith and kept their jobs. They stood in tedious embarrassment before cold altars'. The history of CLS can be understood as a struggle against conscription into the priesthood; instead, Unger says 'we [CLS] turned away from those altars and found the mind's opportunity in the heart's revenge'.

Further readings

Brown, Wendy Bron and Halley, Janet (eds.). 2002. *Left Legalism/Left Critique*. Durham, NC: Duke University Press.

Kairys, David (ed). 1998. *The Politics of Law: A Progressive Critique*, 3rd Edition. New York: Basic Books.

Unger, Roberto. 2015. *The Critical Legal Studies Movement: Another Time, A Greater Task*. New York: Verso.

3

CRITICAL RACE THEORY

Emergence and new lines of inquiry

Kamari Maxine Clarke and Ifrah Abdillahi

The 2008 election of the United States' first black president heralded for many the possibility of a 'post-race' world, but this ideal has been challenged – if not shattered – by the resurgence of overt white supremacy and anti-immigrant populism. Even prior to the era of Trumpism, the post-race dream was problematized by re-focused publicity surrounding the shootings of unarmed black men (and women and children) by police. Increased visibility of these killings, documentation of which has been shared widely and rapidly via social media, contributed to significant social justice mobilizations such as #Ferguson and #BlackLivesMatter. These frontline movements have pushed for a renewed critical confrontation with racism, which has been taken up in mainstream fields of law and justice, as well as the domains of science and health, where it is often reframed and euphemized as 'unconscious bias'. There is a growing recognition that race is complexly embedded in all forms of social inequality throughout the US and elsewhere, which have histories and legacies of white supremacy. Scholars and practitioners have taken up with renewed urgency the desire to understand what race has become in our contemporary world and how political, economic, and historical structures of inequality continue to shape the conditions that produce race. These lines of inquiry, increasingly adopted across disciplines and professional fields, are influenced by a field of scholarship termed Critical Race Theory (CRT).

CRT is a framework for studying the relationship among social inequalities, social institutions, and (in)justice. Since its origins in the United States, the approach and need for CRT has evolved and expanded. It has emerged in relation to the limitations of critical theory and has moved beyond it to analyze and confront how race and racism shapes social order (including law and legal systems), culture (including technical and professional cultures), and power (including formal and informal, at multiple geographic scales). Before turning to a sketch of where these developments in CRT may be headed, we turn below to a more detailed look at the history and evolution of the field. We end with some reflections on the need for a more globalized approach to CRT that takes into consideration the work of scholars positioned outside of the US but who are engaged in critiques of coloniality and imperialism that structure contemporary modernity.

The emergence and evolution of Critical Race Theory

Some scholars, including Latoya Johnson, mark the beginning of CRT in the anti-slavery movement, while others such as Kimberlé Crenshaw trace its origins to the writings of WEB

DuBois – especially his disciplinary critique of history epitomized by *Black Reconstruction in America.* Others point to the influence of Critical Legal Studies (CLS). Traditional approaches to the study of law in society have ranged from Naturalism to Legal Realism to Critical Legal Theory. Legal Realism emerged as an approach to law and legal rules that is necessarily based on judicial decisions derived on behalf of society at large rather than a conception of a supernatural authority (Naturalism). In this paradigm, legal duties and related rights are shaped by judicial decisions and driven by social interests rather than abstract rules. As a philosophical tradition that was popular during the 1920s and 1930s, it departed from conceptions of natural law and more formalistic conceptions of law. While some forms of Legal Realism reduced law to being about court activity and presumed that it was possible to produce a value-free approach to law, American legal realism presumed that law was much more than the mechanical application of legal principles. Rather, it raised questions about the relationship between facts and law, illuminating how legal sciences can actually hide the role of political and moral influences.

In response, Critical Legal Studies (CLS) emerged as a movement in the 1970s as a critique of the American legal realist tradition's preoccupation with jurisprudence. The formative work of civil rights lawyer and Harvard Law School professor, Derrick Bell, who drew attention to the effects of racism within legal thought, was particularly influential. Harvard Law School's only black faculty member at the time, Bell observed that many of the rights fought for during the Civil Rights Movement had not had the desired impact on the lives of ordinary people of colour – and in particular black people – in America. Indeed, legal institutions appeared to respond with intensified de facto segregation (e.g. the war on drugs, the rise of the prison industrial system). Following Bell, a growing community of scholars has challenged the efficacy of law to produce justice, and the concomitant inability of American law schools to overcome inherent injustices baked into their pedagogies. CLS advanced a more radical agenda examining the relationship between domination and legal thought. By adapting critical theory, CLS championed a dialectic approach to critiquing social phenomena. Two central trajectories took shape in this tradition and dominated critical thought for many decades. One involved the assumption that CLS represents a particular understanding of society even as it has internal contradictions. The second approach combined more functionalist methodologies and radical goals. As such, CLS critiqued the perceived formalism and objectivism of American legal institutions and its prevailing liberal vision. It reconceived law through critical approaches to legal reasoning, rights, and conventional approaches to the free market (Tate, 1997: 208). It is in relation to this critique that CRT took shape.

By 1990, Bell resigned from Harvard Law citing the institution's failure to hire other faculty of colour. This resignation left no tenured faculty of colour at the school, and students responded with protest. One of these students, Kimberlé Crenshaw, went on to further shape CRT. For Crenshaw, CLS scholars failed to ground their analyses in the lived realities of race in people's lives, and thus also failed to properly analyze the systemic role of racism in society. And finally, it minimized the transformative power of liberalism (Tate 1997: 229–230). CRT both built on the foundations of critical theory but also moved beyond it by departing from its idealism and insisting on centring the issues, experiences, and injustices of racism that were generally unacknowledged within legal studies. Through Crenshaw and others' leadership, CRT expanded beyond the development of critiques of racial power and white supremacy in lawmaking. A range of subfields emerged that allowed for the re-centring of other variables in the rethinking of racial injustice. Topics taken up by these fields have ranged from critiquing liberalism, to questioning the civil rights movement, to interrogating conceptions of race and race consciousness, to calling for further expansion of CRT to grapple with how larger structures of inequality function in the contemporary period. There is a growing recognition that *all* of the legal, political, and social institutions in the United States (and beyond) – including 'objectivist' fields of science and technology – need to be interrogated with a critical understanding of race and racism.

New directions in Critical Race Theory: future considerations

Today, CRT is well-poised to explore the often-obscured operation of racism within science and technology, although challenges to its reach include the rapid rate of contemporary technological change as well as the historically US-centric focus of scholarship that has not always succeeded to illuminate how racism operates beyond American shores. Recently, there is increased momentum within the health sciences to highlight and address the enduring and pervasive health inequities experienced by racialized groups, particularly Black, Caribbean, and people of African descent. As health inequalities have persisted over time, in different contexts, and across various diseases, there is growing recognition that inequalities that exist around racial constructions of difference contribute to consequent inequalities in health outcomes. Recent scholarship in Science and Technology Studies also emphasizes the importance of unpacking the cultural histories and social assumptions that are integral to, for example, what Rena Bivens (2008) has described as how computer programmers write code and design software and how recent scholars have documented the way that digital technologies are produced and circulated through computer software. Where research on algorithms, thus far, focuses on the (un)intended consequences are access and equality of design and engineering at a society-wide scale, we are beginning to explore how the very construction of new technologies – from concept to consumption – is intimately shaped by micro and macro patterns of racism perpetuated at the individual, interpersonal, group, institutional, and societal scales. In the class of Artificial Intelligence (AI), the 'neural networks' of machine learning are being trained to analyse large datasets and to use algorithms to unearth complex patterns. These algorithms can be used to predict individual-level risk to a given disease based on analysis of variable factors. However, these technologies have major limitations in relation to race and equity. As noted in scholarship across disciplines, the algorithms used to test datasets do not take into consideration underrepresented populations (e.g. racialized groups, LGBTQ, etc.) and thereby create an explicit disadvantage against those groups from the onset. Biases in the design of AI models are embedded in the implicit biases of those crafting the models. Accordingly, there is enormous potential for AI to inscribe discrimination in the very skeleton of this new technology, while simultaneously obscuring discrimination under the aegis of technology. Historians of science and medicine have pointed to how the epistemology of racial inequality has historically been developed, normalized, canonized, and championed within these fields, which points us towards the urgent need for CRT to prioritize critiques of technology in the 21st century. Researchers and practitioners concerned with social equity are beginning to take seriously the implications these technologies can and will have in threatening the health and livelihood of marginalized populations. CRT must continue to question the so-called 'objectivity' of seemingly neutral technologies by drawing attention to the assumptions upon which that objectivity claim is based.

The contemporary moment is shaped by the 'post-truth' and 'alternative-facts' realities of the racist nationalism era. Even as all of the sciences struggle to preserve the legitimacy and authority of scientific evidence, battles to secure the conditions for the good life are raging on many fronts. In Canada, rapidly changing demographics are propelling an uptick in political acknowledgement that racial discrimination is prevalent, despite decades of multicultural policies. Unnamed racial dynamics are receiving never before seen attention through a national anti-racism strategy and millions directed at initiatives promoting equity, particularly for Black Canadians. However race-based population level data is either not collected or is not disaggregated. The lack of data collection or the existence of that data aggregated through the category of people of colour (that include black Canadians with recent African immigrants, as well as East Asians and South Asians) produces profound limitations in mapping inequalities with implications for health and other socio-cultural and political conditions.

While contemporary transformations in our world have been adumbrated by new forms of science and technology, empirical work has played a key role in documenting the impacts and effects of racism and its role in shaping various knowledge modalities. A normative approach to applying an equity lens to health research and practice is to focus on the circumstances in which people are born, grow, work, and age, and to trace how these circumstances are shaped by broader social, political, and economic systems in ways that profoundly impact individual and community health. The question before us is: What kind of approaches to racial inequalities is needed to attend to new developments that are producing social inequality in our world? A CRT that attends to the dual manifestations of discrimination in law and in science opens important opportunities for future scholarship.

The future of CRT would be wisely suited to explore the actual production of new knowledge modalities, including empirical inquiry into the forms of expertise involved, and as Eyal Weisman has referred to it, the location of varied experts within matrices of global power. For example, the Agreement on Trade-Related Aspects of Intellectual Property Rights-Plus (TRIPS+), which promotes scientific and technological innovation while simultaneously restricting low-and-middle income countries' (LMICs) access to life-saving knowledge and medicine is an important domain for understanding the relationship between international treaties and their on-the ground implications. Yet, TRIPS remains an under-examined domain in the global burden of disease and an important site for scholarship on global domains of power facilitated by white supremacy and privilege.

There are many examples of the way that structures of inequality actually contribute to forms of racial differentiation. Consider global policies like the Sustainable Development Goals (SDGs). The SDGs offer boundless opportunities for researchers to take up CRT in the study of globalized process and racial stratification. SDG17's pillar on technology is mainly focused on the global diffusion of green technology but it is another emergent global transformation that has the potential to further entrench the inequitable distribution of global power and resources. By exploring the structures of scientific regimes, how they inspire particular conclusions, or shape particular designs, CRT can continue to examine the emergent institutional developments – especially in the global South – in the production of knowledge and the way that that knowledge is shaped by pre-existing biases and latent colonial inscriptions that adumbrate new relations and possibilities. In furthering our understanding of the role of new 21st century scientific and technological developments, CRT can help us to make sense of precisely how technologies both obscure and exasperate pre-existing inequities. In this way, we begin to grasp how objectivity and truth construct the terms through which knowledge and power play out in daily life in ways that continue to subjugate and exclude black and brown people.

Beyond US-centred applications of Critical Race Theory

By the 1970s, scholars in the humanities and social sciences within the US and outside of the US took up the challenge articulated by scholars like Walter Rodney and Edward Said, and later Eric Wolf to examine the way that coloniality and imperialism shaped foundational domains of knowledge, culture, and power in society. For Rodney, it was the role of colonialism in shaping structures of exploitation; for Said, the role of imperialism in shaping the epistemic and representational violence shaped knowledge about the non-West 'other'. It was not until the 1970s onward that legal anthropologists and legal historians engaged this work in order to examine the way that notions of the 'other' structured contemporary modernity and began to build on subaltern studies scholarly traditions. Guyatri Spivak and Dipesh Chakerbarty or postcolonial studies scholars through the work of Eric Williams, Sylvia Wynter, and later Walter Mignolo and Achille Mbembe were amongst them. From this, emerged *postcolonial legal studies* (See Renisa Mawani, this volume) which sought to expand the relationships between imperialism, culture, power, and the law.

In international law, it was the work of scholars of colour, such as Anthony Anghie James Gathii, Siba Grovogui and more engaged in the development of Third World Approaches to International Law (TWAIL) and others writing within and outside of the law. The works of Eve Darian Smith and Peter Fitzpatrick on laws of the postcolonial reflected a commitment to a new paradigm in connecting race, law, and coloniality. Mahmood Mamdani and Kamari M. Clarke engaged international and national legal processes through its relationship to culture, coloniality, and power, and its place in the production of race. When postcolonial legal studies began to take root through interventions that were increasingly in conversation with other post-1980/90s postcolonial studies scholars, CRT took a different direction, instead focusing on US-based racial inequality in the law. However, attempts to situate CRT approaches alongside the study of the imperial foundations of the law are only starting to take shape amongst CRT scholars.

Conclusion

Critical Race Theory and postcolonial legal studies, combined with TWAIL, converge in their complementary agendas by showing how law is a central apparatus in the construction of white supremacy and racial oppression. We have also shown that science with its roots in particular imperial logics and colonial inscriptions have been key foundations in the construction of race and its structural inequalities. Without an approach to the imperial foundations of law and how it has produced race, and without an understanding of how race is an ideological domain through which coloniality, the settler state, and imperial structures have shaped law's modernity, it will be difficult to understand the complex and far reaching relationships between the state, racism, and the history of imperial legacies that structure race and social relations in the contemporary period. This is a domain for renewed directions in the study of Critical Race Theory.

Further readings

Anghie, Antony. 2005. *Imperialism, Sovereignty and the Making of International Law*. Cambridge: Cambridge University Press.

Crenshaw, Kimberle. 1991. 'Mapping the Margins: Intersectionality, Identity Politics and Violence against Women of Colour'. *Stanford Law Review*. 43 (6): 1241–1299.

Gathii, James. 2011. 'TWAIL: A Brief History of Its Origins, Its Decentralized Network and a Tentative Bibliography'. *Trade Law and Development Journal. National Law University, India*. 3 (1): 26.

Grovogui, Siba. 1996. *Sovereigns, Quasi Sovereigns, and Africans: Race and Self-Determination in International Law*. Minneapolis, London: University of Minnesota Press.

Martinez, Aja. 2014. 'A Plea for Critical Race Theory Counterstory: Stock Story versus Counterstory Dialogues Concerning Alejandra's 'Fit' in the Academy'. *Composition Studies*. 42 (2): 33–55.

Tate, William F. 1997. 'Critical Race Theory and Education: History, Theory, and Implications'. *Review of Research in Education*. 22: 195–247. https://doi.org/10.3102/0091732X022001195.

4 FEMINISM

Donatella Alessandrini

Feminism has never been a singular movement, that much we know. Trans-national, intersectional, and decolonial feminisms have challenged dominant historical narratives, including feminist ones, that have failed to account for the experience of the majority of beings (not only women) inhabiting this world. They have also shown how relations of power, oppression, and domination are contingent on many factors so that struggles vary widely across the world. This entry engages with recent calls for feminists to join other anti-systemic movements, in particular anti-racist, environmental, migrant, and labour rights activists. It does so by focusing on one particular tradition of feminist thought, that of Social Reproduction (SR), aware that both this tradition and the position of the writer, including the perspective put forward, are contingent and partial.

Some forms of feminist thought popular amongst legal scholars focus on sexuality as a key or even the key dimension or vector of gender oppression (Catherine McKinnon being a notable proponent of this view, very influential in the 1980s and 1990s, especially in the US). By contrast, SR scholars argue that feminism begins by acknowledging that our capitalist economies rely on a large reservoir of unpaid labour (traditional 'women's work', for the most part) and also on environmental resources, as well as on social divisions and hierarchies, to produce wealth. These feminists claim that this system, this rather ideological split, is unsustainable and undesirable.

The first section focuses on the separation between formal economic production and social reproduction, and on the gender, sexual, racial, and class divisions that sustain the production of value in the global economy. The second part looks at the role played by international economic law in enabling this separation and the third concludes with some thoughts on provocative forms of action.

Re/production and social divisions

We can understand social reproduction as encompassing biological reproduction, including sexual, affective, and emotional services; the unpaid production of goods and services in the home and within the community; and the reproduction of culture and ideology, which can both reinforce and challenge dominant social relations.

Social Reproduction scholars argue that the separation of formal economic production from reproduction is an essential feature of capitalism. The separation is, however, more ideological than real: capitalist forms of production necessarily rely on the devaluation of reproductive activities. Without the unpaid/partly paid labour of those who guarantee the daily reproduction of the labour-force (i.e., birthing, childrearing, educating, disciplining, feeding, cleaning, tending to the ill, providing sex work) there would be no production at all. It is exactly because this essential labour and the resources involved are devalued that capital can accumulate.

Therefore, the first reason for drawing on the 'social reproduction' tradition is that it highlights the large pool of unwaged labour on which capital relies, problematizing the analytical and political privileging of wage labour that has characterized labour movements and socialist parties for a long time. The second related reason is that it emphasizes the constitutive role of social divisions and hierarchies in processes of capital accumulation. Further, by engaging with and taking on the critique of post-colonial feminist scholars who argued that social reproduction and domestic labour are not universal categories, SR scholars have been able to conceptualize how, in addition to gender and sexuality, race, class, and geography have been used to devalue certain types of labour in order to extract value. The feminism that has taken on post-colonial and anti-capitalist perspectives speaks not only of the importance of the colonial enterprise for capitalism but also of the racial, gender, and sexual divisions that enabled the disciplining and cheapening of the reproduction of the workforce, through interventions on the body of unfree labourers and women.

Even in wealthier capitalist societies, the gains made by some of the better-paid working class were based on gender, sexual, racial, and class exclusions. In the US, throughout the 20th century, the omission of agriculture and domestic workers from the Fair Labour Standards Act – as well as the socio-economic exclusion of women through the so-called family wage, paid to husbands, and the hetero-normative eligibility criteria for social welfare – reinforced gendered and racialized divisions in the workforce. The profitability of big industrial capital also relied on commercial and financial arrangements that articulated an unequal international division of labour, granting certain economies the upper hand (or 'comparative advantage') in high value-added production. When that international division was called into question by later events – such as the oil shock of 1973–1974, which raised the international price of petroleum; the ascendancy of newly-industrialized countries like Hong Kong, Singapore, South Korea, and Taiwan; and the demands for economic redistribution enshrined in the UN Declaration of a New International Economic Order – new arrangements were put in place to re-structure global production in the interest of mobile capital and the states that competed for it.

It is therefore important for feminists and others to see the connections between: the crisis of productivity in economies in the global North; the gradual dismantling of the welfare state in the global North; and the structural adjustment policies required by international financial institutions in the global South. Structural adjustment policies imposed by the World Bank, the IMF, and other entities included privatization of government services and goods, and, importantly, the reduction of public spending. Global North international entities also demanded investment and trade agreements which enabled the fragmentation of production along chains and networks, giving ample powers to foreign investors whilst limiting those of labour, and allowing capital free rein to roam the globe while greatly limiting labour mobility. In today's globalized value chains, we can trace the close relationship between 'high tech' industries in places like Silicon Valley, and labour-intensive and low-waged production in other countries – [see entry on 'logistics and supply chains']. We can even see how, as the NGO 'Global Witness' have shown, the search for minerals and rare earths needed in computing is connected to wars which continue to be regarded as purely regional.

By tracing connections across time and space, a feminism that pays attention to capitalism and to the production/reproduction dynamic enables us to qualify mainstream claims about newness and rupture. For instance, it points to the fact that, whilst more women in wealthy capitalist societies are involved in waged work (and some reproductive labour has been commodified), wages have declined in real terms, and the state has retreated from taking on some of the costs of social reproduction. This has resulted in more, not less, unpaid labour for both women and men, calling into question liberal feminist claims about the transformative potential of women's access to the labour market. Indeed, wealthier households rely on racialized and gendered labour for their social reproduction, thereby participating in global care chains; whilst others may not be able to survive on their combined income – despite official unemployment rates being relatively low. A feminist social reproduction lens also calls into question the claim that immaterial, intangible labour has become preponderant. Material and immaterial labour have always been closely connected. Today, despite the rise of the so-called knowledge economy, many of the world's peoples perform a vast amount of material labour and production, which may be in the informal economy, but is nevertheless integrated with world capitalist production.

Here is something qualitatively different that has happened over the past fifty years, which concerns the way our everyday life is connected to finance, something that may not seem at first glance to relate to feminism, but I would argue it does. Whilst in global North societies finance has become the main locus of revenue and profitability for many firms, including non-financial ones, financial capital has also found ways to extract profits directly from social reproduction. Indeed, as the state is no longer mediating the relationship between production and reproduction according to the old welfare-state era pact between capital and labour, we find ourselves ever more dependent on finance for current and future needs (i.e. we have to pay for private healthcare, education, insurance, pensions etc.). In societies, mainly in the global South, where that pact between capital and labour never existed, the connections between reproduction and accumulation have been re-shaped through sovereign debt, structural adjustment policies, micro-credit, and other mechanisms.

Thus, the separation of production from social reproduction for the purpose of generating profits continues to be an essential feature of capitalist economies, although it is performed differently in different parts of the world. Feminist movements can expose the political work that is quietly done by separating production from social reproduction. One such mechanism is law: feminist scholars have troubled the perceived neutrality of many areas of law, including the powerful domain of international economic law, the second point in this entry.

International law has developed treaties and other powerful legal techniques of colonial and imperialist expansion, domination, and exploitation, since at least the 16th century. These techniques have relied on the construction of powerful racial hierarchies and social divisions. The early Spanish thinker of empire, Francisco Vitoria, for example, asserted the European 'natural law' right to freely trade and evangelize, by force if necessary. Although the law of the post-World War II/post-colonial period did not rely directly on the older colonial language of racial or cultural superiority, hierarchies and divisions persisted, for instance, in the way former colonies, about to become 'developing' countries, were required to abide by the superior economic rationality of those who considered themselves 'developed'. Thus, when the international economic rules of the post-war order – drafted mainly by the US and UK in the early 1940s – were disputed by 'less developed' countries on the grounds that they promoted an unjust international division of labour, their claims were dismissed on the basis of them being political rather than economic. A certain way of seeing the world and counting value was taken as equivalent to economics as such.

This particular way of claiming, organizing, and structuring knowledge about the international order – what matters counted as economic and which ones did not – was to become a powerful aspect of the design of the post-war international legal system. The way in which the international economic and legal fields are organized has changed over time. However, one continuity with this earlier period can be found in how the economic realm of the international legal system continues to be construed as more technical, neutral, and objective than the domain of 'social' policy and social regulation. This view insulates areas construed as strictly economic (i.e., trade, finance, and investment law) from claims made in areas construed as social or political (i.e., socio-economic rights, labour, and environmental regulation), preventing the latter from radically affecting the former. While rights have always been exclusionary – the fact that the separation continues to be performed is a feminist issue. Indeed, the separation between economy and society is as old (or 'modern' if one takes a decolonial angle) and arbitrary as that between male/female, reason/emotion, and nature/culture.

This point can be illustrated by way of a recent example, that of the 2007–2008 financial crisis. The crisis could have offered the possibility of bringing together economic and social regulation by acknowledging the inextricable connections between, on the one hand, production and finance, and, on the other, labour, environment, and socio-economic rights. Instead of seeing these connections, however, the discourse that prevailed was one about separations, starting with that between the so-called real economy and finance – the former seen as the healthy domain where goods and services are produced, and the latter as the realm where excessive speculation distorted valuation in the 'real' economy. Consequently, financial regulation (often limited to reviving some banking regulations dating from the Great Depression) was seen as sufficient to eliminate speculative excesses and return us to a 'healthy' productive system. But the speculation that led to the crisis grew out of financialization processes that, as seen earlier, have been re-shaping the nexus between reproduction and accumulation for the past half century. The complex financial innovations that caused the 2007–2008 financial crisis were built on new ways of profiting from social reproduction: securitized home loans, credit card loans, auto loans, and student debt, including student debt that was then sliced and diced and sold on (that is 'securitized').

Thus, by not acknowledging the important ways in which social reproductive activities contribute to, and are indeed constitutive of, both 'real' production and finance, the mainstream response to the financial crisis did nothing to conceptualize, let alone change, the ways in which production and finance are connected and work to exploit and divide people within and between countries. The result was that questions about what we produce, how we produce, and for what purpose were pushed to the background.

A feminist focus on social reproduction can therefore help us demand that states and institutions take on more of the social reproductive costs (for example by providing publicly funded childcare).

Earlier feminist scholars and activists interested in social reproduction maintained a focus on institutional arrangements as sites of struggles. The demands they placed on institutions, however, were not conceived as blueprints for actions but as experiments meant to expose contradictions in capitalist processes *of valorisation*. A key starting point was and remains the fact that capital relies on unwaged labour, much of it women's work in the household, for the production of value but cannot afford to fully pay for it. The *wages for and against housework* campaign of the 1970s, through which some feminists demanded that the state recognize domestic labour as productive for capital is one such example. It was not an attempt to get this labour accurately measured so that things could go on as usual. Instead, it aimed to reflect on these contradictions so as to generate different ways of producing and being together through collective experimentation.

Today, the demand by some feminist movements that the state provides both a self-determination income and a job guarantee not conditional on citizenship may be seen as an experiment in this tradition of thought. A self-determination income is the demand that the gains from production be shared by all, and not made conditional on nationality, age, care responsibilities, disability, or temporary unemployment; while a concomitant job guarantee is a demand that current 'production' models and jobs that are so destructive of life be replaced by life-sustaining and life-enhancing ones, starting with those in the environment and care domains. Both demands are meant to challenge the conditionalities for inclusion into wage labour, the welfare state, and citizenship of the 20th century. And it is on the refusal to abide by these conditionalities that their potential to provoke alternative valorisation processes rests.

Feminist experiments cannot assume any easy common ground; they can however start from the acknowledgement that spaces, places, and processes are deeply entangled, not in the sense of being simply connected with, but as being made through, one another. And a key dynamic in this complex web of connections is the way in which various gender, sexual, racial, class, and geographical divisions and hierarchies – including those which hide women's largely unpaid work in social reproduction – make it possible for both industrial and financial capital to reap profits around the world.

Further readings

Arruzza, Cinzia; Bhattacharya, Tithi and Fraser, Nancy. 2019. *Feminism for the 99%: A Manifesto*. London: Verso.

Davis, Angela. 1998. 'Reflections on the Black Woman's Role in the Community of Slaves' In James, Joy ed., *Angela Davis Reader*. Malden, Massachusetts: Blackwell Publishing.

Federici, Silvia. 2004. *Caliban and the Witch: Women, the Body and Primitive Accumulation*. New York: Automedia.

5
GOVERNMENTALITY AND SOCIOLEGAL STUDIES

Pat O'Malley

Introduction: from Governmentality to governmentality

The foundations of this analytic are generally recognised to have been laid in Foucault's 1978 essay 'Governmentality', a term that fuses the ideas of government and mentality and that refers initially to a certain mentality of government that emerged in 17th century Europe. This emerging way of governing centred on the view that what is to be governed is itself self-governing, and thus any act of governance must take account of the self-regulating order of things. For example, a whole 'science' of economics has developed over the past few centuries that seeks to understand the behaviour of economies in order that government may manage the economy 'properly'. Thus, the aim of government shifted away from a focus simply on command and obedience, toward regarding the central issue as the optimal harnessing of these self-governing capacities – the 'conduct of conduct' in Foucault's words. With the emergence of Governmentality, Foucault stressed, governments must always address the question of the 'right' balance between governing too much or too little, governing but not disrupting the self-governing capacities of that which it rules.

'Population' and 'economy' are two of the central 'self-governing' categories characteristic of Governmentality, whose more or less contingent emergence Foucault sketches out in his germinal paper. By the end of the 18th century, he suggests, 'the economy' had come to be understood as a natural order governed by its own laws (such as the law of supply and demand), which governments could not avoid taking into account. 'Populations' were understood to have their own characteristic rates of birth, death, increase, and decline. These were 'discovered' through the use of statistical measures and came to be taken-for-granted 'real' entities subject to the expertise of 'economists', 'sociologists', and 'demographers'. But so too, 'individuals' emerged as self-governing entities, whose capacities and potentials likewise had to be taken into account and optimised. Governmentality thus had to take account not only of these and many other self-governing entities, it also had to consider the relations between regulation of organic collective entities and the 'microphysics' of selves: the government of 'each and all' was to be one of its trademarks. In this way, not only did government become more complex as it had to govern at so many 'levels', but in the process had to recognise the limits to its own knowledge and capacities. The absolutist dream of knowing everything and thus governing perfectly by command was displaced by a more *tactical* approach – so that even the sovereign command became only one tactic among many. Likewise, the vision of government being exercised by one omniscient centre was displaced by one in which there are many centres of knowledge and many centres of government and self-government.

In order to investigate and understand this new and complex form of government, new techniques of analysis need to be developed, which give rise to the second meaning of 'Governmentality' – as that analytical approach appropriate to understanding 'Governmentality'. In this respect, Foucault's account of new forms of power emerging from the 17th century was key. Perhaps notoriously, his *Discipline and Punish* (1975) opens with the horrendous execution of the regicide Damiens in which he was literally torn apart and burned to ashes. If this account of the spectacle sears the mind, this was exactly the intent: the execution was staged in order graphically to demonstrate the might of the monarch contrasted with the powerlessness of the errant subject. In a strong sense, it worked on the audience more than the individual offender. Even as this execution proceeded, this form or 'diagram' of power – 'sovereignty' – was being shadowed by an emerging diagram that was to become even more pervasive in the centuries that followed: 'discipline'. In contrast to sovereignty, discipline was to lay hold of the subject as lightly as possible. Illustrated by Jeremy Bentham's Panopticon prison, discipline operated by establishing a norm or standard, and 'examining' the individual offender in order to assess the nature and degree of her deviation. After examination, the offender was subjected to regular but minimal corrective stimuli: it is as though force has been broken down into a myriad of small interventions as opposed to one massive one. In contrast to sovereignty, it could be effected in enclosed spaces such as the prison because the object of correction was not the crowd but the individual. Schools, universities, bureaucracies, factories – almost every modern institution – was to become shaped by this form of power. In turn, Foucault suggested that discipline is a 'technology of freedom', by which he meant several things. First, the exercise of coercion is rendered less visible. Second, discipline was directed at training the 'soul' of the subject – that is, the source of habit. Discipline shaped the subject into habits of obedience with the intended result that the exercise of coercion becomes unnecessary. In a restricted sense, governmentality uncovers an illusion of freedom: the illusion that we can ever be freed from governance, simply because we are subjected to disciplinary interventions from birth. But this is uncovering in a very specific fashion: it is not revealing concealed machinations of the powerful, or of social forces; it is laying out in a certain way the already visible operations of techniques of power.

Yet just as discipline was in the ascendancy, alongside it a second form of power was being developed, variously called 'security', 'risk', or 'government'. A classic 19th century example was the network of sanitation apparatuses including sewerage and reticulated water supplies. From the mid-1800s, a host of social workers descended on the poor to discipline them in the ways of sanitation. It was highly intrusive and had limited success for many reasons, including the cussedness of humans and the difficult conditions under which they lived. A solution was found in public health engineering: if clean water were supplied in such ways that it is more available than dirty water, and if sewerage piped away human waste more conveniently than just throwing it into the street, then the aims of public health could be achieved more effectively, less visibly, and less intrusively. This exemplified 'security' or 'risk' in that it did not seek to correct individuals, but worked on shaping the behaviour of masses or statistical distributions, and it worked largely in situ: special and highly visible institutions did not have to be built. In this way, it involved lesser degrees of domination than discipline. Later, however, the focus on statistical distributions was to take a more coercive turn. By the late 20th century, statistical techniques (often borrowed from insurance) were in place that identified 'at risk' individuals and situations. Pregnant women were advised (sometimes required) to limit alcohol and drug consumption because of risks to the foetus. The overweight have been enlisted into dietary regimes because of risks to their health.

They are expected to apply certain 'techniques of the self' to mend their ways. Smokers, drunk drivers, and a host of others become almost pariahs because of risks to others as well as themselves. In criminal justice, as Lucia Zedner warned 'pre-crime' has emerged as a major issue as those yet to offend are identified as unacceptable risks and sequestered or tagged, and former sex and violence offenders may find themselves publicly identified so that publics may be aware of the risks they represent.

The analytic of governmentality

In all this discussion, an analytical model has been outlined. It might be reduced to basics by saying first that a certain goal of government is empirically identified. We might say that government begins with the identification and definition of a problem (a process referred to as 'problematisation') and develops a preferred outcome that resolves the problem. It may be to wipe out crime in a certain area or city, or it could be to lose three kilograms. At this stage, it is an ideal, but a set of techniques is put in place that are intended to make this mentality real.

In this process of design, a critical feature of the preferred outcome is the envisaging of certain kinds of subjects: a healthy individual, a safe driver, individuals who govern their exposure to criminal predation, and so on. Together, this may be thought of as a mentality of government, or also a *diagram* of power or governance. Either term emphasises that this is a form of analysis which for the moment focuses on the plans or blueprints – not on all the failures, inconsistencies, misunderstandings, and resistances that inevitably make government messy and almost invariably incompletely achieved.

The purpose of this focus on 'ideal designs' is to clarify the overall objectives of instances of governance, what governmental regimes want to make us into and by what means. It is a form of analysis that neither makes a judgment about whether any regime is a 'good' or 'bad' arrangement, nor does it say that this is how things will, in fact, pan out when institutions and authorities of various sorts attempt to put the diagrams into effect.

The focus on 'mentality' in 'governmentality' immediately raises the old Marxist spectre of idealism, which for Marxists was a false mode of analysis focusing on the 'superficial' level of ideas rather than to the material domain. Central to the Foucaultian idea of mentality is indeed analysis of the ways of *thinking* about government – how problems and people are thought about, what solutions to problems are dreamed up, what ends are imagined as ideal outcomes. But the 'governmentality' approach is not intended to suggest that plans and ideals are either coherent or successful – only that we can only recognise or imagine them in certain ways. The analytic of governmentality in this sense is concerned with appearances – the words used to describe problems, the discourses in terms of which subjects are characterised, the categories that are used to explain policies – rather than with any assumed generative substrate such as the productive order or class interests.

However, this does not mean the governmentality studies are merely analyses of philosophies of rule. Governmental mentalities are governmental precisely in the sense that they attempt to shape the conduct of those things, events, and subjects they seek to govern. They are in this sense intensely *practical* – they imagine the world as governable: problems are construed in ways that make them amenable to practicable solutions. Accordingly, such mentalities, or 'government rationalities' (more or less consistent sets of problematics, goals, categories, subjectivities, and so on) are always linked to *technologies* for doing things.

Thus, welfare liberalism, the governmental rationality, that dominated much of the 20th century in Europe, was linked to all manner of practical technologies – such as the risk-based schemes and apparatuses of social insurance – that literally changed the world. In this process, not only is a governmental rationality distinguished from philosophy, but it is intended by its adherents to make real that which was once only thought about. Accordingly, it has been possible for research in governmentality to focus on a very wide array of practices in key institutional areas over the past two centuries, ranging from insurance, child welfare, medicine, public health, psychology, traffic regulation, social work, environmental policy, and urban government.

Governmentality, freedom and politics

Governmentality came to the fore during the period in which Marxist theory lost favour and many critical social theorists were seeking an alternative framework. The intention of Foucault and his colleagues was precisely to undo reductionist thinking that subordinated all manner of developments to mere epiphenomena or effects of some 'real' underlying meta-historic logic whose operations could only be revealed by the theory concerned. In Marxism's case, this hidden 'reality' was the forces and relations of production, which could be expected to resolve themselves into a predictable future. Against this, governmentality centred inventiveness and contingency. Mentalities of rule were systems of ideas and understandings: they could emerge and unfold in unexpected ways.

For example, few Marxist social theorists foresaw the rise to dominance in the 1970s of 'neo-liberal' programs that, rejecting welfare state logics, sought to embrace economic competition and re-establish exposure to risk. And on the other side of the political spectrum, few of the ideologues who advocated neoliberalism foresaw the ways that failures and resistance would cause their plans and programs to morph into new forms and come under the influence of strange alliances. Such an emphasis on contingency and the inventedness of governance is a methodological way not only of stressing that things needn't be as they are, but also that no theory of history can claim a right to govern because it can expose the forces of history. Governmentality interrogates the apparent inevitability both of the present and the future. It is for this reason that (to the frustration of many) governmentality does not develop its own theory and practice of government and refuses to present its own vision of the way forward.

Despite this refusal, governmentality is not merely an analytical approach. It also reflects and advances a certain kind of politics, specifically a response to the student-led revolutions of 1968 and beyond. The analytic was hostile to the kind of politics that suggested that all social ills could be transformed by one grand revolution. It cast doubt on the idea that power could be 'smashed', and liberty and egalitarianism could be installed, for example, by seizing the state or the means of production.

Even the idea that liberty or freedom was the aim of a progressive politics was rendered problematic because freedom itself was not one thing. Rather, governmentality saw freedom itself as always defined by some particular political program. What was freedom to 19th century liberals – economic independence and democratic liberalism – is not the same as it is in left-wing politics. Governmentality doesn't deny the possibility of freedoms, it assumes only that different versions of freedom are seen as produced by variable sets of techniques, subjectivities, and rationalities.

The political question that arises within governmentality studies, therefore, is not 'how should we define or achieve freedom', but how can we understand the 'freedoms' we are offered by governance in ways that help us with another question: 'should we be governed thus?'.

Coupled with freedom, central to most academic theories and political rationalities, almost inevitably is some idea of 'power'. For Marxist theory, power had always been in the possession of the dominant class, and in turn, this was mandated by the nature of the productive order. In most critical sociological theories, power likewise was possessed by some dominant group or elite. While such visions of power were ubiquitous in 1960s' critical sociologies, they all assumed that power was a thing that could be possessed and wielded, for example, in the exercise of 'social control'. Such theories (including Marxism) had great difficulties convincingly explaining how the weak scored significant victories (as in Vietnam) or how states legislated against the interests of dominant classes.

Against this zero-sum view, Foucaultians re-imagined power not as a thing to be possessed, nor as an irresistible force of history, but as a set of strategies and techniques whose success is always in doubt. That is why history is understood as contingent and the future as open. Thus both analytically and politically, for governmentality, power is not monopolized by any class or institution; it is exercised in a non-zero-sum game by a multitude of agencies through many apparatuses, institutions, and architectures. The analysis of power is thus the same as the analytic of government: power is no more than the mentalities of rule and the technologies they deploy to realise their ends.

As this implies, in governmentality, power is not to be thought of merely as constraining: power is also creative. Consequently, freedom and liberty are not to be thought of as natural states that can be restored or achieved by the stripping away of 'social control' or by smashing 'class domination'. Freedom is never the absence of constraint, but is always the product of governance, including self-governance. Freedom is always something to be invented and realised, taking its form from the diverse ways in which governmental rationalities imagine what it is to be 'free' and the ways in which they go about seeking to create it.

For governmentality, therefore, the role of analysis is not to point the way to some new 'true' programmatic politics which will liberate us all. For Foucault and his colleagues, the oppressive potential of such regimes of truth was the main lesson to be learned from the politics of Marxism. Instead, the focus is on destabilising and questioning the present by revealing its contingent formation – its non-necessity. In governmentality, emphasis would be on how that which appears as necessary is to be understood instead as assembled together out of available materials, ideas, practices, and so on, in response to a specific understanding of the nature of the problems to be solved.

In tandem with this, emphasis is placed on the actually existing understandings and constructions of the world that give rise to efforts to change it. In this view, that which appears natural is not to be taken for granted as something – like 'population' or 'the economy' – unproblematically real and just waiting to be discovered. Rather, it is to be regarded as invented, reflecting, or embodying governmental *understandings* of the way things are.

As the arbitrariness and historical contingency of many taken-for-granted categories in the present are made visible, possibilities for change emerge – the analytic gives rise to insights into how things might have been otherwise, and thus how they *could* be different in the future. Thus, while the response of many was to regard the Foucaultian project as pessimistic or nihilistic, its rationale was to invite us not to look for simplistic formulae for freedom for all, but to focus on the implications of all governing regimes – including the implications of each individual's self-government and its implications for others.

Foucault thus broke away from an array of progressive theories that to date had dominated the Left. The Marxist emphasis on class and capitalism was to be de-stabilised not only because of its claim to truth for all, but because his de-centring of power meant that destroying class and capitalism would not eliminate other oppressions. Similarly, in the 1980s, it became apparent that feminism would not eliminate racial and colonial subjection, and so on.

Governmentality's profound analytical scepticism means that it has the potential to operate as an unending critique of government. But while this may be represented as a valuable political resource, it leaves two difficult questions:

The first question is: how is it possible for governmentality analysis to operate in this way without itself claiming to speak the truth, or else, collapsing into total relativism? Part of the answer to this lies in aspects of the analytic already outlined. By focusing on the rationalities of government as outlined empirically by the regimes themselves (in manifestos, rules, manuals, speeches, and so on), it makes no truth claims of its own; it provides only a kind of intellectual grid in terms of which the truth claims, the plans, the techniques, and the subjectifications of governmental regimes are laid out.

The second question is much more problematic: Governmentality de-stabilises the present and necessarily raises the question 'should we be governed thus?' And yet it provides no directions 'forward'. Isn't this a luxury that only comfortable academics can afford? Do politically involved folks not need to act on the present with a vision of the future, in order to address inequality and oppression?

One possible answer is that the governmentality approach is not a complete package but a tool. As Foucault was fond of saying, it is part of a toolbox; it can be deployed in combination with almost any other. Thus, for example, we can point to feminist, postcolonial, and queer scholars who deploy governmentality's insights in exactly this way.

Against this, it could be countered that governmentality leads toward a form of liberal individual quietism: that its logical conclusion is an ethic of the self. Certainly, in his last few years, Foucault wrote much on questions of 'the care of the self' and face-to-face interactions, such as 'truth-telling'. But this interest does not in any way undermine acting in the broader political world in pursuit of wider causes: Foucault also engaged in struggles for prisoners' rights.

To paraphrase Foucault, what the governmentality approach does is to emphasise that all government is dangerous, but also that government is inescapable. Foucault's early work was primarily aimed at refuting or revising Marxism, but it is also a refutation of the then popular anarchist view that all political power is bad. In this light, it is not an analytic that precludes political engagement or the formation of political programs; however, it is one that will always expose its 'host' programs to reflexive critique.

Further readings

Burchell, Graham; Gordon, Colin; and Miller, Peter, eds. 1991. *The Foucault Effect: Studies in Governmentality*. Chicago: University of Chicago.

Foucault, Michel. 2007. *Security, Territory, Population: Lectures at the College de France 1977-78*. Transl. G. Burchell. New York; Picador.

Rose, Nikolas; O'Malley, Pat; and Valverde, Mariana. 2006. 'Governmentality'. *Annual Review of Law and Social Sciences* vol. 2, 83–104.

6
INDIGENOUS LAW
What non-indigenous people can learn from indigenous legal thought

Kirsten Anker

Introduction

Recently, the professors teaching Property at McGill University in Canada met with our colleague Aaron Mills, an Anishnabe legal scholar. We discussed how we might better integrate Indigenous law; not just by including Aboriginal title cases and texts on Indigenous understandings of property, but by situating Indigenous property claims as part of a broader Indigenous constitutionalism, legality or 'lifeway', as Mills calls it. This conversation tells us something about a political moment here in Canada: similar initiatives across the country have followed the 2015 call by the Indian Residential Schools Truth and Reconciliation Commission to require all students to receive instruction in Indigenous law. It also speaks to the current movement in the 'revitalisation' of Indigenous laws and the conviction that non-Indigenous folks can and should be learning about them and should be 'taking them seriously' as laws.

This chapter is written from my perspective as a scholar of European ancestry – and an emigrant from Australia to Canada – about my journey in learning from Indigenous legal thought and practice. There is a huge diversity of lifeways, languages, and legal orders amongst those who are called Indigenous peoples – the *yothu-yindi* duties of children to their mother's clan-group are specific to the Yolngu in the north of Australia, while the *Kaianereko:wa* or Great Law of Peace came from the particular history of the Haudenosaunee confederacy in Turtle Island (North America). Nevertheless, for this introduction, I will allow myself to notice certain commonalities in the ways Indigenous peoples have articulated their laws and legal thought in colonial languages.

- It is difficult to find precise equivalents of the word 'law'. The Yolgnu word '*rom*' signifies appropriate ways of being and doing in the world, while *Kaianereko:wa* has the sense of a good way or 'path' to follow. The Anishnaabe word often used for law, *inaakonigewin*, is said to mean 'a decision or judgment' about what to do. Indeed, the process of identifying Indigenous language terms with a legal lexicon should involve unpacking what is assumed in what *English* speakers mean by law and its cognate terms.
- The 'things' that 'Indigenous law' refers to are diverse. While conventional meanings for 'law' in English centre on the notion of written, positive rules, whether prescribed by a legislator, or developed by courts, Indigenous peoples frequently describe law as coming

from custom, a creator being, dreams and visions, or the land, and being transmitted through stories, songs, paintings, ceremonies, and the teachings of elders. Contemporary Indigenous communities certainly do have centralised governments that produce positive law. However, the diverse forms of Indigenous law make it difficult to assume that Indigenous law is essentially about rules, which – at least in their explicit, positive form – require language that expresses logical or causal relations in general categories. Not all languages even have the grammatical structures that permit this. I like what Cheyenne/Lakota legal scholar Sàkèj Henderson says: 'you *can* talk about Indigenous law in terms of rules, but it would be like reading the flute score of an opera'. Note that this statement also gives pause for thought about the ways in which law in common or civil law traditions are not captured by the idea of positive rules.

- One recurring theme is that Indigenous law is not strictly divided from other domains. European systems have been heavily invested in identifying 'pure' law and separating it from politics on the one hand and social fact on the other (although this is heavily critiqued), just as modern European thought generally is characterised by the atomisation of the world into discrete 'things'. In contrast, Indigenous lifeways are frequently described as holistic: the intellectual, emotional, physical, and spiritual aspects of knowing the world cannot be separated.
- Since everything in a holistic world is related and interdependent, Indigenous law is often said to be centrally about, or to emerge from, relationships. This might mean that maintaining healthy relationships is a key principle, or that legal knowledge actually derives from the pattern of relationships in which we are enmeshed.
- The 'persons' with which we are in relation may also include animals and plants, rocks and spirits. In this way, Indigenous legal thought is embedded in an ontology that is radically different to the one in which liberal legal orders operate, where human subjects dominate non-human objects.
- In a world where potentially everything is alive, or has a spirit, Indigenous peoples developed ways to become open to noticing these spirits by shifting their consciousness. This might happen through fasting, drumming, chanting, mindfulness, intense heat or pain, dreams, or psychotropic substances. Ceremonies can themselves be considered sources of law as relationship.
- Relationships are specific; they are always located somewhere. Law-producing relationships with and within the 'natural' world lead some Indigenous legal thinkers to speak of there being an Indigenous 'natural law' (not to be confused with the European natural law tradition) or of the land being the source of the law. Given that particularity, Indigenous laws may not just be 'culturally specific', but unique to their particular ecologies.
- Not all knowledge about Indigenous law is publicly available. Some legal knowledge might be restricted to certain people within the community, because it is considered sacred. There may also be resistance to divulging knowledge about law given a colonial history in which information about Indigenous peoples has been used against them, or taken out of the control of the community. Conversely, secret knowledge has sometimes been divulged strategically in the hope that governments, courts, or the wider public may be moved to listen to Indigenous claims.

Paradoxically, perhaps, these general introductory observations point to a need to approach Indigenous legal thought through specifics, and this will shape the way I write in the rest of this chapter. My points of departure – three striking features of what I have learnt so far – will be first-person experiences. Other details that I relay have either been published or were shared with me in a context that was neither confidential nor restricted, and indeed where I was encouraged to take this knowledge as simply a good way of living that I might want to adopt for myself. I am grateful to many friends, colleagues, and hosts for sharing with me. While I am sensitive to issues of cultural appropriation, and acknowledge that I may err in this regard, I have understood what has been shared with me in the spirit of a gift to be used respectfully. In particular, I use what I have learnt to notice and understand aspects of my own tradition, and to try to transform those ways of thinking that may be pathological or damaging. The discussion may seem to depart from what is recognisably 'legal' thought; this is what is entailed, however, in unpacking the lifeways of law.

More than words

In my first weeks at McGill, I was excited to attend an event with visiting elders from Saskatchewan. I remember looking forward to an intense learning experience, packed with information about Cree culture. Having read much ethnographic literature, I knew how rich and complex cross-cultural conversations could be. I sat in anticipation after the introductions had been made, waiting for the elder to speak. We remained in silence for what seemed a very, very long time. I was confused. The elder eventually said a few words, and I only remember that they seemed very general and prosaic. There might be many reasons for this; nevertheless, many of the elders I have met have been people of few words. When giving speeches about law, they often say things that might be taken as platitudes, like 'It is important to respect Mother Earth' or 'The longest journey in a person's life is the distance between their head and their heart'.

Can law be made up of such apparently simple words? I was fortunate to have Keptin Stephen Augustine – a Mi'kmaq historian and educator who was a key witness in some Aboriginal rights cases that went to the Canadian Supreme Court – as a guest in several of my classes. He presented the Mi'kmaq creation story: Gisoolg, the creator, makes Nisgam, grandfather sun, and Oositgamoo, the Earth; lightning strikes bringing the first person, Glooscap, from the Earth, who travels in all four directions learning about the rocks and water; the fish, birds, and animals; the fire; and the Great Spirit. Labelled by Keptin Augustine as a constitution, this story was – like the Canadian or Australian constitutions, I figured – fittingly general and relatively brief. However, in my tradition, there is then a great deal of ink spilled arguing about how to interpret those general statements of principle. I asked Keptin Augustine if there were a similar exegetical tradition around his creation story, or perhaps longer 'advanced' versions where the detail was elaborated. He answered that the usual version recounted within the community is actually shorter, as he had added many explanations for our benefit; further, the story is not actively dissected. On the other hand, the layers of detail, for him, came from ever deeper sensory experiences with the various elements in the story: listeners might feel the wind, and smell the herbs or tobacco burning, for instance. John Borrows, an Anishnabe legal scholar, calls this practice of paying attention to a law about how to learn: 'nuh' in the Ojibway language. If someone were to tell you 'nuh', it meant that you were to stop and take in everything that surrounds you, to feel with all parts of your body.

These reflections might make us notice – as do scholars writing about law and performance or legal aesthetics – that sensory phenomena are also important in the way European legal traditions are learnt and comprehended, although these aspects of law are frequently invisible or undervalued. Everything from courtroom design and architecture, to the look of an on-line database and the rhetoric of formal writing, shape the way law works and affects us. There is a striking difference, though, between situating law in the range of artefacts and performances that humans create and enact and situating it in the water of a lake or in a tree. Most non-Indigenous legal scholarship, whether it is formalist, positivist, critical, sociological, or humanist, takes law to be a purely human creation. However, as Anishnabe writer, Leanne Simpson, puts it, land itself can be 'pedagogy' and teach us about interdependence and reciprocity. For many, this is a kind of natural law.

Law from the land

In thinking about what it means for nature (or ecologies) to be a source of law, I suggest that there are at least three ways to conceptualise it. The first is that ecologies provide models for human behaviour. These can run from the literal to the metaphoric: when we pay close attention to the activities of plants and animals within ecosystems, we can make better decisions about the management of natural resources because we can work in synergy with the inherent organising principles of the Earth. But those behaviours might also provide analogies for the social life of humans if they represent some general insight about well-functioning systems. Learning from the land in this sense has been taken up by proponents of wild law, green legal theory, or ecological jurisprudence. While some of this work has been influenced by Indigenous legal thought, to my mind it tends to restrict learning law from the land to scientific observation and does not pay much attention to the importance of the mythological and ceremonial aspects of land-based Indigenous legal thought.

The second way of understanding the connection between law and land is that when places, plants, and animals become storied, they work as legal texts. I remember an exercise we did in my freshman psychology class where we were given a limited time to memorise a list of twenty words, typically with an imperfect recall rate. In a second round, we were asked to associate each word with a point on a route we walked frequently and to create a visual image for each association: this time I remembered each word! I imagine a more complex version of this exercise when I hear about the Songlines in Australia – the tracks traced by Ancestral Beings across the continent as they went about creating landscape features and bequeathing songs and ritual knowledge or law. The link between stories and places is reciprocal: walking the land is a memory trigger for the songs, while the songs themselves are map-like itineraries that can be used to navigate. More generally, when places record the events of mythological time or remind us of stories that include them – the formation of geological features, the invention of the sun, or the creation of species, for instance – they function as a mnemonic for those stories such that the land can be 'read'.

In turn, the stories about the land are a form of jurisprudence. Dene and Saulteaux legal scholar Val Napoleon says that 'stories are good to think with'. A research methodology she developed with Hadley Friedland uses Indigenous peoples' stories as a way into their legal order; they work with communities across Canada to revitalise laws in relation to a range of subject matters from environmental protection to child welfare. In collaboration with community elders and representatives, stories are analysed both for substantive principles, such as what kind of harms may be redressed, as well as for legal processes, such as how decisions are made and carried out. The focus is on stories as an intellectual resource for decision-making, emphasising law as a rational activity, and taking the common law case-based method of deriving legal principles as a model. For example, Alan Hanna analyses the origin story of the Tsilhqot'in people – in which the ancestor Lendix'tcux defeats a moose guarding what becomes Tsilhqot'in territory, pieces of its body turning into a variety of animals – as propounding the source of Tsilqot'in jurisdiction (in part) via the creation of reciprocal obligations through relationship. The creation of animal life was an initial gift to the animals that bound them to Lendix'tcux and his descendants in a perpetual cycle.

It would be a mistake, though, to simply take the landscapes recorded in these stories as a purely symbolic resource for human law-making. The stories themselves often point to something much richer. For instance, when the Haudenosaunee's Skywoman fell to Earth and was aided by birds and animals to survive, she performed a dance of thanksgiving that creates the Earth: it is made 'from the alchemy of all the animals' gifts coupled with her deep gratitude', as Potawatomi biologist Robin Kimmerer writes. Similarly, the stories themselves, I think, are not simply autonomous products of the human mind created for their jurisprudential utility, but have emerged out of Indigenous peoples' relations with their environments. So, the third way of thinking about law in or from the land is that law stories are told in sensuous relationship *with* the land.

Paying attention in the sensory way necessary for deep learning about the land leads to a distinctive form of participatory consciousness that, I argue, is represented in Indigenous tropes of kinship, fusion, metamorphosis, and the spirit or sentience of the world around us. These elements of the lifeways of Indigenous legal thought are particularly hard for those raised in the rationalist, disenchanted modern tradition to grasp other than as metaphoric, as I can attest. Nevertheless, it is a challenge that the current climate and ecological crises beg us to confront, as data set after data set witness the extent to which we are intimately a part of and affected by the intertwined systems for sustaining earthly life. The surging 'rights of nature' discourse may be using a blunt and inapt tool to achieve a balanced human-Earth relationship; Indigenous legal thought offers ways to approach that relationship other than in instrumental and intellectualised terms.

Ceremonies

Ceremonies formalise and intensify the practice of immersing ourselves in a sensory world. Where I live, the Haudenosaunee peoples perform the *Ohen:ton Karihwatehkwen* (the words that come before all else, or Thanksgiving Address) at sunrise or the beginning of a gathering. While the variations are endless, the *Ohen:ton* expresses gratitude to each element of creation, often with the refrain '*ehtho niiotonhonkne onkwa nikon ra*'/now let our minds be one. The first time I heard it in English, my then colleague Mike Doxtator took us on a dizzying journey from an exploding universe to the microbial beings that produce soil, via the biographies of the ancestors of each individual that led to us being present. What immediately struck me was the rapid telescoping of any human concerns or differences into minimal significance given the scale of what is addressed in gratitude. Others have similarly noted that the *Ohen:ton* is a kind of political tool that permits opposing groups to unite in what they all share: their situatedness in the web of life. As I came to hear the address regularly pronounced outside, I realised that it helps me to single out birds, wind, and insects from the background hum of noise to notice the sun hitting my retina, to appreciate the way my stomach contents connected me to plants and animals, and to feel joy in the presence of life all around me.

Other ceremonies in which I have participated offer even more intense connections to creation. The sweatlodge involves both sensory deprivation by enclosing participants in a darkened 'lodge' and sensory intensification through heated rocks and steam, which tends to loosen the body, including its control of deep emotions. In the practice learnt from his Lakota relatives, and which I underwent multiple times, Mohawk elder Tom Cook invites sweatlodge participants to focus on someone or something that is on their mind, and leads four rounds of songs to rise up different aspects of their reflections on, and wishes for, that person or thing. The constant drumbeat represents the heartbeat of Mother Earth, which underneath us itself physically provides cool relief from the steam. The rocks glow, sizzle, and crack with a kind of primal energy. Herbs scattered on the rocks find their way into our lungs and become part of us. While the sweating itself promotes physical health by releasing toxins, the aspects that foster both introspection and direct connection to a range of elements can be used for emotional and spiritual healing.

The lesson I wish to stress is that ceremonies do not just represent a principle of relatedness or thanksgiving; rather they help us to cultivate relations and gratitude. The law that is 'in' them or 'from' them is not, then, only a matter of reading them as texts, but of letting ourselves be affected by them.

Conclusion

As I see it, there are three reasons for non-Indigenous folks like me to learn about – and from – Indigenous legal thought. The first is to recognise the ways in which we have been both blind to and prejudiced against specific forms of Indigenous law, and thus the ways in which colonial legal orders suppressed Indigenous ones. Putting aside the expectation that law is enforced by centralised authority, or that is constituted by specific prescriptions for behaviour, for instance, may help create space within the larger society for the revitalisation of law in Indigenous communities.

The second is that Indigenous legal thought may offer a critique of, and alternative to, dominant ways of thinking about law. For instance, much within modern legal orders is based on a liberal worldview in which autonomous human subjects exercise their agency through contract, property, and so on, within a market model in which non-humans are exchanged for maximum profit. Learning about relational legalities has helped me see the ways in which these traits both misrepresent our interdependence and are damaging in their impacts. Focusing on 'good paths' or healthful relationships in learning about Indigenous law also makes me realise how much of my common law legal education was centred on, as Cree legal scholar Jeffery Hewitt put it to me, dysfunctional people who can't resolve their own disputes.

Third, in the specific case of Canada, our constitution (not the formal document, but the grounds of our political arrangement) derives in part from the historical treaties that colonial officials signed with Indigenous peoples. The agreement reached was understood by Indigenous parties with reference to their own legal orders, in many cases within which treaty is an affirmation of, or commitment to, our existing interrelatedness, indeed, our kinship. To fulfil my responsibilities under treaty, I need to understand the nature of that kinship as constitutional law. As many of the lessons I recounted above suggest that learning process is not so much a matter of grasping the intellectual content of a legal principle, but of engaging in actual relationships (with people, as with 'Mother Earth') so that they may inscribe a way of being lawful in me.

Further readings

Black, Christine. 2011. *The Land is the Source of the Law: A Dialogic Encounter with an Indigenous Jurisprudence.* London: Routledge.

Borrows, John. 2010. *Canada's Indigenous Constitution.* Toronto: University of Toronto Press.

Williams, Kayanesenh Paul. 2018. *Kayanerenkó:wa: The Great Law of Peace.* Winnipeg: University of Manitoba Press.

7

LIBERALISM

Mariana Valverde

The words 'liberal' and 'liberalism' are used for a variety of purposes. A US peculiarity is that people on the left wing of the political spectrum are often described as 'liberals' – whereas in the European parliament the self-described 'liberal' bloc is made up of mainstream to conservative parties that are opposed to the block made up of social democratic and Green parties. And in Australia, the main right-wing political party is called the Liberal Party.

The confusing inconsistent usage indicates that 'liberalism' is not a distinct philosophy. It is best described as an umbrella term under which many projects have historically huddled in uneasy togetherness. Here, I will begin by reflecting on the origins of the terms 'liberal', and then go on to highlight two central ideas of legal and political liberalism (as distinct from liberal economics): the social contract and the 'harm principle'. In the concluding section, I discuss the strong links between the principle that gave us freedom of expression and privacy rights (the harm principle) and the reality of European 19th century imperialism.

The use of the word 'liberal' to denote an approach to politics goes back to Spain in the 1810s. ('Liberal' as a political word first appears in English in 1815). In the aftermath of the Napoleonic invasion, a progressive group of politicians met in the Andalusian town of Cadiz to draw up a 'liberal' constitution. They opposed the Spanish branch of the Bourbon monarchy, which was waiting in the wings to seize power after the French troops were defeated (as indeed happened, to the dismay of the progressive Cadiz 'liberals').

At that time, and for decades afterwards, the term 'liberal' did not have a very specific meaning. The word vaguely indicated an opposition to absolute monarchies and to the power of the Catholic church to monopolize education and ethical thought. Liberals were inspired by the new notion of 'liberty' made popular by the American and the French revolutions. But like many of their counterparts in continental Europe, Spanish liberals knew what they were against more clearly than they knew what they were for.

Britain's 19th century was free from the revolutions and counter-revolutions that shook much of continental Europe as well as other regions, such as Latin America. Perhaps because of that relative stability, it was in the British context that the two primary meanings of 'liberalism' came to be somewhat settled. The first meaning of 'liberalism' drew on Adam Smith's late 18th century 'liberal' political economy. Liberalism in the economic sphere, mainly articulated in Britain, was a vision of a state that would facilitate private enterprise instead of strongly regulating imports, exports, and the prices of basic goods, as was routine in the 18th century. Economic liberalism was also associated with freer trade between nations.

Today, economic liberalism is not a single theory, but an umbrella term that refers to policies that favour capitalist enterprise and the vaguely defined entity known as 'the market'. Liberal economics dislikes high corporate taxes and is suspicious of strong labour unions. Traditionally, economic liberals disliked tariffs and other trade barriers: in recent years, however, some keen defenders of the unregulated marketplace, such as President Trump, have gone back to pre-liberal protectionist tariffs – which goes to show that nobody owns the copyright on the word 'liberal' even just in the economic domain.

For the purposes of this chapter, the meaning of 'liberalism' that matters most is not related to economics, but rather is connected to political and legal theory and policy. Two elements of liberal political-legal theory can be highlighted. One is the view, first propounded by Thomas Hobbes (*Leviathan* 1651) but given its most influential form by John Locke in the late 1600s that political sovereignty lies originally and fundamentally in the individual. The individual is regarded as existing by nature and as autonomous and separate by nature – contrary to many other earlier and later theories of self-hood and society. (In the 1960s, Canadian political scientist C.B. McPherson made the argument that the isolated individual that social contract theory imagines as existing pre-culturally is, in fact, a cultural product of free market societies – that argument is still influential today.)

Political theories that begin with an original isolated individual regard political formations, including governments, not as growing organically out of particular social realities, as medieval thought had it, but rather as institutions born out of rational choices made by abstract isolated individuals. In this view, the basis of political order is a horizontal agreement amongst individuals to give up a chunk of their God-given autonomy to the collective, in order to provide the conditions for market activity and to ensure basic safety and security.

Importantly, Hobbes and Locke did not claim that the 'social contract' was a historical event (or not consistently, at any rate). They believed that a proper government is one that can be understood **as if it were the product of a simultaneous agreement** amongst naturally free individuals to give up a part of their natural sovereignty, pooling those in what would then become the state. Views differed as to how much of that original sovereignty ought to be pooled. Hobbes thought government can only work if individuals give up all their rights, surrendering all their natural sovereignty, whereas John Locke argued that some rights are 'inalienable' (as the US Declaration of Independence has it) and remain with the individual. (The word 'inalienable' is a legal term meaning something cannot be sold or given.) That Lockean view is the source of legal codes that protect some individual rights against the state.

In liberal political-legal theory, the state appears as a rational solution to the chaos of 'the state of nature' – human life visualized in a pure state of pre-legal, pre-political individualism. That the 'state of nature' is 'nasty, brutish, and short', in Hobbes' famous words, is indeed a logical necessity if one assumes complete individualism. (This view is not just a highbrow oddity: many sci-fi narratives take up Hobbes' pessimistic view of human nature as inherently competitive in order to scare us about the future.)

In countries that, like the US, have Locke-style constitutions, the founding assumption is that all citizens, even all residents, have somehow agreed to live under a common government, even if we didn't sign anything: we are therefore obliged to abide by the rules even if we don't like the rules or they're oppressive. Foreigners living abroad thus pose a constant problem for social contract theory, as John Locke anticipated; so do Indigenous people living in settler societies, who feel they were never and are still not part of any overall 'social contract'.

Along with social contract assumptions, the second key building block of political-legal liberalism is the notion that even when living in organized societies with governments, individuals are entitled to a certain autonomy, a sphere of 'privacy' – the 'inalienable' rights invoked in the US Declaration of Independence.

The classic formulation of the idea that there is an invisible boundary around individuals, inside of which they should enjoy 'liberty' (here meaning individual freedom), is that of the 19th century Englishman, John Stuart Mill, in his 'harm principle'. The harm principle is the basis of much constitutional law, including the right to privacy. Mill's argument was that coercive authority should be limited, with individuals being entitled to think and act as they see fit, as long as 'harm to others' does not ensue. This is the principle courts and legislatures have used in various countries to decriminalize morally contentious behaviour such as abortion, homosexuality, prostitution, and marijuana use.

Two points need to be made about the 'harm principle'. First, as American criminal law theorist Bernard Harcourt has argued, 'harm' has come to be interpreted so broadly as to justify the very type of paternalistic laws that Mill opposed. 'Harm to society' is currently used to justify coercive measures against victimless behaviours such as panhandling or sex work. Canadian courts have justified the criminalization of pornography, for example, because even though such laws negate individual freedom of expression, pornography harms 'society', in some diffuse way.

John Stuart Mill would have said that 'harm' needs to be defined more narrowly and precisely, to target visible harm to specific others. Nevertheless, political actors always take terms and principles inherited from the past and do novel things with them. The expansion of the harm principle to encompass vague or remote harms to the social order underlines the key point of this chapter, namely that 'liberalism' is a more capacious and flexible notion than its 19th century pioneers imagined. As Harcourt's noted article 'The collapse of the harm principle' shows, the harm principle, usually thought of as a bedrock of liberalism, can in practice merge with conservatism. However, some liberals who are not as fixated on individual liberty as Mill would say that political liberalism is not about maximizing individual freedom but about establishing a balance between social order and individual freedom. As mentioned before, nobody has the copyright on the word 'liberal'.

Last but not least: it is very important, in today's 'postcolonial' world, to appreciate that Mill never saw the individual liberty that the harm principle seeks to protect as universal. As Uday Mehta and others have shown, the fact that Mill worked in the East India Company and was an active participant in debates about how to run the Empire is not incidental or contingent for liberalism's present legacy. Imperialism and colonialism were intertwined with the harm principle at the moment of the principle's birth. As Mill wrote in the introduction to the "On Liberty" collection of essays:

> It is, perhaps, hardly necessary to say that this doctrine [the harm principle] is meant to apply only to human beings in the maturity of their faculties. We are not speaking of children … For the same reasons, we may leave out of consideration those backward states of society in which the race itself may be considered in its nonage … A ruler full of the spirit of improvement is warranted in the use of any expedients that will attain an end, perhaps otherwise unattainable. Despotism is a legitimate mode of government in dealing with barbarians, provided the end be their improvement.

It is important to appreciate that Mill's imperialist and racist views were not those of the biologically oriented thinkers of the late 19th century and early 20th century. For Mill, the vast gap separating peoples and nations entitled to individual liberty from those that must be treated like children and ruled by wise despots is not a biological one but a cultural and institutional one.

Along with many others in his time (Emil Durkheim, for example), Mill believed that all human societies could be placed somewhere on a single upward line of cultural progress, from barbarian to truly civilized. The notion of multiculturalism would have shocked him, as it would have shocked Durkheim. In Victorian thinking, many of the peoples ruled by the British Empire, especially those in Africa, were seen as savages, as having never achieved any civilizational milestones. But other nations, especially in India and China, were different. They were not savage: as the noted Palestinian intellectual, Edward Said, showed in his famous book *Orientalism*, Victorians including Mill appreciated the thousands of years of written philosophy that India and China had. 'Orientals' were seen by Europeans as at the end rather than at the beginning of their collective evolution. They had once been at the leading edge of world civilization, Mill often commented, but they had become stagnant, decadent. Custom and tradition crushed creativity and innovation. Their peoples had lost the collective desire to innovate and their citizens had lost the desire for individual liberty.

The cultures lumped together under the banner 'Oriental' were represented as having fallen so far into stagnation that only a strong outside leader – a colonial ruler from a civilized nation – could shake them out of their stupor by imposing 'improvement' measures.

Thus, Mill's theory of individual liberty was thoroughly imperialist: for him, the human desire for liberty was not inborn in all humans but is the product of a civilizing process that has succeeded only in a few favoured nations. For Mill, the leaders of these few truly civilized nations had not only the right but the positive duty to 'improve' their colonial subjects. The 'white man's burden', the supposed duty of educated white men to travel from the metropole to the colonies and impose despotic measures to 'improve' the natives, was of course thoroughly racist; but the lines dividing nations and peoples were drawn mainly by means of cultural, not primarily biological, stereotypes. The idea was that at some far-off future, the colonies might become sufficiently civilized – having been 'improved' by wise colonial leaders – as to experience the yearning for liberty. However, any future liberty that might come to be enjoyed in the distant future by post-colonial subjects would (surprise!) look exactly like Englishmen's idea of liberty. No pre-colonial laws and institutions were worth saving. Progress was Eurocentrically defined.

It may seem that Mill's frankly imperialist view of why liberty was suited only for the select Northern European few, with most of the world seen as suited only for undemocratic rule by a civilized despot, is a thing of the past. However, if one looks at press coverage of political events in the global South, especially in Asia and Africa, one can find in politicians' speeches and in journalistic prose more than a few echoes of the notion that peoples in a 'backward state' should not be left to govern themselves using their own cultural and legal resources, but should be guided or even coerced into improvement – no longer by imperial rulers, perhaps, but now by the IMF, the World Bank, and mainstream NGO's.

The separation of the world into peoples who deserve liberty and peoples who are mindlessly stuck on tradition and need to be coercively improved by others did not disappear when the sun set on the British Empire. One example of this type of thinking is found in European bans on the Islamic veil or hijab – the usual rationale for such laws being that Muslims, including Muslim women themselves must have Western ideas of liberty foisted upon them from outside.

All of this suggests that the legacy of liberalism, even when leaving economics out of it and speaking only of legal and political liberalism, is an extremely mixed one. Those who today worry about the fate of what is often called 'liberal-democracy', as if it were one word, might do well to ponder the non- or even anti-democratic dimensions of many liberal ideas and projects.

Further readings

McPherson, Crawford B. 1968. *The theory of possessive individualism*. Harmondsworth: Penguin.

Mill, John S. 1859 [1975]. *On Liberty*.

Mehta, Uday. 1999. *Liberalism and Empire: A Study in Nineteenth-Century British Liberal Thought*. Chicago and London: University of Chicago.

8

POSTCOLONIAL LEGAL STUDIES

Renisa Mawani

Any discussion of postcolonial studies typically begins with the work of Edward Said. In his path-breaking book, *Orientalism*, Said claimed that colonialism was not solely an economic, political, or military project of European domination over distant lands and peoples. Rather, he argued that colonialism and imperialism were also *cultural* formations that were informed by and engendered various types of knowledge. Drawing from British and French cultural archives – including art, literature, and travel writings – Said insisted that the 'Orient' or the 'East' was not an actual place that one could identify on a map. Rather, distinctions between east and west, orient and occident were cultural and political productions of the European imagination, ones that have deeply shaped how we think of the globe, historically and in our present day. British and French colonists, travellers, bureaucrats, and intellectuals created 'the Orient' as a space of inscrutability and inferiority, a fantasy that enabled and ultimately justified the symbolic and material subjection of peoples, lands, waters, and ways of life. Crucially, for Said, orientalism shaped the colonies as much as it did European centres of imperial control. The presumed inferiority of the east informed and fortified claims of western superiority and supremacy.

To be clear, Said was not the first to write critically of colonialism or of colonial power. His predecessors include African and Caribbean intellectuals, many of who were directly involved in struggles for decolonization. For example, Amilcar Cabral, an African poet and intellectual, led decolonial movements in Guinea-Bissau and the Cape Verde Islands. Frantz Fanon, who was born in the French colony of Martinique, worked closely with the National Liberation Front (FLN) and wrote trenchantly of France's colonization and occupation of Algeria. Yet, scholars identify Said as the progenitor of postcolonial studies for several reasons. First, *Orientalism*, which was published in 1978, marks a clear intellectual break from Marxism, which inspired much of the anticolonial literature, including Fanon's *Wretched of the Earth*. In his efforts to analyse culture and knowledge as sites of colonial and imperial power, Said drew upon post-structuralism, most notably, the work of Michel Foucault that was gaining prominence at the time. Second, and in contrast to Fanon who reworked Marx's ideas through colonial struggle, Said directed his attention to the symbolic, representational, and epistemic violence of colonialism and imperialism. He was especially concerned with how European dominance over lands and territory informed western ideas of 'the east', and how these knowledges shaped violence and coercion directed at colonial subjects in British and French colonies. Importantly, Said did not use the term 'postcolonial'. Postcolonial studies was a phrase developed by others and in retrospect to identify the emergence of a new field that encompassed his scholarly contributions as well as those of his contemporaries, including Gayatri Spivak and Homi Bhabha. Following Said, the 'postcolonial' signals not a historical moment or political status but a conceptual shift in how we think of European and American colonialism and imperialism, in their economic, political, and cultural forms.

Said's writings from *Orientalism* onwards have offered alternative frameworks and new approaches for thinking about colonialism. However, many of his ideas have been the subject of critique and continue to be debated today. For Said, orient/occident, east/west were never distinct entities. Rather, they were mutually constituted and always in a relation of inferiority and superiority. Whereas the idea of the Orient was produced in British and French literature, travel writings, and bureaucracy as a space of incomprehensibility and dependency, these writings said as much about the Occident as they did about the spaces under European control. For Said, projections of the Orient as weak and easily dominated, which were widely circulated in novels, travelogues, and other forms of creative writing, helped to maintain the expansion of western power. But according to his critics, Said's emphasis on constructions of the Orient in the European imagination have obscured and even erased the voices of colonial subjects, and more crucially, their many individual and collective acts of resistance against western rule. Colonial subjects, scholars have pointed out, engaged in everyday acts of defiance and insurgency and also planned large-scale revolts against French and British administration, as evidenced in the Haitian Revolution (1790–1804) and the Indian rebellion of 1857.

In *Culture and Imperialism*, published in 1993, Said responds to his critics directly. Here, he clarifies that orient/occident and east/west are cultural productions that shaped how Europeans and also colonial subjects have viewed the world. The power of orientalism, he explains, is not to suggest that colonial subjects passively accepted domination, subjection, and/or violence. Said acknowledges that colonial rule was fiercely resisted and actively contested in the colonies and the metropoles alike. However, his larger project is to draw attention to culture and knowledge as critical domains of colonial and imperial power, an endeavour that has been taken up by socio-legal scholars in important and interesting ways.

Law and postcolonial studies

Orientalism offered a piercing critique of literature and literary studies, the disciplines and fields in which Said worked. Since its publication, many scholars have extended Said's insights to anthropology, area studies, Middle East Studies, history, and more recently sociology. Postcolonial studies have had a much slower and far more gradual influence in law and socio-legal studies. Although legal scholars were writing of law and colonialism in legal anthropology and legal history since the 1970s, few took up Said's theoretical challenges or methodological insights. It was not until the late 1990s that the contours of something called *postcolonial legal studies* began to emerge. In *Laws of the Postcolonial*, Eve Darian-Smith and Peter Fitzpatrick observe that '[e]ngagements between law and postcolonialism have been infrequent if stunning'. This 'lack of engagement is both astonishing and understandable. "Astonishing" because postcolonialism is now the main mode in which the West's relation to its "other" is critically explored, and law has been to the forefront of that very relation' (4). The infrequent attention to postcolonial studies is also understandable, they admit. Like Marxism, feminist legal studies, and critical race theory, postcolonial studies also posed a disruption to foundational ideas of law, order, and authority. Thus, to engage with Said and other postcolonial thinkers would be to challenge the very basis of law and to recognize its colonial origins and violent effects.

In the past two decades, postcolonial legal studies have become a vibrant though eclectic field. Said's insights have been taken up largely by legal theorists and legal anthropologists, and to a lesser extent, by legal historians. Despite this unevenness, socio-legal scholars have expanded and developed Said's work in important ways. Critics of *Orientalism* charged Said with being too focused on the discursive and representational and thus failing to engage with the material effects of colonialism and imperialism. But as key interventions in postcolonial legal studies demonstrate, the material and representational are not easily disconnected from one another, which is a point that Said made early on. For example, jurists and legal institutions produced cultural meanings of racial civility, savagery, superiority, and of who held legal authority and sovereignty. At the same time, specific laws were enacted to exert force on those regarded to be colonial and subjected peoples, destroying communities and cosmologies in the process. These insights – of law as knowledge and law as force – have undoubtedly influenced how we think of western law's relationship to its *others*, but also law's relationship with itself. Said ends *Orientalism* with a short discussion of the U.S. Empire. Although he lived through the 9/11 attacks on the World Trade Centre and the Pentagon and the initial and early U.S. occupation of Afghanistan and Iraq, Said's death in 2004 meant that he did not bear witness to the intensification of orientalism and Islamophobia, not only in Europe, the U.S., and the former white Dominions (Australia, New Zealand, and Canada), but also in China, Myanmar, and India. Orientalism, Said taught us, features centrally in the administration of colonial power. Drawing from these insights, postcolonial legal studies reveals how orientalism (as racism, xenophobia, and Islamophobia) shaped and continues to shape global legalities, imaginaries, and geopolitics today.

Law, identity, and territory

In *Culture and Imperialism*, to assuage his critics, Said ties the discursive and representational focus of *Orientalism* more firmly to questions of geography and territory. 'Neither imperialism nor colonialism is a simple act of accumulation and acquisition,' he writes. 'Both are supported and perhaps even impelled by impressive ideological formations that include notions that certain territories and peoples *require* and beseech domination' (9). For Said, colonialism and imperialism were always rooted in jurisdiction over land and territory. While colonialism was about occupying and settling on land that was owned by others, imperialism for Said was administration from a distance. In the first decade of the 20th century, at the high mark of British imperialism, the empire extended control over 85 per cent of the earth's surface. To accomplish this expansive territorial control, they established different types of colonies for labour, resource extraction, and settlement. Not all colonies, protectorates, or territories were considered equal, however. In the white settler Dominions, which were already occupied by Indigenous nations, Britons and Europeans made territorial claims through legal authority and political sovereignty that continue to be contested today. Canada, Australia, and New Zealand were considered to be racially and culturally most similar to the imperial centre, and were afforded considerably more political autonomy than British colonies such as Hong Kong, Malaya, or India. The supposed dependency of the colonies, most notably India, was reinforced and maintained by a plethora of orientalist writings that were philosophical, political, and legal.

In its earliest formations, postcolonial legal studies focused on questions of law, identity, and territory. Legal scholars expanded Said's formulations of the orient to problematize other continental regions of the globe – Africa, Asia, North and South America, and Australia – in which colonial subjects, Indigenous peoples and their lands and waters were appropriated by European legal control.

One question that has drawn considerable attention is how orient/occident and east/west have been invoked historically and in the present to distinguish between those peoples that supposedly respected law from those that were believed to be lawless. From the 19th century onwards, European powers produced widely circulating narratives of who had law and who did not. The British Empire prided itself on a deep investment in the rule of law. The common law, as unwritten law, was thought to be a law of and for the people and the times. Britons thus viewed themselves as an imperial power that exported law to far off places, a colonizing force that was far gentler and more civilized than its European counterparts, particularly France. More recently, the U.S. has congratulated itself on being the world's so-called promoter of human rights and democratic forms. Following the attacks on the World Trade Centre and the Pentagon on September 11, 2001, representations of law and democracy were used to justify the occupation of Iraq and Afghanistan. In the case of both the British and U.S. Empires, declarations of the rule of law have led to colonial interventions, dispossession, and violence.

To facilitate their competing aspirations of settlement, natural resource extraction, and administrative control, British colonial authorities created legal regimes of various kinds that they imposed on lands, waters, and peoples. Many of these laws were borrowed or at least inspired by colonial rule in other places, most notably India. For instance, the Indian Penal Code (IPC), which was a protracted project, became a blueprint for other written codes that were used to govern criminal activity in the Dominions and colonies. The IPC was adopted in various parts of the British Empire including the Straits Settlements and Ceylon. It also formed the basis for the Criminal Code of Canada. As socio-legal scholars have argued, criminal laws were imposed on Indigenous peoples and colonial subjects who had their own legal regimes and modes of dispute resolution that were not often recognized by British and European authorities. In India, British criminal and colonial laws were imposed alongside personal law systems but always in a vertical relationship of inferiority and superiority. In the settlement colonies of Canada and Australia, criminal laws were used to diminish the territorial and political sovereignty of Indigenous peoples.

Long before the rise of settler colonial studies (see Hunt in this volume), postcolonial legal scholars focused considerable attention on British imperial efforts to deterritorialise Indigenous peoples from their lands, resources, and waterways. Focused on legal technologies such as maps, land surveys, and conceptions of property and ownership, scholars have shown how Indigenous peoples were removed from their lands through legal regimes that did not recognize their histories, territories, or forms of legal authority. In order to produce legal claims to ownership, British colonists cleared the land of vegetation and foliage. Armed with criminal and colonial laws, they also sought to clear the land of its Indigenous nations. In efforts to address the 'Indian problem', British authorities created new legal spaces such as reserves. The reserve system in Canada became a spatial and legal form of racial containment that was exported and adapted elsewhere in the British Empire. The boundaries of law, territory, and identity that were established by British, French, and other European powers continue to be contested today, as evidenced in contemporary struggles over Indigenous sovereignty in Canada, the U.S., Australia, New Zealand, and in decolonial struggles in Palestine, Kashmir, and Hong Kong.

Colonial maritime legalities

Critics have challenged socio-legal scholars to move beyond thinking of the globe in terms of land alone and to expand conceptions of territory beyond a two-dimensional surface. European powers, including the British Empire, often struggled with extending law and order to inhospitable and rugged terrains. The instability of British legal order is evident in the unbearably hot climate of Australia, in the mountainous regions of Jamaica, British Columbia, and the Northwest Provinces (India), and in the estuaries, riverine regions, and ocean arenas that ran through and surrounded island regions. Mountains, deserts, valleys, and wetlands were often ungovernable spaces that defied legal categories and mechanisms of control. Drawing from these insights, postcolonial legal scholars have recently problematized the globe not only in terms of east/west, orient/occident, but also along the coordinates of land and sea.

Said wrote extensively about Joseph Conrad, a mariner turned novelist. Yet surprisingly, he said little about oceans and how the sea figured in Europe's pursuits for imperial control. For some European thinkers (e.g. Georg Hegel and Carl Schmitt), the orient/occident and east/west divides mapped clearly onto land and sea. Whereas Europe was associated with the sea, the orient was associated with land. European naval mastery – whether in terms of Portuguese and Spanish trade and navigation, or Dutch and later British shipbuilding – inspired new forms of law, sovereignty, violence, and territorial control. Socio-legal scholars, including postcolonial ones, have started to think with and about the sea. What Irus Braverman calls 'blue legalities' (see her chapter in this volume) and what I have termed 'oceans as method', directs attention to oceans not only as surfaces of colonial power but also as generative sites from which to think and rethink the relation between law, colonialism, and imperialism.

To be clear, colonialism and imperialism would not have been possible without the technologies of ocean-going ships and evolving techniques of maritime navigation. European powers – including the Portuguese, Spanish, and Dutch – expanded their territorial reach first at sea and then across land. It was struggles at sea, over trade, commerce, and control that resulted in the European production of new global lines. The Papal Bull issued by Pope Alexander VI in 1493 divided the globe into north/south and east/west, to distinguish the regions under Spanish and Portuguese control. Just as no land was ever uninhabited, as Said declared in *Culture and Imperialism*, oceans and seas were also never empty. Indigenous and Asian mariners travelled the sea for millennia, to fish, trade, and to pursue religious pilgrimages. Their seafaring produced new forms of law, order, and authority in the process. Thus, European maritime legal orders were imposed onto aqueous regions that were already sites of existing legal orders.

What makes oceans and islands interesting to think with is that they place into question the land biases inherent in how we think of law. Many watery places – including rivers and ocean arenas – have long resisted and even challenged legal categorization. The high seas, for example, were situated beyond the sovereign claims of empires and nation-states. In some cases, the ungovernability of the sea inspired laws directed at the legal administration of land. The Torrens land registry system is a case in point. Named after Sir Robert Richard Torrens, and used initially in South Australia, and then New Zealand, British Columbia, Malaya, and elsewhere, the system of land registry was inspired by the registration of merchant ships. The Torrens system erased prior histories of land and was therefore a key legal technology in the dispossession of Indigenous peoples. Though its maritime histories are often forgotten, the effects of these colonial land registries remain a site of intense contest and conflict today.

The representational and material forces that Said highlighted in *Orientalism* and then clarified in *Culture and Imperialism* have resurfaced in recent legal studies of ocean regions. To be sure, oceans and ships were sites of colonial and imperial legal and political power. However, they also informed counter-imaginaries and sites of resistance to the orient/occident divide, as evidenced in the figures of pirates and outlaws, and in the pursuits of Indigenous and Asian mariners.

Conclusion

A fundamental critique of postcolonialism has centred on the prefix 'post'. Scholars and activists have asked how one can invoke the term postcolonial when many so many peoples across the world continue to live under violent and oppressive conditions of colonialism, settler colonialism, and imperialism. As I have suggested above, postcolonial studies – inspired by the work of Said – is not a periodization. Rather, it offers a series of conceptual, methodological, and political interventions. These highlight the significance of textual production, the importance of close and critical reading, and the necessity to include culture, identity, and representation in any discussion of colonialism and imperialism, past or present. Socio-legal scholars have developed the field in interesting ways. Many have extended Said's insights to new geographies and territories, taking up his challenge to undo European distinctions not only between orient and occident, east and west, but also land and sea, which itself is a modern colonial invention. The ongoing challenge for postcolonial legal studies is to rethink law as a representational and material force simultaneously.

Further readings

Darian-Smith, Eve and Fitzpatrick, Peter (eds.). 1999. *Laws of the Postcolonial.* Ann Arbor: University of Michigan Press.

Fanon, Frantz. 1963. *The Wretched of the Earth.* New York: Grove Press.

Ferreira Da Silva, Denise and Harris, Mark Harris (eds.). 2017 *Postcolonialism and the Law.* First Edition. London: Routledge.

Said, Edward W. 1978. *Orientalism.* New York: Vintage Books.

Said, Edward W. 1993. *Culture and Imperialism.* New York: Vintage Books.

9
QUEER THEORY AND SOCIO-LEGAL STUDIES

Sarah Lamble

It has become almost customary to begin any introduction to queer theory by noting the impossibility of defining queer and the inherent paradox in doing so. If there is anything that encapsulates queer, it is the refusal of categorisation. Queer thought and queer practice have focused more on boundary-pushing rather than boundary-setting; contesting identities rather than establishing them; transgressing rather than institutionalising social norms. Moreover, there is no singular form or programme of queer theory; queer entails a multiplicity of perspectives, approaches, and objects of study. So, in the very act of seeking to demarcate the field of queer theory, one inevitably risks undoing the queer project itself.

Nevertheless, queer theory can arguably be recognised by its embrace of a particular repertoire of aesthetic styles, critical modes of questioning, and sexual sensibilities. The challenge in mapping out its terrain is to give sufficient character to render it graspable, without pinning it down or domesticating its insubordinate impulses.

Queer as anti-identity

Although queer is now commonly used in popular western culture as an umbrella term to refer to non-normative gender and sexual identities (i.e., lesbian, gay, bisexual, transgender, two-spirited, non-binary, etc.), queer theory and politics originally emerged from a critique of identity itself. In contrast to lesbian and gay 'rights seeking' projects, which sought to recognise and normalize homosexuality in relation to heterosexuality, and to obtain new legal rights for groups defined by their identity, queer aims to question the very idea of gender and sexual identity itself. Reclaiming 'queer' from its derogatory use as a slur, queer politics rejected strategies of normalisation and deliberately embraced a positionality of oddness, abnormality, and subversion. As David Halperin famously described, 'Queer is by definition whatever is at odds with the normal, the legitimate, the dominant … "Queer", then, demarcates not a positivity but a positionality vis-à-vis the normative.'

Resisting the more conventional assumption that identity – including the gay, lesbian, and bisexual identities of the early gay rights movement – denotes an essential truth about one's 'authentic self', queer perspectives see identity as a liminal and variable effect of broader power relations. Thus, queer theory has sought to draw attention to social processes of identity formation and constructions of identity. For example, the very categories of the 'homosexual' and the 'heterosexual', though often taken for granted, are actually a relatively recent phenomenon in modern Western culture. The emergence of these categories marked an important shift from social norms organised around sexual acts anyone might commit (e.g. sodomy) to the social, legal, and medical governance of relatively stable sexual identities (the homosexual, and later others as well). As queer scholars have revealed, the regulatory shift from policing *sexual acts* to targeting and investigating *sexual identities* has worked to produce 'normal' and 'deviant' subjects, and the corresponding enforcement of particular sexual and gender norms. It has also encouraged people to think of themselves in terms of having an 'inner truth', understood as a matter of identity.

By uncovering the ways that sexual identities emerge and become contested sites of power and regulation, queer theory has revealed the ways that identity is central to modern governance. In this sense 'queerness' is a political ethos that works to question and deconstruct identarian logics. Queer is a political positionality or stance in relation to the 'normal' or dominant, but queer thought is sceptical about all identity categorizations, even LGBT ones, and sees identity-based rights claims as limited.

This emphasis on positionality rather than identity means that queer analysis is not reduced to a focus on sexual orientation per se, but rather to a range of subject-locations and formations outside of dominant norms. Queer becomes a point of connection and affiliation for the outcast, rejected, and stigmatised – or what Cathy Cohen famously encapsulated, in the US context, as 'Punks, Bulldaggers and Welfare Queens.' In other cultures, countries, and amongst Indigenous peoples, queer expression takes particular forms that may not be recognized as 'queer' in dominant Anglo-American discourses, but nonetheless challenge gender and sexual orthodoxies. As Cohen has argued, 'If there is any truly radical potential to be found in the idea of queerness and the practice of queer politics, it would seem to be located in its ability to create a space in opposition to dominant norms, a space where transformational political work can begin'. These spaces of resistance arise within and against particular geographic, cultural, racial, and class contexts, and so queer actions and writings are context specific. There is no one-size-fits-all global queer theory.

Queer as politics

What came to be described as queer theory in academic circles initially emerged in the 1980s and 1990s in North America, particularly in conjunction with street-based activism around the AIDS crisis, rising neoliberalism and the ascendency of LGBT 'respectability' politics (such as fighting for same-sex marriage). Its early instigators were not professors, but graduate students, many of whom were active in community organising and street protests with groups like ACT-UP (AIDS Coalition to Unleash Power). In its early expressions in the academy, the queer theory that was initially most recognised tended to be written by white gay men, but its roots and traditions are broader than that, with many of its key influences coming from other radical political strands, including Marxism, feminism, and the Black radical tradition.

It has been argued that the radical gay and trans liberation movements that were manifested in earlier protest events like the 1966 Compton Cafeteria riot in San Francisco, and the 1969 Stonewall Riots in New York City, were the precursor to 'queer politics' – particularly with their multi-class, multi-gender, and multi-racial (albeit sometimes fraught) alliances. These insurgences were led by social outcasts from working class neighbourhoods – ostracised gays and lesbians, gender non-conforming street workers, prostitutes and hustlers, drug users and transsexuals – who fought back against ongoing police brutality and harassment.

Queer theory and politics partly grew out of concerns that the more radical visions of the 'gay liberation' movement were increasingly being eclipsed and assimilated into a more conservative 'equality and respectability' politics. Queer theorists and activists rejected strategies to present lesbians and gays as equivalent to 'normal' straight people and resisted respectability as a condition for securing access to housing, employment, health care, and spousal rights. Queer activists were wary of moves to simply include gays and lesbians within the existing terms of the heterosexual world and its associated social institutions (e.g. marriage, monogamy, the nuclear family, the military). Queer activists didn't seek 'acceptance' or 'tolerance' from 'straight society'; they wanted to dismantle and transform straight society itself.

For this reason, queer politics tended to embrace the provocative, the subversive, the uncomfortable; queer has deliberately taken up strategies of unsettling and disrupting. This was evident in the very names and edgy tactics of many of the organising groups that articulated a specifically queer politic such as Queer Nation; OutRage; PUSSY (Perverts Undermining State Scrutiny); Transsexual Menace; Lesbian Avengers. These groups were comprised of organisers that actively embraced gender and sexual nonconformity: drag queens, leather daddies, radical fairies, butches and femmes, transsexuals and studs. Such groups expressed a discernible queer aesthetic and marked a clear strategic departure from the more politically cautious 'homophile' organisations of the 1950s and 1960s, and the white 'professional' class of gays in the 1980s and 1990s, who in the global North sought to challenge stigma and discrimination through appeals to respectability and social approval. Queer activists were unapologetically anti-assimilationist and anti-establishment.

Queer theory in the academy likewise sought an irreverent, playful, and disobedient approach, resisting academic conventions. In many ways, queer theory marked a breakaway from the more identity-focussed 'lesbian and gay studies,' though the boundaries between the two fields remain porous and overlapping, albeit with arguably distinct aims and politics. Queer theory challenged what later became known as 'homonormative' and 'homonational' tendencies within particular strands of LGBT identity politics and mainstream culture, where, under the conditions of neoliberalism and empire, lesbian and gay identities have been deployed politically and socially in the service of conservative and nationalist projects, such as the global 'War on Terror' or pro-military campaigns in the USA.

Queer as critical ethos

While queer theory and politics cannot be reduced to any particular programme or set of goals, some common themes broadly characterise its ethos. First, queer has a broad commitment to interrogate social processes of normalisation. Queer studies owes a particular debt to Michel Foucault's work, as his understanding of power, knowledge, and subject-formation (both in relation to sexuality but also more broadly), and his critical history of how human beings came to be governed in the West through identity categories, have played a key role in analysing the ways in which sexuality, desire, and intimacy have become key sites of governance and social regulation. Queer theory has taken up questions of governance by questioning taken-for-granted norms and sensibilities relating to gender, sexuality, and desire. For example, in examining the criminalisation of homosexuality, queer theory not only questioned the particular laws which regulate same-sex sexual activity, but also exposed the ways in which wider constructions of deviance work to produce 'good' and 'bad' sexual subjects as well as 'normal' and 'abnormal' sexual practices. Queer theory and politics have tended to push back against the status quo, marking an ethos of what Michael Warner has described as 'resistance to regimes of the normal.'

Second, queer has an interest in questions of sex, desire, erotics, pleasure, and sexuality broadly conceived. Early queer theorists such as Gayle Rubin and Eve Sedgwick argued that the domains of pleasure, sexuality, and desire were not being adequately explored within existing feminist analytical frameworks of 'gender' or 'biological sex' and instead warranted study in their own right. This was party response to particular strands of feminism which were dominant at the time, which tended to neglect issues of sexuality or treat sexuality within a presumptive heterosexual framework. Such queer feminists also pushed back against tendencies within queer theory to prioritise and fetishise white gay male sexuality and culture. However, the relationship between 'feminist theory' and 'queer theory' is arguably synergistic with considerable crossover between the two. Likewise, owing to an attentiveness to questions of power relations in general, queer theory – particularly more recently – has resisted efforts to consider desire in narrow terms. Instead, queer theory has scrutinised the ways in which gender and sexual practices and expression intersect with other social relations of power, such as race, class, and disability, particularly within the fields – critical race studies, Indigenous studies, and 'crip theory'. Queer thinking has also interrogated the ways in which sexual practices and norms are woven through other social formations such as time, place, culture, capitalism, and empire. For example, queer Indigenous scholars have challenged the ways that colonial legacies continue to shape and frame how Indigenous sexual practices and gender expressions are named, understood and governed in contemporary cultures. Rather than focussing on sexuality as attached to identities per se (i.e. how LGBTQ subjects understand themselves), queer has sought to examine the ways that social and cultural norms and practices are infused with logics of sexuality and erotics of power. Queer questions both conventional understandings of sexual identity and sexual practices and considers the ways that sexuality, intimacy, and desire are often present in unlikely or unexpected places.

Third, queer has an interest in transformative social change. Though frequently characterised as a deconstructive practice that largely functions through negation (i.e. critiquing what is, rather than prescriptively saying what ought to be), queer analysis has sought to challenge the status quo not simply for the sake of critique itself, but in order to forge new possibilities, generate new worlds, and enact alternative ways of being. While tending to eschew any kind of programmatic politics which seeks to impose a new normativity over existing ones, queer theory and politics have long been interested in a politics of generative possibility and potentiality, albeit in experimental, tentative, and playful ways – with both utopian and dystopian strands. For example, by playfully invoking non-normative kinds of desire, queer politics disrupt and question dominant power relations (e.g., hierarchies between heterosexual and homosexual, the channelling of desire through scripts of monogamy, marriage, and particular kinds of family formation), and in doing so, create space for non-normative desires to emerge and proliferate. It is precisely the disruptive moment of 'queering' that is arguably most generative, as it opens up space for new possibilities that exceed established norms. In this way, queer theory does not advocate for any particular kind of desire or practice or expression, but rather for troubling normative practices in order to reshape the conditions of possibility. Hence queer investments in transformation are necessarily ongoing and incomplete. As José Esteban Muñoz writes in *Cruising Utopia*, 'We may never touch queerness, but we can feel it as the warm illumination of a horizon imbued with potentiality'.

Queer as practice/method

Queer theory does not depend on a specific domain or object of study (e.g. a focus on sexuality) and can be seen instead as an approach or *way of doing critique*. In moving away from the notion of queer as an identity and towards that of queer as a practice, queer becomes a way of doing something, a mode of asking questions, a way of seeing the world – a verb rather than an adjective or noun. To queer something is to disrupt straight logics, to view the world askance, to engage in practices of troubling. As Carla Freccaro describes, queer performs 'outlaw work.'

Although much early queer theory first emerged within humanities tradition, it can now be found in a wide range of interdisciplinary endeavours, including cultural studies, sociology, geography, history, economics, politics, law, and philosophy of science.

To queer something is to enliven it with a queer sensibility or orientation. For example, in literary contexts, a queer approach might entail a re-reading of an ostensibly 'straight' text for its 'queer sensibilities'. This might involve not so much looking for hidden gay or lesbian characters but rather identifying queer or camp subtexts, or reinterpreting relationships between characters in queer ways. It is a practice of perceiving things differently, or drawing attention to unnoticed patterns, signs or symbols, of reading alternative meanings into dominant frames. It can be a practice which *queers* the boundaries between reading and writing. For example, 'slash fiction' is a popular genre within 'fan fiction', where fans of a popular film or television show invent their own narratives through visual representations or fictional accounts. This might entail taking specific characters and creating imagined plotlines that involve straight characters developing gay romances or taking up queer roles. (Early Star Trek shows, for example, have given rise to many interesting queer interpretations and sequels). Such readings take up the role of fantasy to remake and reshape conventional narratives in queer ways.

The practice of queer can also be said to encompass certain kinds of aesthetic style which reflects its political bent. Queer is often characterised by aesthetics of disruption, contestation, and critique. Queer tends to be edgy, playful, experimental. It is at ease with the contradictory and instable, and dabbles with moments of liminality, crossover, and transgression.

Relevance for socio-legal studies

As it has been taken up within socio-legal studies, queer theory has contributed to the existing critical legal tradition. Rather than trying to simply interpret, improve, or reform the law, queer socio-legal approaches have sought to understand the broader social, political, and historical context for laws and legal norms – both as they are imagined to be and as they operate in practice. If socio-legal studies seek broadly to go beyond a 'black letter law' approach, and instead examine the role of law in society and vice versa, queer socio-legal studies have sought to extend this project to a focus on the socio-legal dimensions of sexuality, sensuality, and desire.

The queer literary tradition has been particularly fruitful in the domain of law, via the practice of interpreting laws and legal cases through a queer lens. Reading a court case queerly might entail looking for the subtle (and sometimes not so subtle) queer subtexts, or the implicit assumptions about sexuality that arise in the presentation of facts, in the analogies drawn between situations, or the metaphors used to explain legal reasoning. Queer analysis might also entail reading a case for what it *doesn't say* as much as for what it does say: seeing where desire is present within a judgement but not explicitly remarked upon; noting where issues of sexuality are conspicuously absent; identifying what silences or absences reveal about broader social and cultural understandings of a given situation or legal debate. There is also a tradition of reading seemingly 'progressive' LGBT rights decisions with a more 'queer eye'; attending to the ways in which such decisions might instantiate new norms and regulatory regimes that reinforce and privilege particular kinds of sexual practices which implicitly work to regulate and castigate others.

Queer theory as applied to socio-legal studies has been particularly attentive to the role that law plays in governing sex and sexualities, both overtly and more insidiously; and demonstrating how policing of sexual and gender identities and practices are central forms of governance and control in contemporary societies. For queer theorists, gender and sexuality is not a 'supplementary' aspect of governance nor an 'add-on' to analysis but is embedded within, and pervasive throughout, everyday legal norms and practices.

Queer theory has also been involved in work that displaces the centrality of law itself; looking at the ways in which legal regulation occurs outside of formal legal institutions such as courts, prisons, and police, and instead attends to the ways in which legal fictions and ideals shape everyday forms of governance.

Queer socio-legal analysis goes beyond the questions previously asked by gay/lesbian rights activists. For example, we might consider the legal issues that arise in relation to LGBT people in prison. A more conventional LGBT rights approach would ask: What are the experiences of LGBT people in prison and how do they differ from 'straight' prisoners? How can we protect the rights of LGBT prisoners? In what ways do current prison laws and policies discriminate against LGBT people? By contrast, a queer approach might ask: What role does gender, sexuality, intimacy, and desire play in the governance of prisons? How is gender and sexuality salient to the operation of power in prisons? In what ways do erotic economies operate in prison and with what effect? How does desire, longing, and pleasure play out in prison contexts? In what ways is heteronormativity part of the institutional life of the prison? Both sets of questions are legitimate and important, and arguably connected. Analytically, however, the first set of questions tends to treat identity – or group populations – as the focal point for analysis, and it often black-boxes and reinforces the liberal individual subject of rights that is central to modern legal systems. The latter set of questions, by contrast, tends to foreground issues of sexuality and governance, particularly in relation to institutional power, and opens space for more systemic critiques (e.g. challenging the prison itself rather than its specific treatment of LGBT people).

On the limits of queer

In the contemporary period, many would argue that 'queer' has become so widely used and so liberally deployed that it has largely lost much of its critical analytical force. Particularly as LGBT+ identities have become a new niche-market for targeted advertising and product sales, queer aesthetics have been likewise taken up in ways that commodify and commercialise its subversive edginess as a form of lifestyle consumerism. In this context, some have argued that the recent proliferation of sexual and gender identities within popular culture and the greater emphasis on 'sexual and gender diversity' is less about challenging social norms, and more about lifestyle branding and 'pink capitalism'.

In an era where queer is more often than not read as identity politics rather than a critique of identity, queer's analytical potency risks being tamed and its insurgent capacities dulled. At the same time, queer theory and politics have invariably found novel ways of pushing back and resisting against the enclosure of the status quo. So perhaps it is precisely within the current contradictory tensions and contestations which will continue to give queer its generative and critical force.

Further readings

Berlant, Lauren; Warner, Michael. 1995. 'What Does Queer Theory Teach Us about X?'. *PMLA* 110 (3): 343–349. https://doi.org/10.1632/pmla.1995.110.3.343

Cohen, Cathy J. 1997. 'Punks, Bulldaggers, and welfare queens: The radical potential of queer politics?'. *GLQ: A Journal of Lesbian & Gay Studies* 3: 437. https://doi.org/10.1215/10642684-3-4-437

Driskill, Qwo-Li; Finley, Chris; Gilley, Bria J.; et al. 2011. *Queer Indigenous Studies: Critical Interventions in Theory, Politics and Literature*. Tucson: The University of Arizona Press.

Jagose, Annamarie. 1996. *Queer Theory: An Introduction*, New York: New York University Press.

Kafer, Alison. 2013. *Feminist, Queer, Crip*, Bloomington and Indianapolis: Indiana University Press.

10
TRANSNATIONAL GOVERNANCE AND LAW
Global security and socio-legal studies

Gavin Sullivan

The last decade has witnessed a resurgence of right-wing populism and the rapid spread of nativist politics across what were once thought liberal democracies. This shift has been accompanied by strongman rhetoric denouncing the perils of 'globalism' and attacks on the multilateral institutions and norms that defined the late twentieth century international order. For Trump, and other authoritarian populists like him in Brazil, UK, Hungary, and elsewhere, 'The future belongs to patriots' ready to defend national sovereignty from 'outside' dilution and 'take back control'. International organisations (IOs) are critiqued, treaties are either neglected or abandoned, and global governance is derided in favour of self-interested state behaviour across key foreign policy areas.

Yet, despite these reassertions of sovereignty from the populist right, global political, and economic problems continue to generate new regulatory instruments and processes that enmesh national and international laws and public and private norms together in novel ways. From climate change to migration, health and financial crises to internet governance, cross-border risks and problems are being governed in ways that transversally cut across national boundaries.

The regulatory tools used to govern transnational problems often bear little resemblance to the formal laws used by states, including international law. Informal 'soft law' measures (like indicators, rankings, and best practice guidelines) and forms of private ordering (such as private contracts and terms of service, standard setting, and dispute settlement) are on the rise, whilst formal international lawmaking is in decline. The expansion of data harvesting within surveillance capitalist societies and correlated growth in algorithmic decision-making is also stimulating new information infrastructures and ways of governing through data that are reordering relations between individuals, states, and global bodies in far-reaching ways.

In the conventional Westphalian worldview, the state is the locus of both national and international law-making. But in transnational governance arrangements, the state is only one actor among many in the norm-making process and may not be the most important or, sometimes, even present at all. Key here is that 'transnational' is not another word for 'international'. Transnational law and regulation is pluralistic and often characterised by friction between multiple or conflicting regimes. It is an area where socio-legal studies can make a real difference in charting how global power is being enacted.

One area where these regulatory changes have been profoundly felt is the domain of global security. The global 'war on terror' that followed the 9/11 terrorist attacks on the US catalysed an international state of emergency marked by visible excesses of sovereign power. From the indefinite detention of 'enemy combatants' in Guantanamo Bay, the extraordinary rendition and torture of suspects in secret black sites around the globe and massive expansion of surveillance powers to 'connect the dots' of terrorist-related information, the US-led war on terror has often been represented as a draconian example of contemporary state sovereignty in action. Yet some of the most profound and enduring regulatory changes made to threats posed by groups like Al-Qaida (AQ) and the Islamic State in Iraq and the Levant (ISIL or ISIS) are transnational in nature and scope. They engage diverse ensembles of actors into novel global security governance processes and generate norms and standards that cannot readily be explained through the traditional framework of national and international law.

In this contribution, I briefly outline three key problem areas where we can observe this transnational security governance in action – (i) Terrorism Financing; (ii) Foreign Terrorist Fighters; and (iii) Terrorism and Extremism Online – before closing by highlighting a few salient points arising from this shift for the socio-legal study of transnational governance and justice struggles more generally.

Terrorist financing

After the September 11, 2001 attacks on the United States, countering the financing of terrorism became a key regulatory concern. Freezing the assets of suspected terrorists, targeting them with sanctions, and listing individuals and groups for providing support to terrorism is a strategy that can produce visible results in the 'war on terror' in ways that invading states harbouring terrorists (Afghanistan) and indefinitely interning suspects (Guantanamo) cannot. This approach was also consistent with the doctrine of pre-emption which began dominating security policymaking in this period. The US National Security Strategy (2002), for example, aimed to 'disrupt and destroy terrorist organisations at an early stage by denying … sponsorship, support and sanctuary'. Techniques of financial warfare and incapacitation were deemed critical towards this endeavour.

Terrorist financing is a paradigmatic trans-boundary collective action problem. If states fail to effectively co-operate in their efforts, there will be gaps in the transnational chain where financing of terrorism or potentially terrorist groups can flourish. Without a consistent approach, the actions of the reluctant few can undermine the governance efforts of the many and enable further terrorist attacks to take place.

The UN Security Council led the way, through the adoption of Resolution 1373 (2001), agreed within weeks of the September 11, 2001 attacks. It required all states to change their domestic legal systems by introducing laws that criminalise terrorism, terrorist financing and support, and to freeze the assets of terrorism suspects and prohibit the provision of funds to them. It also created a counterterrorism committee to monitor implementation, and required states to report steps taken in compliance with it.

Historically, the UN Security Council had adopted resolutions in response to particular and concrete threats to international peace and security that were time-limited and would expire once the threat receded. But through this resolution, and other sweeping measures adopted since, the UNSC has transformed itself into a new kind of quasi-world government – imposing binding legal obligations (or 'global legislation') on all states in ways not tied to specific conflicts that apply indefinitely by default.

UN sanctions have also been radically expanded since 9/11 to target individual terrorism suspects and groups deemed 'associated with' Al Qaida and ISIL. The UN Al Qaida and ISIL terrorist list has been a hugely controversial mechanism of global security governance. It relies on secret intelligence and an incredibly opaque nomination process by states. It fails to provide targeted individuals with an effective remedy or access to the underlying material to challenge their listing. This list is not merely a sanctioning tool of the UN Security Council. It also puts an array of transnational regulatory arrangements into motion and is the conduit for new forms of security governance to develop. The UN expert group responsible for administering this listing regime, for example, routinely make far-reaching recommendations to the Security Council to legislate on issues such as border control practices, biometric identification, and data interoperability standards to enhance implementation of the list.

Norms aimed at terrorist financing have also been translated and strengthened by a range of transnational regulatory bodies. The Financial Action Task Force (FATF) was originally set up by the G7 group of wealthy nations to combat money laundering, but after 9/11, its mandate was extended to include terrorist financing. The FATF is a powerful transnational standard-setting and policy diffusion network. The key regulatory instruments it uses include: the 40 FATF 'International Standards on Combatting Money Laundering and the Financing of Terrorism & Proliferation', the detailed interpretative guidance on how these standards should be applied, best practice guidelines, and a regular 'mutual evaluation' process (akin to peer review) to monitor compliance and identify high-risk jurisdictions. Whilst these standards are technically non-binding, non-compliance can have powerful and coercive effects. Jurisdictions identified as deficient are subjected to corrective Action Plans and publicly identified as high-risk, effectively rendering them pariah states in the global economy until they introduce regulations that meet FATF standards.

Not-for-profit and charitable organisations have been singled out by the FATF for being at particular risk of 'terrorist abuse'. This has facilitated crackdowns on human rights defenders and restrictions on NGOs and civil society organisations by repressive governments under the guise of countering terrorism. Humanitarian, peacebuilding, and development organisations now have to comply with norms aimed at terrorist financing control as condition of their funding. This has securitised their work in far-reaching ways and led to some aid agencies withdrawing from conflict zones under de facto control by listed groups (such as in Somalia). Money transfer businesses crucial for transferring remittances from diaspora communities to family members in countries like Somalia have also been severely restricted as a result of terrorist financing norms and the 'de-risking' practices of large banks. In such ways, counterterrorist financing norms are disproportionately targeting Muslim communities.

The FATF requires states to create Financial Intelligence Units for collecting and analysing 'suspicious action reports' from financial institutions, placing banks on the frontline of transnational norm implementation. This privatisation of global counterterrorism governance has spawned a massive industry in compliance and risk mitigation software. Data companies like World Check compile more than 400 sanctions and watchlists into a database for financial institutions and others to do due diligence checks to 'Know your Customer'. They also create their own listings of individuals and organisations that World Check deem to be risky or associated with terrorism, based on publicly available sources (including online news stories, court records, and blogs). Individuals removed from formal sanctions lists (like those of the UN Security Council) often remain on the World Check database and are left unable to open a bank account or work as a result. In these ways, public and private governance, and formal and informal norms work together to control risky financial flows.

Foreign terrorist fighters

After the Islamic State declared a worldwide caliphate in Syria and Iraq in 2014, around 12,000 foreign fighters from more than 80 countries travelled to join the fight with them. By 2015, this number had swollen to almost 30,000 foreign fighters from 100 countries. Since the caliphate's collapse, the key security issue has become about identifying and controlling the movements of these people and governing the threat they pose as they seek to return home. The 'foreign terrorist fighter' threat is another paradigmatic transboundary security problem: a globally diffuse risk difficult to counter using conventional tools of national and international law.

In response, the Security Council adopted resolutions that introduce sweeping measures to stem the flow of people travelling to and from conflict zones to fight, train with, or support terrorist groups. Like earlier Council measures, Resolutions 2178 (2014) and 2396 (2017) are 'global legislation' that impose binding obligations on all states to change their legal frameworks. Yet, these measures go much further by putting far-reaching transnational data infrastructure programs into motion that require new forms of information exchange, public-private collaboration, and data-driven governance.

Commercial airlines, for example, are required to provide Advance Passenger Information to states for algorithmic analysis and states must 'intensify and accelerate the exchange of operational information' across a range of different areas and formats. Systems for collecting and analysing Passenger Name Record from the aviation industry and biometric data must be developed to identify terrorists and 'foreign terrorist fighters'. Watchlists and databases of known or suspected terrorists are to be constructed for screening all travellers and conducting 'evidence-based traveller risk assessments'. All this data should be shared 'responsibly' with other states and organisations, like Interpol, and widely distributed amongst law enforcement, border security, customs, military, and intelligence agencies in ways that comply with international human rights and the rule of law.

This governance is extended by the *Global Counterterrorism Forum* (GCTF), which defines itself as an 'informal, a-political, multilateral counterterrorism platform'. The GCTF was designed by the US as an expert-led and action-orientated arrangement, structured around a number of transnational thematic working groups. It produces best practice guidelines to assist states and others to meet their security governance obligations, channels technical expertise to support technical capacity building across different areas (from watchlisting to countering violent extremism), and promotes 'flexible partnerships' between a range of public, private, and civil society actors. Despite their 'soft law' status, informal GCTF norms often shape more formal international law-making processes like those of the UN Security Council.

The 'foreign terrorist fighter' has also catalysed swathes of other informal best practice initiatives and collaborative governance programs that overlap and compete for dominance. Concerned by the clear threats that the resolutions and practices mentioned posed to the protection of human rights, in 2018 the UN Office for the High Commissioner for Human Rights produced a guidance document for states with best practices on human-rights compliant responses to the 'foreign terrorist fighter' threat. The UN Counter-Terrorism Committee similarly produced the *Madrid Guiding Principles on Foreign Terrorist Fighters*, consisting of 52 'practical tools' to help states stem the flow of FTFs. The guiding principles are non-binding, but like the GCTF memoranda that many of them build upon, they have been endorsed by the Security Council – which bolsters their legitimacy and power as transnational norms.

The *Countering Terrorist Travel Programme* (CTTP), launched in 2019 by the UN Office of Counter-Terrorism, globally coordinates capacity building to collect and analyse aviation data. It aims to share targeting rules between states, standardise protocols for transnational data exchange, and disseminate best practices for stopping terrorist travel. The CTTP works closely with a wide range of UN agencies as well the International Civil Aviation Organization, the International Air Transport Association, and Interpol. One of the main aims of the CTTP is to provide states with expertise needed to establish dedicated Passenger Information Units for the collection and algorithmic analysis of passenger data. They have also made targeting software (*goTravel*) available to states to support automated data analysis.

Together, these measures constitute an increasingly thick web of transnational norms and regulatory obligations, both formal and informal in nature. A diverse array of actors and institutions – from the Security Council and other UN agencies and organisations like Interpol and international air transport organisations, civil society organisations, security think-tanks, radicalisation experts, national states, and private bodies – are all immersed in this legal work and competing to exert influence. Building and interconnecting new information infrastructures for transnational data-driven governance to identify and stop 'risky' travellers before they board planes or cross borders is widely seen as the key way to effectively counter the 'foreign terrorist fighter' threat. This regulatory work is constructing new forms of transnational surveillance and movement control that apply to all travellers, altering the ways borders are enacted and building capacity for pre-emptive security to proliferate around the globe, with potentially grave consequences for the protection of rights.

Terrorism and extremism online

The sophisticated use of the Internet by ISIL to recruit fighters from other countries to Syria and Iraq made terrorist use of the internet and online extremism an urgent issue of global security concern. As the former UN Secretary General Ban Ki-Moon put it, 'The Internet is a prime example of how terrorists can behave in a truly transnational way; in response, States need to think and function in an equally transnational manner'. But both states and international organisations are limited in what they can do to prevent online extremism because much of the Internet's infrastructure and data is administered by the world's privately-owned Internet platforms. Social media platforms used to argue that it wasn't their responsibility to regulate online extremist content, but rather the responsibility of governments. But with such phenomena as the livestreaming of terrorist attacks, powerful states (UK, France, and Germany), regional bodies (EU), and intergovernmental organisations (G7) have pressured platforms to proactively regulate online terrorism and extremism and take Internet content moderation more seriously. As a result, online terrorism is now subject to novel transnational governance arrangements involving private platforms, states and IOs, using mostly private norms and regulatory techniques along with machine learning algorithms and other digital technologies to 'clean' the Internet.

In June 2017, for example, four of the world's largest tech companies – Microsoft, YouTube, Twitter, and Facebook – launched the *Global Internet Forum to Counter Terrorism* (GIFCT). The GIFCT was set up by the tech industry as a voluntary self-regulation initiative to disrupt the promotion of extremist propaganda online. It also administers a program of knowledge-sharing and technical collaboration between larger and smaller platforms and formulates best-practices for transnational governance in this area. The EU was threatening to regulate and fine platforms for failing to remove extremist content at the time of its inception. The GIFCT was advanced by platforms to offset this move by showing that a 'for-industry, by-industry' approach works better. But at the time of writing, the Forum is being reconstituted as an industry-funded independent organisation.

Each of the platforms involved have different policies, Terms of Service, and other private norms for defining and enforcing the removal of offending content. YouTube governs violent extremism using policies on hateful and violent content that prohibit material deemed 'shocking, sensational, or gratuitous'. In addition, it relies on the US Foreign Terrorist Organisation list and UK List of Proscribed Terrorist Groups or Organisations to remove 'content intended to recruit for terrorist organizations'. Facebook has its own 'Dangerous Individuals and Organizations' policy that regulates material from 'any organizations or individuals that proclaim a violent mission', which include far-right groups such as the English Defence League and so goes much further than any government proscription list. They also use their own internal definition of terrorism to govern terrorist content, which has been criticised as too broad and imprecise by UN human rights experts.

All of this private regulatory activity is taking place within a context where there is no internationally agreed definition of terrorism or extremism. So, each of the large platforms is effectively endorsing and exporting particular domestic definitions of terrorism around the globe, generating a private transnational legal ordering that is reshaping how terrorism and extremism should be governed.

Algorithmic governance and other digital technologies are critical in this regulatory space because of the immense global scale in which platforms have to moderate online content. 98% of the videos YouTube removes for violent extremism, for example, are detected by machine-learning algorithms. But how such algorithms associate scraps of data to infer 'terrorism' or 'extremism' remains opaque, presenting complex accountability and governance problems. When YouTube began using machine learning to detect extremist content online, thousands of videos documenting human rights violations and potential war crimes in Syria by independent news agencies and bloggers were automatically deleted. When those affected complained, YouTube said their algorithmic processes for detecting extremist content were new and they advised users to add metadata to their content so the algorithms could learn to differentiate terrorist propaganda videos from news coverage of violence more effectively. After the 2019 terrorist attack in Christchurch, New Zealand, was livestreamed on Facebook and circulated on Twitter, both platforms were roundly criticised for their ineffective response. Yet the key problem was that their algorithms for detecting terrorist content had been largely trained on the basis of how Islamist users behave, not far-right white supremacists, so platforms failed to spot it.

Another crucial governance technology used by the GIFCT to regulate online terrorism and extremism is the Hash-Sharing Database. A hash is a unique digital 'fingerprint' of an image or video file. Because the same files have the same hash, hash databases can quickly identify duplicates online and automate their removal. There are more than 200,000 unique hashes of 'known terrorist images and videos' in the GIFCT database. It was used by Facebook after the Christchurch terrorist attacks to remove 1.5 million online videos of the attacks within 24 hours. As a result, this database has been widely touted by the GIFCT platforms as an effective regulatory instrument. Yet how offending content is classified, included, shared, and removed from the database remains poorly understood. Because it removes all copies of flagged content across all platforms and jurisdictions in which the platforms operate, there is a clear risk that legitimate online expression may be indefinitely and globally deleted without any form of redress by being included in the hash database.

The GIFCT is not the only transnational governance platform active in this space. In 2019, after the Christchurch attacks, France and New Zealand convened the Christchurch Call to Action Summit. The summit, attended by numerous states and the world's leading Internet platforms, adopted the *Christchurch Call*. The Call is an informal 'action plan' in which states and platforms pledge to work collaboratively and commit to a range of voluntary rules to eliminate terrorist and extremist content from the Internet. It operates as a multi-stakeholder initiative or transnational public-private partnership, with no binding powers or enforcement mechanisms.

The Call has rapidly grown since inception and currently includes 50 states, 8 leading online service providers (including Amazon, Facebook, and Google), the Council of Europe, UNESCO, the European Commission, and an Advisory Network of more than 40 civil society organisations. As with the GIFCT, the Christchurch Call is imbricated in extending new forms of privatised governance on the Internet. This transnational governance is enacting and reshaping rights (e.g., to freedom of expression) through the technical design choices and user agreements deployed by platforms to regulate online security.

Conclusions

The landscape of contemporary transnational governance is dynamic, diffuse, and exponentially expanding. Powerful actors are increasingly opting for faster and more flexible regulatory solutions to cross-border problems than multilateral diplomacy has traditionally provided. And as new actors are enrolled into transnational norm-making processes, the scope of this governance is broadened even further. The foundational principles that have defined national and international law (e.g., 'the state', 'sovereignty', 'international authority', 'human rights', 'public/private') no longer work in the ways they once did. There are huge stakes involved in these transformations: the strengthening of global hegemonic powers, deepening economic immiseration and inequality, the erosion of rights and freedoms, the emergence of new forms of domination, and the diminishing capacity to hold the powerful accountable through legal means. Yet this reordering is also an opening for socio-legal studies to chart and critique new architectures of transnational power and thus contribute to emergent justice struggles around the world.

Many of the shifts in transnational security governance outlined above aren't being enacted through traditional formal laws but through a diverse array of other regulatory devices. These include best practice guidelines, global legislation, lists and databases, forms of standardisation, transnational data exchange networks, data infrastructures, Terms of Service, and algorithms. Empirical socio-legal scholarship is uniquely placed to follow how these devices are reassembling social, political, and legal relations. Rather than engaging in abstract debates about what law and authority *is*, socio-legal studies can redefine the terms by showing *how* transnational legal ordering and power are unfolding in practice. Informal norms (like those generated by the FATF and GCTF) are reshaping formal laws and interacting in complex ways. Platforms, 'action plans', and other flexible regulatory initiatives are increasingly being used to govern global problems instead of more traditional forms of organisation. Private actors (like banks, airlines, and Internet platforms), once reluctant followers of counterterrorism laws, are increasingly being enrolled into doing frontline security governance work. New forms of security expertise and novel assemblages of actors are empowered through these shifts and older forms of authority are recomposed, rather than replaced. And as in other regulatory fields, the prolific spread of algorithmic decision-making and predictive analytics is radically transforming the way security is done. Mapping how norms circulate and are embedded, unmasking the operations of global power and articulating the possibilities for justice is something socio-legal studies can do very well. The stakes have never been higher.

Further readings

Ali, Nathanael Tilahun. 2018. *Regulatory Counter-Terrorism: A Critical Appraisal of Proactive Global Governance.* Abingdon: Routledge.

Easterling, Keller. 2014. *Extrastatecraft: The Power of Infrastructure Space.* New York: Verso.

Rodiles, Alejandro. 2018. *Coalitions of the Willing and International Law: The Interplay between Formality and Informality*. Cambridge: Cambridge University Press.
Roele, Isobel (forthcoming). *Articulating Security: The UN and Its Infra-Law*. Cambridge: Cambridge University Press.
Sullivan, Gavin. 2020. *The Law of the List: UN Counterterrorism Sanctions and the Politics of Global Security Law*. Cambridge: Cambridge University Press
Zumbansen, Peer (ed.). 2020. *Oxford Handbook of Transnational Law*. Oxford: Oxford University Press.

PART II

Sites of engagement

11

AGRICULTURE, LAW, AND THE STATE

Matthew Canfield, Amy J. Cohen and Michael Fakhri

Agriculture is the domestication of plants and animals. Today, the most pressing political questions about agriculture are questions about the production and distribution of food. Food, in turn, is a boundary object. It is both a market commodity and a basic need, straddling the boundaries of the public and private. Contests over food thus serve as key sites for socio-legal inquiry about national and international law and institutions, for example, how states and nonstate actors construct legal rules that govern market exchange, allocate the world's natural resources, regulate safety and risk, and articulate ideals of economic and social development and the common good.

In this entry, we briefly trace how states and other actors have used legal rules and institutions to govern the provisioning of agriculture-based food and to shape national and global power since World War II. In sum, from the 1950s to around the 1980s, governments used law to stabilize and protect (select) domestic markets. From the 1980s to 2008, by contrast, powerful nations and international organizations instead prioritized a legal regime designed to maximize global trade. Since the food crisis of 2006–2008 – when the prices of staple cereals skyrocketed in response to rising energy prices and competition for agricultural land – conflicts over global agriculture have led development planners to ask how agricultural trade can serve the needs of the people who lack food security including by promoting a market-led smallholder-driven agricultural sector in the Global South. We suggest that today the increasing global integration of food supply chains alongside changes in biotechnology have produced new sources of legal and political authority and new struggles over the future of agricultural accumulation.

From the 1950s until 2008, development planners largely neglected agricultural development. Third World governments and international institutions alike shifted resources away from agriculture towards industrialization. In the 1950s, many former colonies exported a small number of primary commodities; as such, they were particularly vulnerable to fluctuating world prices in ways that followed imperial trade patterns. The question that Third World governments put to each other was whether they could redistribute global power and wealth towards the world's poorest countries by creating their own industrial sectors.

In the West, a legal regime of state intervention facilitated this effort at industrial transformation. When, after World War II, the Bretton Woods international monetary system tied world currencies to the U.S. dollar, the United States created an export subsidy when it sold wheat abroad in negotiated prices. It cleverly named this policy 'food aid'. Between 1956 and 1965, US food aid comprised a stunning one-third of world trade in wheat. Food aid enabled the United

States to protect domestic producers while cultivating Cold War allies among newly independent countries. Many of these countries were agrarian societies, self-sufficient in food grains, but increasingly imported subsidized grains to provide people cheap food and encourage the growth of industrial labour. Along with food aid, the US also encouraged these countries to adopt new 'Green Revolution' technologies – high-input practices of industrialized farming – meant to boost production and ameliorate hunger. However, in many places, the Green Revolution served to exacerbate rural inequalities, while also creating a global market for Western-based agrochemical firms.

During this period, both the Third World and First World constructed and used multilateral trade institutions to work out the relationships among national sovereignty, food security, and markets. These institutions included the International Trade Organization, the United Nations Conference on Trade and Development (UNCTAD), the Global Agreement on Trade and Tariffs (GATT) (which excluded agricultural products from trade liberalization), the Food and Agriculture Organization (FAO), and various international commodity agreements. This global legal regime facilitated the interlocking of national markets through policies of supply management that typically included production controls, price supports, and import/export subsidies. It ended, however, in crisis when, in 1972–1973, the Soviet Union purchased 30 million metric tons of grain from the US (three quarters of all grain on the world market) to feed its people. This sale, along with spikes in international oil prices, meant that the price of wheat more than tripled, compelling developing countries dependent on imports to borrow money from banks. And because in many developing economies, urbanization outpaced industrialization, the number of people who needed to purchase imported wheat outnumbered those who could afford to do so. 'Suddenly', Harriet Friedmann recounts, 'billions of people were defined as "food insecure"'.

In the 1980s and 1990s, powerful states, international organizations, and agribusiness corporations used these international institutions to reconfigure agriculture as a question of trade liberalization. Developing countries wanted a regime that made it easier for them to export their agricultural goods to developed country markets. Agribusiness corporations in the West, which had benefited from decades of government subsidies, had vastly increased in power and transnational scale; by now, they wanted a regime that prioritized exporting products from Western markets to the rest of the world. European countries which had adopted protectionist policies had amassed their own surpluses and began to compete intensely with the United States for export markets. As competition amongst more powerful countries increased, export subsidies could no longer masquerade as aid.

In 1986, the US, the European Community, and other wheat exporting countries began negotiations to include agriculture within formal legal trade agreements that culminated in 1994 with the World Trade Organization (WTO) Agreement on Agriculture. Food security, the United States argued in these negotiations, is 'best provided through a smooth-functioning world market'. After the 1970s, powerful states no longer described food security as a moral imperative to provide food for their citizens but rather reinterpreted it within a globalizing market-based framework.

A new international legal regime replaced state intervention with neoclassical economics. Countries in the West stressed technological innovation coupled with strong protection for intellectual property rights such as patented seeds and other forms of biotechnology. For developing countries, conditions imposed on loans in the 1980s and 1990s required governments to limit agricultural subsidies and increase interest rates. These 'structural adjustments' had disastrous effects on developing countries. Rather than smooth-functioning markets, the first decade of the 20th century witnessed riots in numerous countries around the world that reflected food shortages and price spikes – recurring instability precipitated not only by climate change but also by global markets that facilitated food price speculation and the effect of the global biofuel industry on food prices.

This 2006–2008 global food crisis challenged the reigning liberalizing ideology for global food provisioning. Transnational social movements, such as the *La Via Campesina* and the International Planning Committee for Food Sovereignty, demanded more democratic forms of global food and agricultural regulation. Their relentless pressure since the signing of the Agreement on Agriculture in 1994 led to the reform of the UN Committee on World Food Security (CFS) in 2008, which was transformed from a largely ineffectual intergovernmental body into an innovative, multi-actor platform that describes itself as the 'the foremost inclusive international and intergovernmental platform for all stakeholders to work together to ensure food security and nutrition for all'. The CFS has been hailed by observers as a potential exemplar for participatory global governance due to its inclusion of social movements in decision-making processes. Similarly, across the global food chain, multi-stakeholder governance arenas have proliferated. Yet questions remain about whether these new forms will democratise food chain governance and improve the livelihoods and welfare of small farmers in developing countries.

Consider, for example, 'value chain agriculture' – an increasingly dominant multi-stakeholder development model meant to alleviate rural poverty through creating profitable smallholder-driven agricultural economies in countries throughout the Global South. In 2008, the World Bank catalysed this paradigm shift when it proposed that small farmers too can learn to participate in 'the "new agriculture" of high-value products' – a marked change in its policy.

Since 2008, the emergent dominant political and legal ideal is neither state-managed agriculture nor free trade focused on the liberalization of national markets as much as a 'third way' where global value chain proponents and policymakers focus on lead firms as much as nations. They consider, for example, how agribusiness firms operate as governance bodies by setting quality and performance metrics, distributing profit and risk, and creating patterns of access and exclusion so that new sources of value are both created and shared. In this regime, critical questions of law and regulation often emerge less through either state intervention or deregulation, but rather through new voluntary or contractual codes of conduct, third-party certification schemes, and other forms of non-state private standards. Non-governmental organizations setting standards for 'good' seafood or 'good' lumber and paper products are examples of this new type of agricultural governance.

Conflicts around the production and provisioning of food have thus generated a contested legal landscape, with competing and overlapping regulatory networks, including those created by social movements and agribusiness corporations as well as by international organizations and states, all of which now seek to influence the governance of global agriculture. These networks illustrate a partial denationalisation of authority as a range of actors struggle to set rules and standards over investment, markets, and food safety. Whether these new forms of governance will prove democratic, equitable, or collaborative is a question for future empirical research. For example, one arena of transnational conflict is over the regulation of agricultural biotechnology. Disputes about the safety of genetically modified organisms (GMOs) and other technologies have grown increasingly acute as powerful global actors, including agribusiness corporations, propose them as a solution to climate change. While the United States has taken a permissive approach to biotechnology regulation, Europeans and much of the rest of the world have taken a much more precautionary approach, prompting prolonged conflicts in both the World Trade Organization and the European Union. These conflicts reveal not only how culturally contingent understandings of scientific authority shape regulation, but also the potential limits of global cooperation.

This entry has traced how conflicts over agriculture and food have long served to construct but also to challenge dominant legal rules of exchange, development, and distribution among humans. Today, these conflicts are increasingly about human relationships with the natural world. Agriculture presently contributes almost one-third of total global greenhouse gas emissions. This is a result of a century pursuing highly mechanized industrial agriculture and the spread of an increasingly meat-heavy diet. The future of agriculture and food now cannot be analysed apart from climate change and sustainable economic development. While powerful actors continue to promote industrial agricultural development, insurgent social movements have promoted ecological forms of agricultural production. Ultimately, the struggle between these competing models will be resolved through regulatory arrangements that have profound impacts for the future of transnational governance and law.

Further readings

Canfield, Matthew. 2018. 'Compromised Collaborations: Food, Fuel, and Power in Transnational Food Security Governance'. *Transnational Legal Theory* 9 (3–4): 272–287. https://doi.org/10.1080/20414005.2019.1571772

Cohen, Amy J. 2015. 'The Law and Political Economy of Contemporary Food: Some Reflections on the Local and the Small'. *Law and Contemporary Problems* 78: 101–145.

Fakhri, Michael. 2014. *Sugar and the Making of International Trade*. Cambridge and New York: Cambridge University Press.

Friedmann, Harriet. 1993. 'The Political Economy of Food: A Global Crisis'. *New Left Review* 197: 29–57.

12 ANIMALS

Irus Braverman

In March of 2017, legal scholar Steven Wise argued in front of a New York state court for the freedom of two captive clients after filing *habeas corpus* petitions on their behalf. Habeas corpus petitions are something defence lawyers often do to demand the release of their clients from detention. But Wise's clients were different. They were chimpanzees.

Wise was working with the Non-human Animals Rights Project (NhRP), which deployed legal arguments to convince courts to grant legal personhood to certain animals, arguing that their battles were the modern-day equivalent of the legal moves by which Black slaves once had to be granted legal personhood. Wise's campaign did not succeed: in June 2017, an appeal court ruled that chimpanzees 'are not legal persons who have a right to be free'.

But the NhRP's years-long efforts to use the courts to expand the scope of legal personhood have been taking place in a broader context. Indeed, a growing number of legal practitioners and scholars are increasingly concerned with the legal and ethical treatment of animals. Some of the questions they are now asking include: what is the role of non-humans, and non-human animals in particular, in the making of law? How does law make animal life? And how can we overcome the bias of modern legal systems toward humans – what is often referred to as 'anthropocentrism' and 'species-ism,' and which is tightly interwoven with other 'isms' such as racism and colonialism, and their present-day legacies?

Law and society scholarship have tended to marginalize such animal law related inquiries, or to reduce them to the field of 'animal rights' and think that they are irrelevant as such to anything that humanists might be concerned with. But I have been arguing that socio-legal scholarship, and also political activism aiming at using law to further justice, could benefit from a serious consideration of non-human life. To do this, it is necessary to move beyond the individualistic foundation of most existing animal laws in contemporary western societies to contemplate, instead, ecological traditions that consider non-human networks, populations, and ecosystems within a comprehensive land ethics that ties them all together. The term 'land ethics' was coined in 1949 by Aldo Leopold, who is widely considered the father of the modern ecology movement.

Looking at the law through the lens of the non-human animal can therefore open up fundamental questions that speak to the very meaning of being human, and also to the ethical and political concerns that emerge in the project of governing human and more-than-human life. By acknowledging the myriad ways of being in the world, more-than-human legalities can extend the advocacy-oriented scholarship on animal rights to highlight the larger question of how both animality and humanness are deeply embedded in various legal constructions, and, reciprocally, how law is acutely relevant in the construction of animals and other-than-human beings.

In the remainder of this brief chapter, I will first briefly describe the traditional approaches in animal law of animal welfare and animal rights, and then go on to present a novel, ecologically oriented approach that does not reproduce the liberal category of the individual with rights advocated by animal rights scholars, nor the utilitarian discourse of the animal welfare movement.

Animal law is taught in more than 90 law schools in the United States and is also offered at law schools in many other countries, from Portugal to New Zealand. Most of these courses are based on the animal rights paradigm, which remains inescapably anthropocentric, to the detriment of life forms that look less like humans than the chimpanzees with which this article started. Animal rights, as an approach, developed as a critique of the older animal welfare, anti-cruelty approach. Tom Regan, an influential animal rights writer, explained the difference between the welfare and the rights approaches this way: whereas welfarists advocate for larger animal cages, animal rights proponents push for dismantling the cages altogether.

Peter Singer, a leading animal ethicist, has argued that the capacity to 'enjoy life' should determine the attribution of moral value to both human and non-human animals. He would thus limit legal rights to sentient beings or, more narrowly, to beings who can be described as self-aware – a position that has been rejected as species-ist by many other animal rights advocates.

A key debate in the area of animal law has been whether animals may be considered property. Zoo animals, for instance, are said to be owned by the zoo, but some animal rights activists have questioned whether wild animals can be turned into property.

My own research on animals in different countries has explored the variability of ownership models. I have suggested that the zoo's ownership of its animals is different from the ownership of pets by private individuals, which in turn differs from the ownership of wild animals by the state, or the ownership of domesticated animals by farmers. Additionally, certain arrangements have been emerging lately that question traditional ownership regimes to allow less conventional possibilities – for instance, in the form of a 'public trust' or 'common ownership'. In such situations, the ownership is blurry and almost becomes something like stewardship. The common ownership model is prevalent in wildlife conservation practices. For example, the United States Fish and Wildlife Service owns, on behalf of the American people, certain endangered species, such as black-footed ferrets, red wolves, polar bears, and Gila trout, and the Brazilian government owns all specimens of the golden lion tamarin, wherever they may reside. Along similar lines, since the 1970s, Switzerland, Brazil, India, and Slovenia have enacted laws that place limits on the human use of animals as property. This type of legal reform has been motivated by a combination of animal welfare concerns and animal rights perspectives.

Beyond animal welfare and animal rights, new perspectives on non-humans and the law are emerging that are influenced by ecological approaches. Unlike the more traditional western legal systems, such approaches do not draw sharp lines that separate humans and their rights from the rest of nature. Indeed, environmentalists have long-criticized the animal rights movement's exclusive focus on the rights of individual animals who belong to certain species (and not others). By contrast, environmentalists are mainly concerned with biodiversity, relations among populations, and the adaptability and resiliency of ecosystems. For that reason, an ecologist or conservation biologist might be in favour of killing, what they often refer to as 'culling,' certain invasive species, a move that would be adamantly opposed by animal rights advocates and, in certain instances, also by animal welfarists.

The fundamental tensions between the individual approach and the ecosystem paradigm are exacerbated by the rapid environmental and climate changes that we see today, in an era referred to as the Anthropocene, a term coined by Paul Crutzen in 2000. The point of referring to the time since the Industrial Revolution as a new geologic era is to underline that human activity has, for the first time, caused major disturbances to global physical processes, both in the oceans and on land.

As a result, scientists, legal scholars, and philosophers alike have been stressing the importance of bridging the gap and of drawing the connections between animal rights and ecosystem protection.

The convergence of animal welfare, animal rights, and ecological perspectives has resulted in proposals for 'compassionate conservation' as well as in constitutional attempts to establish 'rights of nature' (rather than just animal rights). The argument in the constitutional context is that the traditional legal classification of the world that divides persons from owned objects no longer makes sense if we want to protect vital aspects of life on Earth. In the words of Mari Margil, from the Community Environmental Legal Defence Fund: 'to make progress in this area, we must break away from legal structures that were never intended to apply to nature, such as legal personhood, and establish a new structure that addresses what nature needs. Perhaps we can call this framework legal naturehood'.

In 2006, the first law recognizing the right of nature was enacted in Pennsylvania. Two years later, Ecuador enshrined the rights of nature – described from an Indigenous perspective as Pachamama, the Quechua word for 'Mother earth' – in its Constitution, and Bolivia soon followed suit. Some have suggested that it is not a coincidence that many of these constitutional reforms took place in countries with very strong Indigenous political and cultural movements. This also goes to highlight the important links between environment and justice.

The evolution of socio-legal thinking on animals and nature has raised fundamental questions about law itself, by exposing the anthropocentric and individualist bias of traditional legal doctrines. But while legal reasoning and legal practice continue to be overwhelmingly anthropocentric, there are some indications of change. Legal geography, for example, is a socio-legal tradition that engages space, matter, and corporeality and is therefore a particularly apt field for staking out the central concerns of more-than-human legalities. Through pursuing a posthumanist, more-than-human, multispecies account of law, socio-legal scholars could point the way out of the restrictive logic of the individualistic and human-centred liberal rights framework.

Many laws classify animals into fixed categories, such as pets, farm animals, and wild animals. Initial efforts to extend legal personhood to some of these categories have depended on their perceived similarity to humans. In effect, writs of habeas corpus were drawn up for chimpanzees and other apes, but not for crustaceans or other invertebrates, for example. Indeed, the less visible and less human-looking life forms have largely been ignored both by animal welfare and animal rights activists. This is why focusing on such forms of life could present important challenges to conventional legal thinking about human-animal relations. Along these lines, legal scholars who work in the ocean context have been arguing that legal thinking has been far too land-centric, and that law itself might be re-imagined from the standpoint of the ecologically crucial oceans – an intellectual change that is occurring in many disciplines and has often been referred to outside of law as 'the blue turn.'

I have been arguing that it's time for a blue revolution in law, too.

The move towards ocean legalities or blue legalities is in keeping with the ecological perspectives that are arguably broader and more inclusive than traditional animal welfare and animal rights approaches. This move will simultaneously lead to recognition that we live in a mixed and messy multispecies society, and will urge us to make way for an animal turn – alongside a broader non-human turn – in law.

Further readings

Bradshaw, Karen. 2020. *Wildlife as Property Owners: A New Conception of Animal Rights*. Chicago: The University of Chicago Press.

Braverman, Irus. 2015. *Wild Life: The Institution of Nature*. Stanford: Stanford University Press.

Braverman, Irus. 2013. *Zooland: The Institution of Captivity*. Stanford. Stanford University Press.

Braverman, Irus and Johnson, Elizabeth R. (eds.). 2020. *Blue Legalities: The Laws and Life of the Sea*. Durham: Duke University Press.

Deckha, Maneesha. 2020. *Animals as Legal Beings: Contesting Anthropocentric Legal Orders*. Toronto: University of Toronto Press.

13

ARTIFICIAL INTELLIGENCE AND PUBLIC LAW

Jacob Livingston Slosser

There is a common – and all too pervasive – misapprehension that law is fundamentally on the backend of progress. The idea is that without some major efforts at reform and constant legislative action, the law will continually be catching up to social and technological innovation. There are, of course, many circumstances where this is overwhelmingly true (particularly in the realm of changes in social attitudes), but to ascribe an inability to adapt to new circumstances to the law's long-established nature is a mistake that risks passing over legal remedies that are well understood and thoroughly developed. Furthermore, this deceptive assumption risks compartmentalizing any remedies aimed at a new technology into its own fragmented discipline (law and technology) and isolating them from well-established principles of law. There is no better example of this fear getting the better of legal thought and legal research than the case of Artificial Intelligence (AI).

A ubiquitous consequence of this fear is the all too common conflation of the many uniquely tooled versions of AI into a single risk. Though this entry cannot cover all these tools in detail, it will avoid throwing anything to do with Artificial Intelligence into the same bag by distinguishing between the different *fears/risks* of AI and AI's central *functional definitions*.

Functional definitions differ from purely academic definitions in that they place the importance on the differences in practical functions, rather than in an abstract typology that would differentiate by dissecting the meaning of 'intelligence', for instance. This short entry will take for granted that many of the applications of AI are not truly 'intelligent' in the same way that might be understood in popular science publications, news articles, or philosophical discussions, nor need they be to be effectively and lawfully introduced.

Part I briefly covers these definitions while Part II uses the right to explanation as an example of how AI confronts the law and presents ideas towards their reconciliation. This short entry thus presents a foundational argument for public law principles to be championed for their ability to foster AI implementation, whilst protecting hard won fights for equality, fairness, and justice. It argues that far from being antithetical to a human-centred ethics, the introduction of different AI systems can serve to reinforce those ethics through established law.

Part I: what do we talk about when we talk about AI?

For the purposes of this entry, we will have to simplify what we mean by AI. In this sense, machine learning, deep learning, and similar terms will be reduced to a set of processes that we will conflate into the placeholder of AI.

AI, in its simplest form, is a classification task. It takes a dataset as input and classifies it in a way to make decisions or determine an outcome. For instance, imagine a government body deciding on whether or not to pay an unemployment benefit to an applicant. There will be a number of binary (true or false) sub-decisions to be made that lead to a final outcome. The order in which these sub-decisions are made and the relations between those sub-decisions is what makes the process *algorithmic* – in that it follows a set of repeatable instructions to a result. This is not a quality limited to machines. The simplest form of this is what is called a rule-based or expert system, where the rules determining the outcome of the sub-decisions are provided by experts in that field; in our example, all of the criteria that lead to an assessment of a person's eligibility for unemployment benefits. This again doesn't differ significantly between a machine application and a human administrator. Those sub-decisions might include relevant factors like: time spent unemployed, sector of work, average income, number of dependents, marital status, etc., each with a relationship to a final outcome based on rules as defined by, in our case, statutes and regulations and employment lawyers.

Where things get more complicated (and contentious), is where these rules are not provided by experts but computed by an algorithm from a set of previously existing examples (e.g. previous applicants for an unemployment benefit), called a training set. In one type of algorithm, this training set is assumed to be innately correct as it is classified. The algorithm then uses the examples to compute a set of rules that can reconstruct this classification to a high degree of precision. There are hundreds of different types of algorithms, prescribing how to best determine features from the example data and how they determine the outcome. Among the most well-known and often used are: decision trees, random forests, and neural networks, each with their significant benefits and drawbacks, covered generally in section II.

As there are a number of different approaches to AI in general, as well as in law, we can also make a distinction between the *internal legal applications* (those that are applied within the community of legal professionals) and *external applications* of AI (those that either apply the law or that invoke public law principles). Within the law, there are the applications that make it easier or more efficient for a lawyer to research and argue a case (e.g. automated text analytics, predicting legal outcomes, or modelling legal argumentation), a judge (e.g. assessing recidivism risks of potential parolees, suggestions for sentencing), or perhaps a legislator (automated statute design). Then, there is the use of AI in the application of law outside the legal profession but still within fields covered by public law: a police officer provided with predictive information of crime activity in different parts of a city; a city employee allocating resources for the care of homeless; or our example of an administrator deciding on unemployment benefits. In all of these settings, there exists a spectrum for how integrated the AI is in a decision-making process. We can characterize this spectrum as going from the fully human to the fully automated. The fully automated end, though not realized yet, is not far off. Most applications of AI discussed here, being used now and likely to be commercialized in the near future, are somewhere in the middle, as hybrid decision-making process.

Given this hybridity, the main challenge for AI lies in the gap between human and machine understanding. In some forms of AI (and often the better performing and therefore more widely appropriated form), there is not only a question about the correctness of the data from which an algorithm trains, but also a lack (or sometimes an impossibility) of understanding of how an AI determined an outcome. This is the dreaded and oft-mentioned 'black box' problem. In some applications of AI, it is impossible to understand the 'why' of outcomes. These leads to (at least) three prominent fears.

The first is about control, responsibility, and agency. If a judge is given a score by a computer of 8 and 2 for two defendants, regarding their risk to society, it is not all that likely that the judge will give the riskier defendant a lesser sentence. The fear here is that even in a hybrid system the judge (or other decision-maker) is merely a rubber stamp for a machine that hasn't been designed or regulated as a fully autonomous decision-maker. The second fear concerns dignity. This is the fear that as the spectrum swings farther to the side of automation, human experience is reduced to a dataset rather than lived experience and subjective voice. The third fear, and one currently reflected in practice, is the fear of bad data. This is not just the fear that data is incorrect but, like the first fear, even correct data can be used in a fallacious and discriminatory manner. This has occurred all over the world, from police departments to google searches to differential pricing to courtroom sentencing. Not only can these algorithms develop rules that identify proxies for groups legally protected from discrimination, but when used, the algorithms can reinforce those same proxy associations, hence worsening the outcomes for those already marginalized on the basis of race, sex, gender, religion, among others.

Part II: explanation

These fears have led to numerous calls from around the world for ethical guidelines for data transparency, new algorithm-specific legislation, the outlawing of black box algorithms, or even the outlawing of automation itself. Many of these guidelines call for a transparency that hopes to address the concern of *why* a decision has come out the way it has. Did the judge give the 'riskier' defendant a harsher sentence just because the algorithm said so? Was the score transparent to the defendant? Did they have a right to challenge it? Was the score based on erroneous or discriminatory data?

This want of an answer is enshrined in public law in the form of a general right to explanation of a decision, sometimes called the duty to give reasons. The judge in her sentencing of a criminal, or the administrator in their decision about an applicant for benefits, are bound by a legal duty to explain the reasons that were weighed to produce their result. Jurisdictions vary in how and in what granularity this is expressed. This is a long existing and well-established principle. The 'why' question has existed long before algorithms. And yet, we do not require intimate knowledge of the neurons in a judge, nor the psychological history of an administrator, if looking for explanations of human decisions. Explanation is not different now that it is algorithmic. The benefit of using established legal principles is that their meanings have been dissected and thoroughly applied in case law.

The general right to explanation exists in almost every jurisdiction in Europe, nationally and supra-nationally in the form of the European Convention of Human Rights and the General Data Protection Regulation (though not always *explicitly* stated). Other countries have similar norms, though often not enshrined in statutes. Overall, there is a long-standing tradition of explanation as a core function of legitimacy and due process in liberal democratic regimes. In general, explanation should require: the legislative grounds on which a decision rests, the legally relevant factors of the case, and the connections between those facts, i.e., the discretionary or interpretive elements in weighing the relevant factors. It is this third element that is often regarded as the major infringement of black box systems. However, as it is now, at no point would a decision-making body be required to describe the decision at the level of granularity at which a black box system obfuscates this information. In no way is a human decision-maker required to account for all the inputs that may have been explicitly or implicitly used to consider the application in all its biological, social, and psychological complexity. Yet, there remains this anxiety.

One constant call from ethicists is to keep a human in the loop. This makes sense, but only if implemented in a specific way. The mere presence of a human in a system does nothing to protect against the fears listed above. Both humans and machines are after the same goal: to classify new input into an already understood category and produce explainable and consistent results. Learning from historical cases and reproducing their language in new cases by connecting legal outcomes to given fact descriptions is not far from what human decision-makers would do anyway. Whenever a caseworker/judge/etc. is attending to a new case, he or she will seek out former cases of the same kind to use as a compass to indicate how the new case should be decided. The difference between the human and the algorithm is that algorithms tend to be more rigorous than humans. Humans respond more organically to past cases because they have a broader horizon of understanding: they contextualize their task to a much richer extent and can therefore adjust their decisions to a broader spectrum of facts. Algorithms, on the other hand, operate without such context and can only relate to explicit texts. Hence, they cannot evolve in the same way. This limited contextualization of algorithmic 'reasoning' will create a problem if all new decisions are drafted on the basis of an algorithm that reproduces the past and if those drafts are only subjected to minor or no changes by the human collaborator. To avoid this and protect against the dangers of bad data, the suggestion is to build-in a type of Turing test for AI decision-making. Unlike the original Turing test, concerned with discerning the intelligence of a machine by reference to human intelligence, this test would be focused on outcomes of good (i.e. legal) explanation.

This could be done in the following way: a certain percentage of the entire case load – say 10% – could be given both to a human and to an algorithm. Both the human and the algorithm would produce a decision draft for the same case. Both drafts would be sent to a human judge. In this set-up, the human judge would not know which draft came from the algorithm and which came from the caseworker (and formats for issuing drafts could be formalized so as to reduce the possibility of guessing merely by recognizing the style of the drafter's language), but would simply proceed to finalize the decision based on which draft was most convincing for deciding the case and providing a satisfactory explanation. This final decision would then be fed back to the machine-learning algorithm – for fresh learning. The Turing model keeps a "human in the loop" and does so in a way that is systematic and meaningful because it takes a specific form: they are built around the idea of human-AI collaboration, safeguarding against the fear of a non-human-centred decision-making process.

Conclusion

It is not to say that ethical restraint isn't prudent or that the introduction of automation should be given free rein. However, such an approach as outlined, will be optimal for providing working conditions in which AI, in the long-term perspective can grow into a means for assuring *better* detection of hidden biases and other bureaucratic deficiencies than the system does now. This approach may help allay the fears of the black box and allows for a greater scope of review of rubber stamp occurrences by being able to compare differences in pure human and pure machine decisions by a human arbiter. The opponents of automation might envision that we are sacrificing hard won ethical principles for the sake of the 'progress' of automation. This is not a misguided fear, but it is also not a new one. The upside of detailed hybrid-AI suggests that the fight against discrimination, bias, and inequality and the aim of justice might be humanity's best bet to refine an already non-transparent human system. As our world continually shows us, it is likely easier to fix code than it is to change someone's mind.

Acknowledgements

This work was produced with the support of the Carlsberg Foundation Postdoctoral Fellowship in Denmark project CoLLage: Code, Law and Language, grant number: CF18-0481 and the Independent Research Fund Denmark project PACTA: Public Administration and Computational Transparency in Algorithms, grant number: 8091-00025B.

Further readings

Eubanks, Virginia. 2018. *Automating Inequality: How High-Tech Tools Profile, Police, and Punish the Poor.* New York, Picador: St Martin's Press.

O'Neil, Cathy. 2017. *Weapons of Math Destruction: How Big Data Increases Inequality and Threatens Democracy.* New York: Broadway Books.

Palmer Olsen, Henrik; Livingston Slosser; Jacob; Troels Hildebrandt, Thomas. *Forthcoming.* 'What's in the Box? The Legal Requirement of Explainability in Computationally Aided Decision-Making in Public Administration', working paper version available at: http://ssrn.com/abstract=3402974.

14
CAPITALISM AND CAPITAL

Bryant G. Garth

Law and Society as a field and set of scholarly practices does not typically address abstract terms such as capital and capitalism. Both terms have an accumulated history, a politics, and a developed and engaged scholarly literature. To make this brief essay concrete, I will ask what a Law and Society perspective might bring to these abstract topics.

This essay will make five basic points. First, it will suggest that studying lawyer/agent/brokers is essential to see the role of law in relation to capitalism (and other societal changes). Second, it will suggest that law follows and legitimates power through societal changes, including the rise and spread of capitalism in Europe and into imperial relationships. Third, law and lawyers work with persons in power under capitalism to legitimate and work with the basic idea that individual merit determines success under capitalism, which justifies the great inequalities that modern capitalism produces. Fourth, there are still major differences in law, lawyers, empire, and capitalism among different states, with important consequences. For present purposes, one can compare the 'Rhine capitalism' associated with France and Germany on the European Continent with 'Atlantic Capitalism' associated with the United Kingdom and especially the United States. One could add to the comparison what may be emerging as Chinese capitalism, but I will not try to do so here. Of course, the developments of each 'model' relate to interconnected histories and developments elsewhere. Finally, the essay will discuss how lawyers reveal the forms of capital valued in increasingly unequal capitalist societies and invent new forms of economic capital that contribute to further inequality.

Markets and ideology

Without trying to produce a precise definition, capitalism is generally characterized as a dynamic form of social organization traced to the rise of economic markets, relatively 'free competition' in goods and services, and wage labour – all replacing relatively stable feudal relationships. One key to capitalism is the supporting ideology that the competition is open to all on the basis of individual talent and drive. Great Britain was the first home of identifiable capitalism. The rise of the European empires from the 15th century into the 20th century, including especially the British empire, is closely associated with the rise of capitalism.

Lawyers and capitalism

The relationship between law and capitalism can be examined by concretely looking at the activities of lawyers. That does not mean studying the 'legal profession' as such. Instead, this approach focuses on lawyers as agents/brokers who link capitalism, law, and existing power structures. As with respect to other social phenomena, lawyers are important less as initiators of major social and economic changes than as agents who shape institutions in ways that build the importance of law and lawyers. The historical rise of capitalism is one example.

A focus on lawyers as brokers reveal the stories of continuity neglected in accounts of the rise of capitalism. The 'break' from feudalism to capitalism was more complex than frequently posited. As recently noted by Jodhka, Rehbein, and Souza, 'The capitalist transformation is a real revolution but a revolution is not a creation out of nothing. It entails socio-economic mobility, separates social structure from the division of labour and creates a whole new range of professions for all social groups. *But it does not abolish older inequalities; it only transforms them and makes them invisible*'. This observation applies both to Europe and to the European colonies, including the United States. An examination of the role of lawyers provides insight on how elites could adapt and profit under a new and dynamic social order.

The legal profession grew up by offering highly sophisticated legal arguments that could be used to serve different individuals, interests, and groups, including commercial groups, and that also could provide technical and seemingly neutral solutions that took into account hierarchies of power. As Kantoricz famously observed (1997), lawyers provide neutral rules that both legitimate and serve power. The perceived neutrality of rules produced by judges, bureaucrats, lawyers, and legal scholars – even if tilted to power – means that the powerful will at times be forced to be accountable to the law. But in exchange their social and economic positions gain legitimacy.

Lawyers are key providers of legitimacy for those who reap the profits of capitalism. They offer the rules that protect traditional forms of property and invent rules for new forms of property such as intellectual property or property embedded in complex financial arrangements. The law protects an evolving status quo because lawyers as power brokers succeed in part by embedding social and economic power in the law and in a legal establishment connected to that power. When change is managed through law, it ensures that the relatively privileged are protected.

This historical role of lawyers and legal scholars developed in the medieval period when the legal profession came into existence (1140–1230). The discovery of Justinian's Digest of Roman law late in the 11th century and the establishment of the University of Bologna and other new schools of law built and built on the prestige of Roman law. Ambitious scholars with the necessary resources and talent used their tools to 'attach themselves to the elite classes that ruled Western society' and from which they typically came. This prestigious and sophisticated set of universals matched with a growing number of opportunities to put it to work.

The imperial powers, culminating late in the nineteenth century, also brought law to legitimate imperial and colonial relationships – at home and abroad – through the same process. They co-opted local elites into the law and therefore strengthened and legalized local elite positions along with empires ostensibly based on western-produced rules. The local elite/lawyers became brokers between centre and periphery, and embedded local power into the local law. They ensured that imported capitalism was not inconsistent with relative stability in social structure even though the new basis for legitimacy was equal economic opportunity, which in theory threatened some of the familial and other bases of power.

The general argument here is that law was part and parcel of the development of capitalism in three major ways. First, the social capital embedded in the law protected the position of feudal and other groups faced with dynamic markets and new economic mobility. Second, lawyers accomplished this in part by bringing emerging economic groups into the evolving legal establishment associated with capitalism. Third, legal scholars and policy-makers erected a whole legal apparatus around the legitimating doctrine of individual responsibility for success or failure. To quote from Jodhka, Rehbein, and Souza again, 'capitalism claims to be egalitarian and fair', which translates into 'a complex of formal equalities that populate the constitution and all the legal codes and contracts of mandatory application'. That is to say, 'There is a whole complex juridical and contractual order designed precisely to show that people, so diversely apportioned by life by their belonging to a class, are treated as equal'.

I will not speculate about whether law was essential to the development of capitalism because, for example, of the contribution of legal rules and institutions to legal certainty. I suspect that this idea is best seen as part of a legal history written by and for lawyers. What we can say is that the social construction of capitalism as we know it owed much to the ability of capitalists to draw on law and lawyers and on lawyers to ingratiate themselves with emerging capitalists along with traditional social elites. That social construction and its relationship to what existed before also produced differentiated capitalisms both at the core and the periphery of the empires that emerged with capitalism.

Different capitalisms

Lawyers and law, for example, play somewhat different roles in two competing forms of capitalism – Rhine capitalism associated with Germany and France (and influential elsewhere) and Atlantic capitalism associated with Britain and the United States. The former is linked to stronger state power and elite law in the service of state power, while the latter is associated with more emphasis on private markets and elite law especially in the service of private power. These are the two models featured in the 1990s debate about 'capitalism versus capitalism' (Albert, 1993). We could also talk about 'crony capitalism' or 'oligarchical capitalism' and we would find variations on the leading lawyers in the service of the power structure (Dezalay and Garth, 2010; Prevezer, 2017). The two capitalisms discussed here historically represented different models of state, market, and economy. Each has influenced the other as well.

At the time of the New Deal in the United States and into the 1960s, lawyers in the United States became closer to state capital since, under the influence of the Cold War, social and racial inequalities undermined the position of the US against the Soviet Union. Similarly, Britain emphasized the welfare state under the influence of the Labour party and concerns with inequality. Influenced by moderate Socialist movements in Britain that helped produce the welfare state, India, for example, sought after independence to develop a moderate socialist economy. The comparative global prestige of state involvement in the economy helped strengthen the role of Rhine capitalism and those who mimicked it such as South Korea, Japan, and even Latin America under the guise of the 'developmental state'. US law and development in the 1960s and 1970s even supported the role of law and state that emphasized the strong leadership of the state and the lawyers and economists that supported it. The highest prestige lawyers were those with strong roles in the state. Even in the Unites States, the Washington, D.C. corporate lawyers oriented toward the state became more prestigious than the Wall Street lawyers.

The neo-liberal revolution came after the end of the Cold War and the rise of US-led globalization. It began as an alliance of monetarist economists in Chicago and businesses and financial interests responding to a kind of crisis of the state blamed on high taxes, regulations, and expensive social programs characteristic of the US version of a welfare state. Elite lawyers at the time in the US were largely on the side of the welfare state, but they adapted to the new orthodoxy. The neo-liberal economists acted globally. They built their credibility in part by helping to fight against socialism in Chile and encouraging Chicago-educated economists in Chile – the 'Chicago Boys' – to join Pinochet and apply the recipes of the monetarists. Reagan was the first US President on board with a neo-liberal capitalist revolution identified first with Pinochet and prior to Reagan in place in Britain under Margaret Thatcher.

This revolution was re-exported globally. There were many importers and exporters especially economists and economist-dominated organizations such as the IMF and the World Bank, but lawyers were also part of the story of a legal revolution accompanied by a retooling of elite lawyers. As in the United States (and in the earlier rise of capitalism), lawyers were not at the forefront initially of the neo-liberal revolution. As noted, the movement outside the United States was led by economists including the famous 'Chicago Boys' in Chile and the 'Berkeley Mafia' in Indonesia. The first legal allies were typically the small corporate law sector modelled on US corporate lawyers – and whose clients were typically multinational corporations based in Europe and the US.

Legal elites, as noted above, had been linked not so much to corporate power but to the state, leading families, and legal education, and they were not part of this movement initially. The law firms, to be sure, expanded and began to serve privatized local businesses, embedding corporate power in the legal field; but also a significant number of lawyers found a place in the new order through the emerging human rights movement that in many ways had its 'take-off period' in Chile in the late 1970s. This focus on rights helped to make neo-liberalism more legitimate and also kept an important place in and around the state for legal elites and the social capital they embodied. It helped spread the neo-liberal version of Atlantic Capitalism versus the versions of Rhine Capitalism, with law and legal brokers facilitating and legitimating the switch and profiting from the global spread of corporate law firms once limited to the United States and much later the UK. These changes can be seen in Europe and throughout Latin America and Asia at least. Vauchez and France, for example, have suggested that the neo-liberal influence in France now leads lawyers in the state to retire into corporate law positions rather than positions connected more to public service. Corporate lawyers in elite firms play a major role in inventing new forms of property to enrich – according to law – their clients.

The neo-liberal revolution has also brought with it a privatization of education and a further de-legitimation of social and family capital in favour of meritocratic attainment through education. This was consistent with rhetoric that has been part of capitalism from the start. The law plays a more or less active role according to this enhanced ideological trend by focusing on equal opportunity, mainly meaning formal equality of opportunity versus favouritism to family and friends. Anti-discrimination law exemplifies this orientation. But it does not make up for the advantages and disadvantages associated with accumulated privilege.

Under this same logic of supposed free competition, this focus on merit and achievement could mean that those who win in the economic marketplace should be at the top of the social hierarchy. This perspective concedes that some start the competition with more resources, and therefore have advantages, but the idea is that there will still be winners of the lottery who can therefore rise to the top. Lawyers have helped keep this myth of opportunities to rise to the top salient.

Capitalism, different forms of capital, and class

But under whatever form of modern capitalism we examine, the winners of the economic lottery do not necessarily occupy the leading political and social positions. Here is where it is important to talk about capital other than purely economic capital. To properly understand capitalism and its legitimating ideologies, the impact of capital beyond economic capital must be understood. To quote Jodhka, Rehbein, and Souza again, 'Economic capital is easy to understand, because capitalism is about economic capital'. Cultural and social capital are also essential to acquiring social status. The cultural capital of learning, degrees, and familiarity with markers of sophistication combines with the social capital of relationships – built in the family and in institutions of the privileged, including schools at all levels. Even if someone wins the economic lottery against all odds, there is a social glass ceiling.

The positions of cultural, economic, and political leadership – the trappings of the upper class – are much less open to them than to those who from birth and privileged education accumulate the less obvious forms of capital. Consideration of cultural and social capital further makes clear the differences between the ideology of capitalism and the actual opportunities it presents. From this perspective, put provocatively by Jodhka, Rehbein, and Souza, 'This mechanism of reproduction is in many regards safer than in feudal society, where inequality was openly visible … Getting rich [in today's world] means to have "made it". The everyday as well as the scientific criteria of classification confirm this view. But economic improvement precisely changes nothing in the class order'.

Lawyers, capitalism, and class order

Lawyers both contribute to and replicate capitalist class relations. Exactly the same rhetoric about equal opportunity long associated with capitalism characterizes the legal profession. Equal opportunity to attend a law school is becoming more essential to the legitimacy of the legal profession. The role of family connections is becoming less immediate in favour of meritocratic criteria that tend to reproduce social hierarchies. Elite law schools are stocked mainly with the children of the elite. The law graduates are also channelled increasingly into corporate law firms, both serving and legitimating big business and oligarchs from around the world. The partners of the law firms tend to be those from or effectively mimicking the cultural capital of the elites – from golf and country clubs to Harvard reunions and non-profit boards. The cultural capital of the corporate law partners – mainly white males with elite degrees and wives at home – also provides a social cache that helps to integrate the nouveau riche into the legal and social order. The vast inequality of legal resources that serves big business, the nouveau riche and older social elites is another way that law and lawyers make capitalism and class structure seem normal or even invisible.

Further readings

Albert, Michel. 1993. *Capitalism vs. Capitalism*. New York: Basic Books.

Brundage, James. 2008. *The Medieval Origins of the Legal Profession*. Chicago: University of Chicago Press.

Jodhka, Surinder S.; Rehbein, Boike; Souza, Jesse. 2018. *Inequality in Capitalist Societies*. Abingdon: Routledge.

Prevezer, Martha. 2017. *Varieties of Capitalism in History, Transition and Emergence New perspectives on institutional development*. Abingdon: Routledge.

15

CENSORSHIP

State control of expression

Sida Liu and Di Wang

Censorship suppresses speech, writing, and other forms of human action involving communication and information flow. It is a complex social system for controlling and prohibiting ideas and practices that are considered unacceptable or threatening to existing powers. Censorship is often conducted by a government with legal force, but not necessarily so. Private organizations and communities can censor the speech and action of their members and peer groups, and individuals can practice self-censorship too. From the trial of Socrates in Athens in ancient Greece to the 'burning of books and burying of scholars' (*fenshu kengru*) by Qin Shi Huang, the first emperor in Imperial China, the history of censorship goes back for at least two thousand years. Yet censorship is also an imminent topic in the 21st century, especially since the rise of the internet and social media. When every person becomes a 'netizen' with a public voice in cyberspace, the struggles over information control and censorship are increasingly intense and pervasive.

From a socio-legal perspective, censorship is both an instrument and a culture. As an instrument, it provides the government a legal or administrative tool for managing information flow in the public domain. As a culture, it permeates social life and shapes the behaviour of individuals, social groups, and public and private institutions. Arguably, censorship is more likely to be observed under authoritarian and repressive regimes, but even in democratic contexts, it still widely exists. Some speech, writing, or visuals can be more 'extreme', 'radical', 'offensive', or 'discriminating' than what the government or the mainstream public is willing to tolerate, for a variety of reasons, such as political ideologies, religious beliefs, gender and sexual norms, racial stereotypes, demands of counterterrorism, and so on. The instrumental and cultural elements of censorship go hand in hand, as no government censorship, regardless of its scale or intensity, could be sustained without social support from at least some of its citizens.

Why does censorship exist? It is because the object or person being censored presents a potential threat to the state, the dominant class, race or gender, or the mainstream society. A short essay like *The Communist Manifesto* can inspire revolutions and social movements across the globe over two centuries. Photos of the Holocaust or the Cultural Revolution can reveal the atrocities caused by a supreme political leader to the whole world. Videos of police shooting or other violence by law enforcement officers can go viral online and undermine the legitimacy of the legal system and state power. To prevent such threats from occurring, laws and regulations are promulgated to authorize state agents and social groups to remove information from the public domain and give sanctions to the people who produce such information. These rules contain principles and procedures that are designed to promote certain aspects or accounts of social life while undermining others. In effect, censorship rules draw social boundaries between permissible and forbidden forms of social action, including but not limited to speech and writing.

In addition to legal regulation, the enforcement of censorship also relies on social disapproval. While governments and corporations may employ censoring practices, it requires politically and economically powerful social actors, who can mobilize sufficient material and discursive resources, to sustain such a censorship system in society. From the imprisonment of British suffragettes in the 1900s to the execution of 'rightists' in Communist China in the 1950s to the trial of the Pussy Riot band in Russia in the 2010s, dominant social actors have long invested in the suppression of public communication and demands. This is especially true for the suppression of progressive or revolutionary ideas. Embedded in a favourable social environment, censorship laws are formidable governing tools for the state and useful ammunitions for oppression in public discourses.

Like many other forms of social control, censorship is a continuous process of state-society interaction. It requires what Erving Goffman would call a practice of 'impression management' by the state or dominant social groups. This is a systematic form of social control that ideally only allows information officially approved to be put on the 'front stage' of the public domain and visible to the general public. A government can be overt about what is permitted and what is not in speech, books, news articles, videos and podcasts, and other media content and, in the meantime, mobilize its legal system and bureaucracy to punish violators. In authoritarian contexts, some of these sanctions can be perverse. For example, in 2013, China's Supreme People's Court and Supreme People's Procuratorate jointly issued a judicial interpretation, which states that individuals can be charged with defamation if the online 'rumours' they create are viewed by 5,000 internet users or reposted more than 500 times, and they can even be prosecuted if the rumours cause 'damage on the national image', 'adverse international effects', or 'other serious harms to social order and the national interest'. If used abusively, a law of this nature would become an intimidating weapon for the state to criminalize alternative public voices and progressive demands for social change.

In this age of online and offline information surveillance, scholars and activists are increasingly at risk, as demonstrated recently in the 2017 trial of Turkish feminist scholar Pinar Selek and the 2018 death of Saudi Arabian journalist Jamal Khashoggi. Some of these persecutions are blatantly illegal, whereas others are justified by censorship laws. To be sure, not all censoring regulations are with bad intentions or negative consequences. For example, the bans over pornography or hate speech arguably are beneficial to society if administered properly. Nevertheless, a society's tightening of information control is often seen as a sensitive indicator for the erosion of political freedom because it gives the state more coercive power and increases the possibility of power abuse in practice.

Censoring practices can suppress a piece of information by deleting it, blocking or delaying its dissemination, or promoting alternative accounts of the same incident or idea. A censorship system usually has an evolving list of tabooed ideas and issues set by written rules or hidden norms. The boundary between permitted and forbidden ideas can be ambiguous and porous, which gives officials and administrators who enforce censorship laws a great deal of discretion in everyday practice. For instance, a sympathetic censor would make a provocative social media post read-only rather than deleting it. An aggressive censor, by contrast, would not only delete the post immediately but also check other posts by the same author to make sure all potentially risky information is removed from the public eye.

In the early 21st century, the technologies of censorship are rapidly evolving with the advancement of information technology. Automatic term detection, for example, is now utilized to block certain issues from being posted on the internet and to monitor emerging issues in online public discussions. Compared with traditional censoring methods like burning books or banning newspapers, these advanced technologies not only empower the censors but also make self-censorship a constant concern and an everyday practice for anyone who seeks a voice in the public domain. It is often the case that, even before a content gets attention from the censors, it is already pre-screened by the author to avoid unwanted attention.

Technology is a double-edged sword: it can greatly facilitate information flow and, at the same time, enable more powerful and pervasive forms of censorship. But technology cannot function without human agency. Human censors are key actors in detecting potentially risky issues and deleting posts and threads that bring the targeted issues into public discussion. Furthermore, human censors can not only block information but also channel it. In China, for example, the state uses a large number of civil servants and media workers to police the internet. When the content of a social media post is too widely spread to be fully deleted from the web, these civil servants and state-run media platforms would join forces to promote the state's official perspective on the issue and weaken alternative voices that can potentially mobilize people to think critically about the state. These state-sponsored or voluntary 'patriotic' internet users play a complementary role to the official internet censors in policing cyberspace.

Furthermore, state-sponsored censorship can use the market to manipulate the public by increasing the barriers of accessing unbiased information and diverse representation of the social reality. The censors often control means of information distribution, such as media networks, social media platforms, and internet service providers. For example, in 2018, the Sinclair Broadcast Group, which owned 193 local TV stations across the United States, ordered news anchors at some Sinclair-own stations to read very similar commentaries warning of the dangers of 'fake news' echoing Donald J. Trump's rhetoric against his critics. Although Sinclair was not a state news agency, it was criticized for its close ties to the Trump Administration, which had been disrupting press freedom and the access to free information. In other words, the state-market coalition is key to the enforcement of censorship, even in a liberal and democratic society.

Censorship is not merely a top-down phenomenon. It includes self-censorship and horizontal censorship and control over other people's views and actions. During the Cultural Revolution in the 1960s, millions of young Chinese responded to Chairman Mao's call against 'feudal traditions' and formed Red Guard groups around the country to destroy symbols and materials of China's pre-communist past. Red Guards also reported on each other and on their family members or superiors at work on accounts, real or made-up, of violating Maoist values. A contemporary example is the 'If You See Something Say Something' campaign of the U.S. Department of Homeland Security, which asks everyone to report on suspicious activities in the public. In effect, it not only reminds people that they are constantly being watched but also encourages them to censor and monitor each other, implicitly targeting marginalized groups such as immigrants and people of colour. When the principles and procedures of censorship have become pervasive in society, it can program cognitive barriers in people's perceptions. Over time, it constitutes a hegemony that not only blocks individuals from censored topics but also directs them away from processing or even seeing when presented with censored materials.

But ordinary people are not totally powerless when facing state censorship. Resistance to censorship has existed as long as the history of censorship itself. Hidden poems, for instance, were a popular form of political resistance in East Asia and some other parts of the world. In the age of the internet, resistance not only take traditional means such as poems, satires, or open street protests, but also high-tech forms like cartoon emojis and online videos. As the #MeToo movement spread worldwide since 2017, some Chinese feminists converted the words 'Me Too' into 'rice bunny' (*mi tu*, also a story-telling robot designed for kindergarteners) by the similarity of pronunciation in Chinese and designed rice bunny cartoons for online distribution. The cute animal cartoon shielded a strong feminist voice with internet language and market interest and greatly facilitated the mobilization of the #MeToo movement in China.

To conclude, censorship is a legal and political instrument for the state and dominant social groups to suppress the flow of information in the public domain. Yet, it is also a cultural hegemony that shapes the everyday lives of ordinary people. While legal boundaries of speech, writings, and other forms of information flow are a necessity in most societies, censorship often does more harm than good, especially to marginalized people and groups who struggle to make their voices heard. A culture of self-censorship, if institutionalized in a society (authoritarian or democratic), can also become a toxic environment for innovative ideas and alternative voices to emerge and survive. To guard against the potential harms and evils of censorship, therefore, requires not only reasonable laws and sympathetic law enforcement officials, but also a culture of societal tolerance and moderation toward dissidents and alternative opinions.

Further readings

Deibert, Ronald; Palfrey, John; Rohozinski, Rafal; and Zittrain, Jonathan (eds.) 2008. *Access Denied: The Practice and Policy of Global Internet Filtering*. Cambridge, MA: The MIT Press.

Wang, Di and Liu, Sida. 2020. 'Performing Artivism: Feminists, Lawyers, and Online Legal Mobilization in China'. *Law and Social Inquiry* 45(3): 678–705. https://doi.org/10.1017/lsi.2019.64

Yang, Guobin and Wu, Shiwen. 2018. 'Remembering Disappeared Websites in China: Passion, Community, and Youth'. *New Media and Society* 20(6): 2107–24. https://doi.org/10.1177/1461444817731921

Planet Netsweeper https://citizenlab.ca/2018/04/planet-netsweeper/

16
CITIES AND URBANIZATION

Antonio Azuela

When we talk about cities and urbanization terms such as 'chaotic' or 'messy' are often used. When seeing city life in this way, it is difficult to see in law anything other than an unfulfilled promise of order, or an unrealistic illusion. This entry's aim is to show that, for good or for bad, law is far more present in urban life than people think, specifically in how urban space is produced, appropriated, and, especially, fought over.

The rules that affect the way in which cities take shape, taken as a whole, appear as incoherent and illegible. But it is not necessary to know all the laws; when a particular question arises there is always a way to find an answer that furthers order and political action. To orient ourselves in the complex world of laws shaping urban life, one must first recognize that urban legality consists of rules and principles that emerged at different times, sometimes in remote times. Thus, urban legality is an accumulation of layers, somewhat like the layers found under the surface of the earth, although different layers can at any time suddenly break through the surface and become unexpectedly relevant.

Before describing some of the layers that make up urban legality, it is necessary to highlight a feature that characterizes all legal rules: their ambiguousness. Those who elaborate legal rules, usually legal professionals working in a particular political context, may make a big effort to control future effects, but unexpected social results are always possible. This is not only because a rule can often be interpreted in conflicting ways; it is also because the social effects of particular rules are often different from and even opposite to what the framers intended. This has happened in the case of legal rules central to modern cities, such as rent controls or zoning rules or the legal empowerment of neighbourhood associations. The ambiguousness of legal rules is the first feature that must be recognized by anyone seeking to resort to local law to pursue a political aim.

Of all the legal notions that exist in and shape cities, the oldest one is that whereby the municipality is seen as a corporation. In many countries the word 'city', in legal contexts, refers strictly to the municipal corporation. Here 'corporation' does not refer to the for-profit businesses that make up today's capitalism, but rather a grouping of the inhabitants that has the status of a legal person – a person distinct from the citizens and endowed with the capacity to represent the whole. Today, the city as corporation does not exist by itself; this legal form now is part of the complex legal architecture of the nation-state. Although globalization sometimes complicates this architecture, generally speaking, it is constitutional law that determines the relationships between the municipality and legal entities and powers that exist at other scales. Sometimes it seems that higher-order powers are less democratic than local ones, which is why it is often said that municipal government is closer to the people; but often it has been local government that has blocked progressive, democratising moves within city governance.

A second historical legal layer, one that characterises the modern city, is the regulation of public health. During almost two centuries, a huge number of technical standards have developed, many of which are unintelligible to the lay person, but which are based on simple ideas, such as air flow into residential rooms, running water, and appropriate methods of waste disposal. These rules are not purely technical; they also draw moral lines on urban space to set out what is an acceptable space. The effects of these rules include the stigmatisation of those who live in places that fall below the legal standard, as if unhealthy conditions turned inhabitants into criminals. One can take a critical attitude toward this stigmatisation without being indifferent to the ill effects of unsanitary spaces, especially housing. If today's cities are more liveable than those described by Charles Dickens or Frederick Engels in 19th century Britain, it is because public health rules have been imposed.

Thirdly, the legal institution that is most closely associated with the modern city is urban planning, carried out by a professional corps dedicated to thinking through the city's possible futures. From the beginning of the industrial era, cities have been undergoing constant change, and as a result, debates about the future, both short-term and long-term, have been central in and to urban societies. We all usually have an image in our heads of the city we desire to see in the future, but also, we have to recognise that future-oriented urban processes always have certain material and institutional conditions. Thus, citizens have to start by knowing who actually elaborates various urban plans. The authors of plans are usually experts who often work at some distance from the experiences and the needs of ordinary citizens. This is widely recognized and for that reason, democratic regimes open up spaces of participation in which people's needs and experiences can be expressed and can become visible to the authorities. Devising participatory mechanisms, and especially choosing the right scale for participation, is one of the most difficult challenges for urban law today. In many places, citizen participation takes place primarily at the neighbourhood level, which can obscure city-wide needs and makes it difficult for citizens to think collectively, at the city-wide scale, about ensuring that the city is for everyone.

A key mechanism within urban planning is zoning. Zoning consists of distributing land uses (residential, commercial, industrial, educational, etc.) and also governing the intensity of each land use, by imposing restrictions on what landowners can do with and in their properties. Zoning is an interesting legal technique because zoning rules cannot exist without zoning maps. Zoning became popular in the early years of the 20th century, not coincidentally at the same time as the free-enterprise ideology of the supremacy of property also gained hold. That is why there were legal and judicial debates about the extent to which it is legitimate for private property rights to be limited by local authorities. Almost everywhere the conclusion was that it is legitimate for public authorities to restrict private property rights when they come in conflict with the public interest. (For example, that is why it is illegal virtually everywhere for a homeowner to open a bar in their living room).

However, zoning rules too can have a variety of effects. They are used not only to protect society from the chaos that would ensue from the arbitrary use of private property, but also to maintain physical and aesthetic distances separating wealthy from poor areas. Socio-economic and racial segregation have often increased due to zoning rules, which is why zoning rules are often hotly debated. These arguments cannot be settled through legal doctrine. Local experience and a knowledge of the practical effects of zoning rules are necessary, since it is in practice that we see the effects of the interests and the ideologies that are at work.

Although zoning rules and zoning maps may seem to be offering a very definite image of what is and is not allowed in each urban space, in fact in every jurisdiction there is a system of permits for new developments. Since it is impossible to predict beforehand every initiative that might arise in the future, zoning and other planning laws often contain mechanisms to adjust and even to overturn the rules that were originally adopted. The art of urbanism is not so much in setting out in advance a plan for the future but rather in the city's capacity to decide whether each new proposal is or is not in keeping with the general idea of the city embodied in the original plan. This is the context in which the most acute urban conflicts arise. Today, with financial capital playing an increasing role in the economy, opposition to 'megaprojects' is an important manifestation of social resistance to the imposition of capitalist priorities on urban space. However, it is also true that many projects that can benefit everyone, such as transportation and other infrastructures, and some that primarily benefit the poor, such as social housing, may have some negative effects on other groups, which have to be weighed against the collective benefits. The only generalization one can make is that planning work should aim at a fair distribution of the negative effects as well as the benefits of urban development.

The fair distribution of positive and negative effects can take place across a number of issues. An important one is the increase in land value that is caused by new infrastructures, or that is the result of changes in planning rules that allow for more profitable uses of certain plots of land. There is a consensus amongst economists that increases in the value of land that result from external changes become an unearned income for the owner. However, in most cities, increases in the value of privately owned land result from collective effort and collective processes. Therefore, cities should consider mechanisms for recuperating some of the surplus value of privately owned land, as an important tool to finance public goods and services and to partially compensate people for the inequalities that a free-market urban economy generates.

The foregoing discussion does not apply, however, in the case of urban development that takes place outside of the law. Although in countries in the global North there are also illegal developments, sometimes large ones, the phenomenon of illegality is massive in the so-called global South. In regard to this issue, it is first notable that words such as 'slums' or 'informal settlements' are used to encompass very heterogeneous realities. It is crucial to appreciate the diversity of informal urbanization processes, because a site that begins as a settlement outside of the law can become a liveable and agreeable urban space, even in the absence of formal programs to regularize ownership. The stigma carried by the word 'slum' and the negative connotations of the word 'informal' create obstacles for dwellers who want to be recognized as citizens with rights. The stigmatisation is particularly acute in post-colonial contexts, in which informal settlements have historically served to isolate and marginalize ethnic and racial minorities.

Although the problematic word 'informal' covers a large variety of situations, people often talk about informality or illegality as if it were a single, univocal thing. The only thing that the areas called informal settlements have in common is how law produces their situation, in that it is state law's own definitions that create 'informality' and illegality. The language of law is so powerful that people often believe it refers to a naturally existing condition. What is outside of the law thus can appear as 'against nature'.

If law, as stated earlier, is inherently ambiguous, being outside of the law is also ambiguous. Most definitions of 'informality' focus on the absence of property titles, which creates the impression that a property title is a basic condition of a good economy, and even of citizenship itself. However, law and society research has shown that the political and economic effects of lacking property title do not depend on an abstract definition but rather on the concrete political context.

There are also types of informality that do not pertain to property and raise different dilemmas. For example, there can be public-health informality, which produces social stigma but can serve to hold the state accountable for providing all citizens with the minimum physical conditions of urban life. There is also urban-planning informality, whereby changes are made to create and maintain streets and public spaces. There is also environmental informality, in that some neighbourhoods have a greater impact on the environment than others. But even in these latter situations, legal arguments can be made to protect the residents. The recognition of housing as a human right, in international law as well as nationally and locally, has meant that even when an environmental problem suggests that relocating a community would be desirable, the relocation can only be justified if the residents have participated in the resulting solution. Clearly, problems associated with informality cannot be addressed through any general formula. They must be addressed from the local point of view, with legal solutions based on the self-organization of the inhabitants.

Finally, it must be recognized that urbanization is a process that is not found only in cities or affects only cities. First, the climate crisis has shown us that urban processes have to be seen in a broad territorial context, since urban living has many impacts on natural resources, and the ecological footprint of wealthy urban dwellers must, in particular, be recognized. There cannot be urban justice if a few megacities absorb huge quantities of resources and energy without minding the consequences, especially the effects on the rural populations. Secondly, life in rural contexts is increasingly similar to urban life. Even in small communities, residents feel entitled to running water, electricity, and other services. The urbanization of the countryside is a social and territorial transformation of great weight, but national legal frameworks have not yet grappled with this issue.

Some important issues have been left out of this account of cities, urbanization, and law, such as rent control and the use of expropriation. If these issues were addressed, however, the conclusion would be the same: namely, that although the laws in effect in cities represent the interests of the dominant classes, quite often there are legal conditions and legal tools that can encourage resistance and local creativity.

Further readings

Fernandes, Edesio and Varley, Ann. 1998. *Illegal cities: law and urban change in developing countries*. London: Zed Books.

Fischer, Brodwyn, McCann, Bryan and Auyero, Javier, eds., 2014. *Cities from scratch: poverty and informality in urban Latin America*. Durham, NC: Duke University Press.

Valverde, Mariana. 2012. *Everyday law on the street: city governance in an age of diversity*. Chicago: University of Chicago Press.

17

CITIZENSHIP

Engin Isin

Of all categories of person, citizenship is one of the most complex yet enduring. Its European history is often traced back at least to ancient Greeks and Romans and their concepts are still strangely familiar today. Citizenship generally governs the relationship between a polity and its peoples by dividing them into citizens and non-citizens with differentiated rights and obligations. If people are caught in various categories of non-citizenship (e.g., migrants, refugees, slaves), life becomes precarious with few (if any) rights. Yet, citizenship affords various rights that enable citizens to enjoy them with relatively few obligations – except of course when citizens are in second-class citizenship categories. Citizenship has then been a game of domination and emancipation that brings into play the struggles of those who want to protect certain privileges and the struggles of those against being caught in either second-class or non-citizen categories. This entry provides a critical perspective on citizenship to describe it as a performative game rather than membership as is often symbolised by a passport or identity card.

There are five distinct but overlapping senses in which citizenship is a performative game of domination and emancipation. (1) Citizenship involves political and social struggles over who may and may not act as a subject of rights (i.e., to have a right to be recognized as a person before the law); (2) these struggles feature not only citizens but also non-citizens as relational categories of person; (3) citizens and non-citizens are not homogenous and unified categories as they include members of different social groups making rights claims; (4) people become citizens by exercising, claiming, and performing rights and duties; and (5) when people become citizens they creatively transform its meanings and functions by 'making rights claims'.

The first sense in which citizenship is performative concerns the issue of struggle. Although citizenship, for legal purposes, refers to membership in a state, a critical perspective considers citizenship as performance. Who may and may not act as a subject of rights is determined by ongoing political and social struggles over not only the content of rights but also who are or are not entitled to them. If citizenship is a struggle over the subjects of rights, this struggle creates scenes in which social groups contest their similarities and differences and 'play for rights'.

If we recognize that since the 18th century, first in Euro-America and then across the world, the nation-state became the dominant type of polity, we must acknowledge how citizens were divided from non-citizens. The characteristics of a social group signified as propertied, adult, male, 'rational', white, Christian, heterosexual, and able-bodied became the dominant universal characteristics in the modern state. These characteristics created various other social groups as subjects with limited (or without) rights: the poor, young, women, irrational, racialized, non-Christian (Muslim, Indigenous, Jewish), ethnic minorities, queers (lesbian, gay, bisexual, transgender, questioning, intersex), and disabled people were regarded as incapable of fulfilling the obligations of citizenship and hence acting as citizens.

It was only in the 20th century that the struggles for new civil, political, social, and sexual rights started to significantly challenge this through emancipation struggles. These struggles began unravelling the dominant characteristics of citizenship, and the unravelling is still under way. We belong to an age when some claims to full citizenship remain as precarious and subaltern as ever. Youth, children, disabled, Muslims, the poor, the racialized, women, queers – in any combination of these – are making rights claims to citizenship, so that they might cross from their precarious positions to at least more liveable positions.

The second sense in which citizenship is performative concerns how social groups struggling for rights bring into play both citizens and non-citizens. The social groups mentioned above occupy different subject positions ranging from citizens (insiders) to non-citizens (strangers, outsiders, and aliens). We have already suggested that citizens are those subjects whose entitlements or privileges are instituted as civil, political, social, and sexual rights. Often, some people who, while may be accepted into citizenship in the future, are still considered strangers in the sense that they have not yet fulfilled requirements to act as citizens. Modern examples are women (in 19th century Britain) and African-Americans (in 20th century America), and Indigenous peoples (in settler colonial societies such as Canada and Australia) as social groups that were deemed strangers because dominant social groups considered them as lacking capacities. Yet both women and later African-Americans were also understood as indispensable parts of the polity that fulfilled functions for it (reproduction for women as mothers, labour for 'blacks' as slaves first and then as workers). By contrast, those social groups who are deemed outsiders to citizenship may find themselves as migrants and refugees struggling for the right to be present in a polity at all.

The line between strangers and outsiders is often drawn through disposability or deportability: strangers (such as slaves in the American South) are sometimes accepted as indispensable, whereas outsiders may become indispensable too but are seen as remaining outside disciplinary reform and thus sequestered. Jews were ascribed to this category in most of European history and Indigenous peoples in Euro-American history. Straddling between these subject positions provides the dynamics of citizenship as play of domination and emancipation rather than a static institution. Those treated as aliens to a polity typically face total rejection if they are seen as continuing to pose a threat when remaining in the territory. Categories such as enemy combatants or terrorists are examples of persons regarded as irredeemable and whose citizenship can be subject to revocation or denaturalisation.

The categories of citizens, strangers, outsiders, and aliens are neither static nor impermeable. A huge variety of social groups move through or across these positions. The boundaries between citizens, strangers, outsiders, and aliens are dynamic and permeable precisely because they are objects of social struggles. As we have already mentioned, people identify with or are ascribed to various social groups and constantly traverse the boundaries separating citizens and non-citizens. Moving across these positions or breaking down the boundaries between them involves struggles over rights.

Governing ourselves as political subjects means exercising rights that we may have (e.g., the right to vote) and claiming rights that we may not have (e.g., the right to same-sex marriage). It also means making claims for or against rights that others do not have as members of multiple social groups.

So far, we have discussed citizenship as though each polity develops its own distinct and independent conventions of citizenship. Yet, many social groups, even those defined as nations, exist across not only social boundaries but also borders. This gives rise to numerous performative contradictions in citizenship. If the sovereign state is the dominant form of political organization of the modern age, it is because it became dominant through a history involving colonialism, imperialism, and nationalism. Euro-American empires around the world have left indelible marks on polities they colonized and have imposed various citizenship arrangements on them. These arrangements erased or displaced already existing political and social relations between and amongst various social groups and replaced them with forms of imperial and colonial citizenship. These indelible marks of colonialism have not only inscribed names – such as the Americas, Africa, Asia, and the Middle East – but also citizenship arrangements whose legacies remain today. Thus, the complex and enduring history of citizenship is not merely European but a global history.

So, the citizen and non-citizen arrangements that obtained in each polity were disseminated across the metropole and the colony, creating spatiotemporal arrangements of citizens and their others (strangers, outsiders, aliens). The colonial subjects were described for their tribalism and were seen to have never understood or developed the characteristics of citizenship and its ostensible unifying powers. If colonial subjects were deemed to have understood citizenship, it was because it was 'borrowed' in post-colonial sovereign states. This makes studying citizenship in colonial and post-colonial states a vexed activity. Is the description 'what citizens and non-citizens perform in making rights claims' adequate to study citizenship in colonial and post-colonial states?

The third sense in which citizenship is performative is that it is enacted through 'making rights claims'. As mentioned above, making rights claims concerns both the content of rights and what people are doing when they perform citizenship. Making right claims is a concept that contemporary citizenship studies inherit from social movement studies. Social movement studies, in the 1970s and 1980s, used the language of rights to articulate the injustices people suffer and to show that people resist the denial of rights to them and others. Performative citizenship signifies both a struggle (making rights claims) and what that struggle performatively brings into being (the right to claim rights).

There is a difference between claiming to be a subject of rights and making rights claims. This is essential for understanding performative citizenship. The distinction enables us to study how people stage creative and transformative resistances and articulate claims against domination and the injustices it precipitates. Their focus is not only on the exercise of rights and duties as they exist but also on claiming rights and duties yet to come. As Karen Zivi writes, 'To approach rights and rights claiming from the perspective of performativity means, then, asking questions not simply about what a right is, but also about what it is we do when we make rights claims. This distinction emphasizes that, when performing the right to claim rights, there is a putative condition of equality between citizens and non-citizens. This means that citizens and non-citizens are both performing the universal right to claim rights, but in different ways.

The fourth sense in which citizenship is a game of domination and emancipation is that when making rights claims, people are effectively saying, 'I, we, or they have a right to …' whether they have that right in law or not. When people make rights claims, they both reference and cite these conventions, and yet the performative force of their claims often exceeds or breaks them, as philosopher Judith Butler puts it. Yet, when there are clearly enabling conventions, making rights claims may still exceed those conventions.

There is a constitutive tension between the right to claim rights as a universal right and making claims for rights. When people perform acts of citizenship, they play out these two tensions. This is what Judith Butler describes as a performative contradiction, which arises when a person or a group is excluded and de-authorized and yet makes claims about belonging. This is an important reason why the object of analysis in researching performative citizenship is acts and how they transform norms and conventions.

The fifth sense in which citizenship is performative is when through struggling for their rights, the rights of others, and the rights to come, people constitute themselves as citizens. Crucially, what makes citizenship performative in this sense is not only that it involves iterating or exceeding conventions about what people may and may not do, but also that people often resist these conventions and transform them by applying principles such as equality, justice, liberty, emancipation, and solidarity. These principles enable or motivate people to struggle over rights by traversing the boundaries of social groups and borders of polities. By doing so, citizens and non-citizens, with or without rights, assume responsibilities towards each other, traverse boundaries and borders, transform themselves and others, the rights under which they make claims, and the rights to which they make claims.

Further readings

Butler, Judith. 1997. *Excitable Speech: A Politics of the Performative*. London: Routledge.

McNevin, Anne. 2011. *Contesting Citizenship: Irregular Migrants and New Frontiers of the Political*. New York: Columbia University Press.

Nyers, Peter. 2019. *Irregular Citizenship, Immigration, and Deportation*. London: Routledge.

Rygiel, Kim. 2010. *Globalizing Citizenship*. Vancouver: University of British Columbia Press.

Tully, James. 2014. *On Global Citizenship: James Tully in Dialogue*. London: Bloomsbury.

Zivi, Karen. 2012. *Making Rights Claims: A Practice of Democratic Citizenship*. Oxford: Oxford University Press.

18
CLASS AND ECONOMIC INEQUALITY

Mariana Valverde

The word 'class' is often used very loosely. Sometimes people say that on top of racism there is also 'classism', meaning discrimination against low-income groups or indigent people. That makes some sense as a political statement, but it creates conceptual confusion because race and class are not quite on the same plane. If used more precisely, 'class' can be very useful in drawing attention, not so much to quantitative economic inequality, but rather to the qualitative and structural factors that place different groups in completely different economic situations. 'Class' is most useful when used to refer to structural positions, which often correlate with income but not always (one can be a very well-paid worker or a struggling business person).

A second key aim of this entry is to draw attention to the crucial difference between **income** inequality and **wealth** inequality. As the hugely influential work by French economist, Thomas Piketty, and his collaborators shows, income-based numbers conceal the rapidly rising inequalities between those who, whatever their income in a particular year, possess wealth (often, inherited wealth), compared to those who may have good jobs but have little or no wealth to pass on to their children.

The third and final part of the article will briefly explain current efforts made, mainly by the UN, to measure and understand inequality in broader terms, through the 'Human Development Index'. This ambitious approach considers factors that go beyond both wealth and income, including the so-called 'birth lottery' that largely determines whether one has access to good-quality public goods and services, and to common goods such as clean air or a fair legal system.

Class: how is a key 19th century concept relevant today?

It was Karl Marx who in the 19th century developed a systematic theory of 'class' which remains influential today despite the decline of strict Marxism. He famously argued that 'all human history is the history of class struggle', meaning that in every historical period and in every type of society, a fundamental distinction exists between those who own and control the 'means of production' (whether that be land, in agricultural societies, factories, as in industrial capitalism, or financial and/or digital capital, central to today's economies) and those who may have more or less personal property but do not own the kind of resources – the 'means of production'– that can be put to use in creating new wealth.

It should be noted that for Marx no resources or goods are ever in and of themselves means of production. Thus, land only becomes a means of production if it is rented out or used to grow crops for the market, for example, but if the land is left to revert to nature then it is not a means of production. Similarly, if one uses a car for personal use only, then it is not a means of production, but if one's car is used as a taxi, then its economic nature changes.

In Marx's analysis, how much income one has is not what determines class position. A landowner might be going bankrupt, but, especially in societies where land is the key productive resource, that person is part of the ruling class and has interests that are directly opposed to the peasants who have to hire themselves out as field hands. Similarly, a software engineer who is given 'stock options' along with a salary will be partly an employee (through the salary) but partly a capitalist. The stock options – which amount to a bet that the stock of that company will be worth more in the future than at the time the options were created – create a vested interest in the prosperity of the business, and to that extent put the employee on the side of capital. Class interest is structural. It has nothing to do with one's political views.

In the 19th century, when Marx did his work, the division between capital and labour was far sharper. Today, however, many wage workers also own some small portion of capital, usually through a pension fund or a small personal investment portfolio. Indeed, in many countries, workers' pension funds are major players in global financial markets, something Marx could not have imagined. The investment activities of workers' pension funds blur the distinction between capital and labour. But although the distinction between those who own capital and earn profits or dividends and those who do not and must therefore work for wages has become blurrier in the present day, that does not mean that class does not exist or does not matter. A particular person might be both a worker and a capitalist, but the two categories remain not only distinct but in tension. The tension between the two becomes particularly visible when workers' pension funds make investment decisions (such as buying up privatized infrastructure) that result in jobs being outsourced or lost and/or wages being cut. Today, it might not always be possible to label particular individuals as either capitalists or workers, but class and class interests are still important – just as race still matters even when mixed-race people are numerous.

Income vs. wealth

In many countries, especially in the global North, the shocks of the Great Depression and World War II, together with the rise of Labour and social democratic parties, resulted in a major historical shift towards greater equality in the post-1945 period. Income tax rates were far higher in the period from the 1950s to the 1980s than they were before or after, and many states invested heavily in public goods and services (from universal old-age pensions to subsidised public transit to public schools and universities) that allowed even those with lower incomes to enjoy an adequate standard of living and opportunities for their children. Since the 1980s, inequality has increased sharply, especially within countries. Cities, in particular, have seen economic polarization – that is, rich neighbourhoods and very poor neighbourhoods increasing in number. It is a common sight in both the global North as well as the South to see rich people ensconced in luxury vehicles whizzing by sidewalks and parks where homeless people are panhandling or trying to sleep.

Economic inequality is partly a matter of income. In Canada, for example, by no means the worst country in the world in terms of economic inequality, the salaries of senior bank managers are an astounding 200 times larger than the salaries of junior bank employees.

One commonly used way to measure inequality within a country or a region is named after an early 20th century Italian economist called Corrado Gini – the Gini coefficient or index. This focuses not on income itself but on the distribution of income. The index is always a number somewhere between 0 and 1, where 0 represents a jurisdiction where everyone earns exactly the same and 1 represents the opposite extreme, that is, a hypothetical society where one person makes all the income. South Africa is famous for having the world's worst (that is, highest) Gini index, at 0.62 (which proves that one can change race politics without making much of a dent in economic inequality). Brazil had virtually the same highly unequal distribution of income as South Africa in 2001, but successive progressive governments and some fortunate economic circumstances combined to steadily lower it significantly, such that in 2013, it had been reduced by almost 10 per cent, to 0.51. The Gini data are collected yearly and can be very useful to track changes; in this instance, one can anticipate that with the extreme right-wing government that Brazil now has, the Gini numbers will start going up again.

The Gini numbers are certainly one useful data point about a society or a city, and make it relatively easy to use the calculation to see change over time, but it is inappropriate to imagine that the Gini numbers give the whole story about economic inequality. For instance, the violent dispossession of Amazonian peoples by Brazilian agro-capitalist interests, which have included the deliberate setting of numerous forest fires, is unlikely to be reflected in income data.

The large-scale, multi-country historical database of income and wealth that French economist Thomas Piketty has gathered and analysed reveals that in recent decades the most striking inequalities concern not income from one's work – the main data used in Gini index calculations – but rather wealth, including both inherited wealth and wealth accumulated in their own lifetime by individuals (such as the tech industry stock options mentioned above, or, more commonly, homes). Those who are born to parents who own 'good' homes and other wealth (stocks and bonds, rental properties, holiday homes) have a huge and arguably unfair advantage over those who, perhaps with the same education and even working at the same job, do not have the benefit of inherited (or as they say in France, 'patrimonial') wealth.

What makes wealth inequality particularly problematic for anyone who values equality is that it is not independent from income inequality, such that income inequality ratchets up the prior structural effect of wealth inequality. Piketty's vast trove of data shows that young people whose parents have more wealth are likely to stay in school longer and receive a better education leading to better jobs than their peers. Similarly, young people with wealthier parents are far more likely to be helped to buy property, especially residential real estate, in many countries is the most important form of family wealth. The two forms of economic inequality thus ratchet each other up.

Piketty and other analysts who have documented the vicious circle of economic inequality suggest that to begin minimizing the growing economic inequality of our time, governments have to increase income taxes on higher salaries (income tax rates for the top groups sometimes reached 75 or 80 per cent in the 1950s/1960s but have fallen sharply in most countries in recent decades.) Governments also have to take measures to tax wealth separately from income (for example, through raising inheritance taxes and taxing financial transactions). Piketty's data shows that wealth inequality has taken many countries in the global North into levels of inequality not seen since before World War I; to minimize economic inequality effectively, governments have to target both income and wealth.

Beyond money: the Human Development Index

Apart from differences in income and wealth, there are great differences (especially between countries) in the common goods that are provided for all for free or with minimal fees – schooling, health care, potable water, electricity, old age pensions, public transit, etc. African migrants trying to reach Europe are not only seeking decently paid jobs but also want to live in a country with good public education and government-funded health care. Facts about publicly provided goods and services are not generally included in economic inequality calculations. But they are included –alongside such hard-to-measure 'goods' such as a fair justice system, a working human rights apparatus and clean air and water – in the UN's 'Human Development Index' or HDI.

The HDI index goes beyond economics, and so need not be discussed here at length; but for many social justice activists – as well as for the philosophers Amartya Sen and Martha Nussbaum – who pioneered the 'capability' approach to international development – this recent UN concept represents a useful holistic approach to measuring differences in quality of life. However, while the HDI is very useful for comparing countries to one another – and for measuring whether a country has managed to improve conditions for its people over time – it is less useful for documenting internal inequalities within a country or a city.

Conclusion

In conclusion, economic injustice has many aspects and takes many forms. The Marxist concept of class remains useful today, despite having been designed primarily to understand 19th century industrial capitalism, because it draws attention to people's structural position in the economy, rather than focusing on how much money people earn. One's structural class position creates certain vested interests, that is class interests, that are relatively independent of one's political views. Many people today have contradictory class interests (as in the example of workers who have a vested interest in the success of capital markets through their pension fund), but that just makes class analysis more complex; it does not make it irrelevant.

We also discussed the limitations of economic inequality calculations that rely only on income, such as the Gini index, and reviewed key findings from work showing that wealth inequality is a growing problem, and one that interacts in a vicious circle manner with income inequality. Finally, we drew attention to efforts made to go beyond money and try to measure inequality in a more holistic manner, such as the UN's HDI.

From all of this, it becomes clear that if action is going to be taken to reduce the frightening levels of economic inequality that exist now, citizens in every country need to pressure governments at all levels to take strong measures to counteract the vicious circle that is the relationship between wealth inequality and income inequality. Citizens also need to agitate to ensure governments provide the common goods and services that are necessary for a decent life but are not bought and sold, from clean air and water to a functioning and fair legal system.

Further readings

Marx, Karl and Engels, Friedrich. The communist manifesto (many editions).

Piketty, Thomas. 2014. *Capital in the twenty-first century*, transl. by Arthur Goldhammer. Cambridge, MA: Harvard University Press.

Sen Amartya. 1999. *Development as freedom*. New York: Alfred Knopf.

19
CLIMATE JUSTICE

Usha Natarajan

We live in a time of ecological change. Climate change is the best known of numerous interrelated environmental transformations currently taking place. Each year, human activity emits one hundred times as much carbon dioxide into the air as volcanoes do, causing average global temperatures to rise. Even more perturbing is a mass extinction of species for the first time in 66 million years, and the biggest disruption to Earth's nitrogen cycle in two and a half billion years. While our planetary history does evidence cyclical recurrence of such events, in the past, nature played the dominant causal role, whereas this time, colossal injustice is at the heart of the matter. The lifestyles of today's wealthy few are sustained not only by increasing economic inequality but the systemic collapse of ecosystems. The richest 20 per cent consumes 80 per cent of the world's natural resources and generates over 90 per cent of its pollution and waste. For most people, the poorer 80 per cent, their grossly unequal access to natural resources is compounded by also bearing the brunt of the pollution generated by the rich.

Climate change is but the most extreme example of such inequity. The United Kingdom, with a population of 60 million, emits more greenhouse gases than Egypt, Nigeria, Pakistan, and Vietnam combined with a total population of 472 million. One state in the United States, Texas, with a population of 23 million has a deeper carbon footprint than the whole of sub-Saharan Africa with total a population of 720 million. And the 19 million people in New York state emit more than the 766 million people living in the 50 least-developed countries. As an international lawyer dedicated to studying global inequities, this is the widest disparity I have ever encountered. Inequality on this scale is not happenstance, it is systemic. Globalised capitalism ensures the profligately wealthy not only stay rich but get richer. As the impact of climate change is borne by others, there is no immediate incentive for the rich to change their behaviour. The poor contribute very little to climate change but are on the frontlines of environmental harm because of their vulnerable geographic locations, lack of resources and regulatory capacity to protect themselves, ongoing extraction of their natural resources and labour to fuel an unequal global economy, and a systemic transfer of pollution from the rich to the poor. For these reasons, most people writing about climate change turn to the language of justice and look to the law for redress.

International law plays a key role due to the cross-border nature of the problems. The climate is in perpetual interaction globally with species extinction, the spread of deserts, the loss of forests, the destruction of the ozone layer, and the increasing toxicity of air, water and land. So, what have international lawyers done about it? We have brought to bear our treaties, principles, and expertise to create the specialized field of international environmental law. Nevertheless, more greenhouse gases have been emitted in the last 30 years than in all of human history. Why? The two most frequented critiques are that the rich and poor cannot agree on how to protect the climate (usually called the North-South divide) and a lack of political will (people have other priorities).

Do unequal economies and political expediency mean lawyers are not to blame? We tend to point to politicians and corporate actors for preventing us from making good laws and, when such laws are made, failing to abide by them. Conveniently, this approach allows our disciplinary expertise to proliferate despite extant failures. Our very inability to regulate climate change has generated many climate law jobs. It may be the case that past mistakes stimulate the search for better legal solutions but, after three decades of unmitigated failure, we also need to consider whether international lawyers may be contributing to the problems we claim to solve.

International law's understanding of climate change originates from the stirrings of environmental consciousness in Western societies in the 1960s and 1970s. In the United States, these decades witnessed the passage of the *1963 Clean Air Act, 1972 Clean Water Act* and the establishment of the Environmental Protection Agency in 1970. From these domestic developments stemmed international law's engagement with environmentalism. An important point was the *1972 Stockholm Conference on the Human Environment*, traditionally identified as the genesis of international environmental law, and followed by the *1992 Rio Earth Summit*, treaties including the *1992 United Nations Framework Convention on Climate Change (UNFCCC)*, and legal concepts such sustainable development and the principle of common but differentiated responsibilities. Today, the legal specialisation of climate law continues to evolve.

While ostensibly international, this type of environmentalism is the product of the Western development experience, and not necessarily self-evident to or shared by most of the world. Through centuries of colonialism, genocide, slavery, apartheid, and racial discrimination, the global North systemically looted the natural resources and labour of the global South to fuel Northern industry and wealth accumulation. Northern understandings of economic development were eventually universalised through the decolonization process, by conditioning Southern independence upon a commitment to industrial development. Post-colonial states did adopt varied stances on the spectrum between capitalism and communism, including non-alignment, but any society that dared to disavow industry altogether was denied sovereignty, as evidenced most clearly by ongoing struggles of tribal and Indigenous peoples for self-determination. Against this historical background, the onset of international environmental law was greeted with ambivalence by many in the South: Northern desire to globally regulate the harmful consequences of industrial development came too close upon the heels of the South finally achieving a degree of economic freedom, raising fears of 'environmental colonialism'.

Southern states reacted to climate change by insisting on legal recognition that the rich caused climate change and should shoulder most of the responsibility for fixing it. Indeed, international law principles such as sustainable development and common but differentiated responsibilities articulate precisely this. But international climate law is unable to turn these principles into action by getting the rich to change their behaviour. Under the umbrella of the *UNFCCC*, climate law initially allocated strict and clear emission reduction targets to rich states through the *1997 Kyoto Protocol*. However, the United States – the world's biggest emitter – refused to participate and this demotivated other rich states from meeting their own targets. Instead, the US led a movement to replace *Kyoto* with the *2016 Paris Agreement*, which allows all states, rich and poor, to volunteer their own aspirational targets. Thus, in three decades of international climate law, not only has more greenhouse gas been emitted than ever before, but the United States has made a concerted effort to erode legal acknowledgement that climate justice matters.

Alongside its (albeit unsuccessful) attempts to govern greenhouse gas emissions, international climate law is also the main terrain where climate justice struggles are articulated, negotiated, and eroded. For instance, in the lead up to the *UNFCCC*, in 1987, the Brundtland Commission introduced the concept of sustainable development. At the time, it had a relatively straightforward meaning: Development that meets the needs of the present without compromising the ability of future generations to meet their own needs. Sustainable development kept economic development within the absorptive capacity of natural ecosystems, acknowledging systemic limits. In line with this definition, climate scientists estimated the maximum change in global temperature that could sustain human life and calculated the remaining emissions quota to remain within this safe zone. Their calculations indicated that between 1990 and 2050 the world should reduce its overall emissions by 50 per cent. In line with the principle of common but differentiated responsibilities, the United Nations advised that rich states reduce their emissions by 80 per cent and poor states by 20 per cent by 2050. Not only were these responsibilities ignored by many rich states: in the decades that followed the concept of sustainable development lost all trace of its original meaning. It is used in such a broad variety of contexts today that it is routinely exploited for the purposes of obfuscation, confusion, and double-speak.

Both the regulatory and symbolic aspects of climate law continue to matter a great deal to states and peoples. Developing states were cognisant that the shift from *Kyoto* to *Paris* not only meant a shift from mandatory to voluntary emission reductions, but *Paris* also entailed a sharp turn away from the principle of common but differentiated responsibilities. The United States argued that 'emerging' economies such as China and India could no longer be considered poor states for climate purposes. On this basis, they argued that sharp differentiation between the legal responsibilities of rich and poor states should be done away with, and instead all states should choose their own voluntary targets. In the years leading up to Paris, 'emerging' states responded that there were many ways to account for economic growth without giving up on climate justice. India proposed a recalculation of emissions based on per capita emissions in the past, present, and future. China suggested targets based on carbon intensity per gross domestic product. Mexico and Indonesia added that recalculations could also include projected increases of gross domestic product. Brazil recommended that rich country commitments be based on proportionality between historical responsibilities and current greenhouse gas concentration levels and related temperature increases. Bolivia united aspects of all these approaches, dividing available emissions based on a set of indexes that considered historical responsibility, per capita ecological footprints, technological and economic capability, and sustainable development levels. Calculation on these bases allocates more than two-thirds of the remaining carbon budget to the global South, with the global North needing to radically reduce emissions because it has already used up most of its carbon budget. Ultimately, the United States defeated each of these proposals as a pre-condition for its participation in the *Paris Agreement*, rendering a regulatory and symbolic setback for climate justice.

In the wake of spectacularly unsuccessful attempts to reduce greenhouse gases, international climate lawyers today increasingly focus on trying to get rich states to help poor states adapt to climate change through transferring technological and financial aid. Rich states routinely block attempts to develop legal mechanisms for climate redress. For instance, reparation claims could potentially be developed based on state responsibility for transboundary harm and the resulting loss and damage. But rich states prefer the language of aid and largesse to that of causation and responsibility. By channelling their contributions through global funds, rich states control the distribution of finance and technical know-how for climate adaptation. To make matters worse, regular assurances of scaled-up funding for climate adaptation and resilience are echoed in international meetings, yet rich states are yet to deliver on the bulk of it.

By now, it should be evident that international law is structured in a manner that makes climate justice claims difficult to articulate and maintain, let alone address. As we have seen, climate struggles usually transpire in the context of international environmental law, which aims to protect the natural environment. However, the fossil fuels that cause climate change are regulated under a different legal regime. International law characterizes fossil fuels as natural resources, which are governed on the principles of free commence and maximizing efficient exploitation, through a mixture of private international law (international business transactions) and public international law (international economic, trade, and investment law). International law has long-functioned to protect the private economic sphere in ways that promote accumulation and thereby benefit the rich, whether in the global South or North, and further globalised capitalism. The very origins of international law lie in doctrines put forward by European imperial actors to allow private actors from the North to exploit natural resources in the South. The drive to economically develop in this exploitative vein has remained a propelling force of our disciplinary evolution ever since.

Indeed, the power of private law over the public imagination overshadows the climate regime itself, which in both *Kyoto* and *Paris* turns to economic incentives for climate solutions. The 'green economy' and 'green growth' are our supposed salvation, as if a blind faith that capitalism could solve the very problems it creates. This contradiction is epitomized in ostensibly virtuous calls for more efficient use of fossil fuels, instead of tackling the problem of overconsumption. Such approaches ultimately exacerbate climate change through by making more fossil fuels available for unlimited (albeit more efficient) usage. Indeed, many so-called green solutions, from biofuels to electric vehicles, from carbon offsets to carbon trading, are creative ways to fuel economic growth but do not stand up to scrutiny when it comes to stabilizing the climate. Whether we like it or not, economy and ecology are inextricable, as are the chains that link consumption and waste, as are natural resources and the environment. International lawyers' ability to compartmentalize them produces convenient regulatory schizophrenia that allows environmentalism itself to be captured by pro-market forces in the untrammelled pursuit of economic growth.

The separation of international environmental law from other aspects of public and private international law makes it impossible for most of the world to adapt to the changing environment in humane ways. To take the example that perhaps provokes the greatest fear within rich states, it is estimated that hundreds of millions of people will be displaced from the global South as a result of climate change. International law currently addresses population movement in such limited ways that it cannot conceive of a shift in our planet's habitable zones. Refugee law provides international protection only for a very narrow category of people who face specific extreme forms of discrimination (only 0.34% of the world population has refugee status). Everyone else crossing an international border is dependent on the discretion of host states for permission to remain. Such a legal framework wilfully obfuscates mass-protracted displacement produced habitually by dominant development patterns, which are so systemically destructive that they foment conflict, insecurity, inequality, drought, famine, desertification, environmental toxicity, and of course climate change. Yet attempts to acknowledge in the *Paris Agreement* that those displaced by climate change need protection, including extreme cases where entire sovereign territories are submerging, were staunchly vetoed by rich states.

Climate change confronts international lawyers with the systemic injustice we help create and maintain. Climate change is not an emergency that we can react to via specialized legal regimes, extreme measures, and extrabudgetary resources. It has been a long time in the making and is the inevitable consequence of an economic system that protects the rich and powerful from the harmful consequences of their actions. International law helps to structure a system of climate apartheid, where the privileged few can treat everything and everyone else in the world as superfluous, residual, and akin to waste. Against such a system, the language of justice reminds us to take responsibility for the suffering we inflict, reminding us that it is neither desirable nor possible to separate ourselves from our environment.

Further readings

Alston, Philip. 2019. *Climate Change and Poverty: Report of the Special Rapporteur on Extreme Poverty and Human Rights.* UN Doc A/HRC/41/39 (25 June 2019) https://www.ohchr.org/Documents/Issues/Poverty/A_HRC_41_39.pdf

Alam, Shawkat; Atapattu, Sumudu; Gonzalez, Carmen; and Razzaqu, Jona (eds.). 2015. *International Environmental Law and the Global South.* New York: Cambridge University Press.

Khoday, Kishan. 2014. *Environmental Justice: Comparative Experiences in Legal Empowerment.* UNDP. https://www.undp.org/content/dam/undp/library/Democratic%20Governance/Access%20to%20Justice%20and%20Rule%20of%20Law/Environmental-Justice-Comparative-Experiences.pdf

Natarajan, Usha and Dehm, Julia. 2019. 'Where is the Environment? Locating Nature in International Law' *TWAILR Reflections*, 3 (3 August 2019) https://twailr.com/where-is-the-environment-locating-nature-in-international-law/

20

CORPORATIONS

Bhavani Raman

Corporations are commonly identified today with private business corporations. Law school courses on corporate law focus on the governance and finance of private for-profit corporations. However, the corporate legal form has been and is still used for many other purposes: charitable organizations, universities, municipalities, and boroughs. Public bodies such as transit or housing authorities are also often, indeed usually, incorporated. A historical overview of the socio-legal aspects of the corporation is useful because, at each given juncture, the legal form of the corporation has crystallized debates on how collective life should be organized. From its beginnings in ecclesiastical organization and the medieval university to its later incarnation as a vehicle for colonialism, industry, finance capital, and urban redevelopment, corporations have been used to facilitate many aspects of political and economic life.

The word 'corporation' dates to late Middle English and derives from the Latin *corporare*, 'To combine in one body'. The word *corpus*, can refer to a human body, a religious body, or a body politic, or a body of law. Incorporation endows collectives with the attributes of a person before the law. But unlike other collectives, informal associations for example, a corporation is more than a sum of its members. A corporation can borrow money, own assets, and pass them on in perpetuity. It is an 'artificial person;' it enjoys a separate legal personality. An individual can only transmit property to future generations through the machinery of wills and estates. Corporations, however, do not have to do anything about their property when members die. This separate legal personality also secures the corporation's autonomy from interference.

Although the legal tools for pooling resources and organizing collectives can be traced to ancient Rome and European guilds, it was the European medieval church that made the legal governance of collectives the enduring ground for theorizing and debating artificial personhood (*persona fictive*). Over the 13th and 14th centuries, academies, monasteries, cathedrals, and bishoprics sought incorporation to protect their autonomy from kings and secure their property in perpetuity. The largely ecclesiastical orders had remained autonomous from the parish-bishopric structure despite papal centralization. Now they made substantial claims to self-government. Incorporation posed somewhat of a threat to papal authority and in the debates that ensued, various meanings of incorporation were proposed. For Pope Innocent IV (d.1254), the corporation was one unitary entity, *universitas*, composed of its physical members. But a more capacious interpretation of the artificial legal person grew out of litigation about the locus of authority in the church. Jurists preoccupied with where God had implanted authority were by no means certain that the Pope enjoyed exclusive authority.

Further new theories of the artificial person also emerged when the corporation served as a model to theorize political entities, including the state itself, and municipalities. English boroughs began to be attributed with a legal personality in the 12th century. A key move was the incorporation of the city of London, in 1327, with the heads of the merchant guilds as the members of the corporation. The Corporation of the City of London survives to this day, ruling over a tiny area around the Bank of England and having its own police force, quite separate from the Met. In the 14th century, Italian jurists known as the Commentators imagined the city as a corporation, a city-*populus*, simultaneously as a body of many human beings *and* an abstract unity. Thus imagined, the city-*populus* was an entity that acted through its physical members but was quite distinct from its component humans. Baldus (1327–1400) was an Italian jurist who treated the sovereign kingdom as a corporation, an entity that outlived its mortal members. Baldus was famous for proposing the theory of the King's two bodies – the mortal person, on the one hand, and an abstract legal entity (the Crown in English law) on the other. In an important book, *The King's Two Bodies*, Ernst Kantorowicz points out that this second aspect expressed the perpetuity of the *persona ficta* and hence contributed to the growing legal theory of corporations at this time. It imagined the corporate body not as a collection of humans living in the same place and time, but as a 'body politic' not subject to ordinary human mortality.

This history of political and religious life shows that the corporation was not invented for conducting private commerce. Indeed, the legal act of creating a corporation to secure collective property and special powers in perpetuity was generally justified in the name of a higher purpose *other* than private gain. In the late 18th century, William Blackstone made this point in his chapter on 'Of Corporations'. Blackstone defined bodies corporate or corporations as artificial persons who maintain a perpetual succession. The rights and immunities granted to those individuals who were members of the body corporate were preserved entirely and in perpetuity for and in the collective, even as individuals died. For Blackstone, the primary purpose of the act of incorporation was 'the advantage of the public'.

By Blackstone's times, however, the corporate legal form had become the primary legal form for all manner of enterprise beyond churches and universities. It was used for settling and governing overseas colonies, building canals, and organizing finance – but still all in the name of a higher purpose. Only now the 'greater purpose' was not just associated with spiritual or learning collectives but was inseparable from commercial activity. Merchants and the Trading Corporation had played an important role in bringing about this change.

From the late 15th century, both monarchs and merchants turned aggressively to the corporate form to consolidate their powers in alliance and competition with one another, and extend their power to far-off lands. What made these proto-global corporations feasible, politically speaking, was the argument that commerce was a greater good, not just a benefit to the individual – a good that would benefit all including home governments. The monarchy that authorized these trading corporations was expected to also reap benefits from these speculations. Trading corporations also secured autonomy or jurisdictional rights over spaces (forts, cities, trade routes) and the right to protect these privileges with arms, special flags, forts, and sometimes their own coinage.

From among the trading corporations, the joint-stock company emerged as a truly consequential innovation. The great European trading corporations of the 16th and 17th century (the Dutch East India Company, the Hudson's Bay Company, the East India Company) were examples of Joint Stock Companies. The joint-stock company sold shares to individual shareholder-investors, so its members-shareholders were not necessarily merchants themselves. This distinguished the joint-stock company from a merchant guild or a partnership where those who invested in the activity also managed the corporation. As a separate legal entity, this type of company was managed by professionals through the structure of committees who directed its operation.

The Dutch East India Company (*Vereenigde Oostindische Compagnie*; VOC) is an excellent example of this innovation. The VOC was founded by a government-led amalgamation of several smaller Dutch trading corporations in 1602. Beginning with a monopoly on the Dutch spice trade, it quickly expanded its interests in other commercial activities including a little acknowledged slave trade in the Indian Ocean. Its ability to do so came from its decision to pool capital by issuing bonds and shares to the general public, becoming the first publicly listed company. In 1657, the British East India Company explicitly copied the VOC structure.

But such arrangements also forced legal questions: where did authority lie in the corporation? On whose behalf was the Company acting? Could it act like a state and exercise executive power? Powerful criticisms arose from those who feared such 'corporate sovereigns' were diluting the monarch's power and also from those on the opposite side, advocating parliamentary sovereignty. Yet corporations enabled rapid European colonization and often argued that their expansion was, in theory, on behalf of the relevant monarch. Recall also, that along with these shifts in commercial life, the place of the monarch also shifted in the grand age of commerce. Theorists of the state refurbished legal corporate life when they theorized the body politic as based on a horizontal 'social contract' among individuals rather the divine right of kings. Philosophers like Thomas Hobbes argued that a polity produced by and through the consent of its members would be an all-powerful entity that would last indefinitely and ensure the security of individuals. Thus, periodic crises and anxieties notwithstanding, the corporations became a dominant legal form.

Corporations attracted both plentiful private capital and state investment by the early 18th century. Because a corporation enjoyed a separate legal personality and could exist in perpetuity, it could offer a way to amortise costs over a long time while its members or shareholders were exempted from being personally liable should things go wrong. The corporation soon became a solution to solve financial crises and facilitate large investments. The nationalization of Britain's debt, for example, established the Bank of England, as a corporation, in 1694. This act meant that the British Government would be perpetually solvent. Corporation also proliferated in two specific sectors: infrastructure (such as public works, canals) and finance (insurance and banking). Together with the fruits of empire, the legal form of corporations stoked the fires of industrialization.

In Britain and its imperial territories, corporations facilitated the speculative investment of surplus capital overseas in railroads; large canals (the Suez); and transport, agriculture, and finance projects in India and Egypt. The legal standardization of the for-profit corporation enabled an expansion along global supply chains forged by imperial shipping and transport routes merging industrial capitalism with imperialism. At this time, the corporation became the preferred mode for organizing *private for-profit* enterprise.

The landmark case fought in the United States, *Trustees of Dartmouth College v. Woodward* (1819) is often cited for the creation of limited liability private corporations. The case was a dispute over whether a college charter was a private contract between the King and the trustees outside the purview of governments. It influentially defined a large number of corporate entities as private corporations, separate from public corporations. *Dartmouth* was not the last word on whether state legislatures can modify or repeal charters. But it paved the way for the corporate legal form to be used to defend collective entities that prioritized the interests of *their members* rather than the greater good.

In the years that followed *Dartmouth*, Blackstone's higher (public) purpose of the corporation was firmly relocated in the domain of private contract. Two interesting attributes of the corporation followed from this theorization. It was commonly agreed that the corporation as *persona ficta* would ensure that natural persons could protect their property from arbitrary seizure by government. But it was also increasingly popular to argue that the corporation was entitled to the same constitutional protections as a natural person. Another landmark case in the United States, *Santa Clara v Southern Pacific Railroad* (1886) opened up the possibility that a corporation be granted equal protection under the 14th amendment of the United States Constitution. Ironically, the amendment was envisaged originally to protect the civil rights of former slaves. In the present, the US Supreme Court has declared that corporations even have freedom of religion – a finding that allows family owned corporations not listed in a stock exchange to discriminate, say against LGBTQ+ folk whose lifestyle is abhorred by the religion of the corporation's owners.

The resignification of the corporation not only as a private entity, but the redefinition of the greater good as free association and free capital outside the purview of government, did not of course mean the end of the civic or state-funded corporation. The colonization of Eurasia proceeded through state-funded corporations in China and Japan that built railroads and infrastructure. On their part, urban municipal corporations thrived in the industrial age. However, they also changed shape. The municipal corporation traditionally was a property owner and undertook public functions such as maintaining order and welfare. For centuries, these were interdependent roles endowed by Royal Charter to the municipal corporation. In the United States, as Hendrik Hartog's study of New York City shows, these two roles came into conflict in the 19th century. State legislatures compelled municipal corporations, especially wealthy ones like New York City, to stop accumulating revenue-generating assets and only regulate and tax them. More recent debates on the public role of municipal corporations might be dated to these 19th century disputes.

Other public bodies which thrived in the industrial age are statutory corporations. Statutory corporations are derived from a legal tool, the trust, which sets up a fiduciary relationship in which a 'trustor' transfers assets to a 'trustee' to benefit the third party, the 'beneficiary'. The trust can be a business entity, a non-profit or charitable concern or a public statutory corporation (such as today's 'conservation trusts'). The trust was used for a variety of purposes. Public trusts were recognized in the early 19th century as distinct from charitable trusts to facilitate the accountability of governments. Ports and other transportation facilities (e.g. toll bridges) were built by means of the legal machinery of trusts from the 1870s, all around the British empire. Trusts were also used for urban redevelopment. After a plague outbreak in Bombay, the Bombay City Improvement Trust was created in 1898, on the lines of the Glasgow Improvement Trust, which had overseen the management of land use and urban housing in Glasgow. In some parts of the British Empire, local governments were able to combine the English legal category of trust with somewhat similar categories from Islamic or Hindu law. A key legal term here, for example, is the 'waqf', used to convert property, personal or religious, into a kind of trust. This legal tool is widely used today in places like Beirut for urban development purposes, even though the waqf is normatively understood to use assets to further religious goals. Like the European medieval church, here too the corporation's legal form blurs hard distinctions between the secular and religious.

Conclusion

Many contemporary innovations, such as public-private partnerships, are not neo-liberal inventions. Instead, these entities are built from the same legal tools that gave us trusts and corporations (those two categories of course not being very discrete, as has just been shown). Many current debates over the corporation are animated, often unknowingly, by far older discussion about the corporate 'person'. But most importantly, the legal history of the theory of corporations decentres the private business corporation as the primary model for understanding collective economic and social agency. The history briefly sketched here reveals a complex account of how the corporate legal form has facilitated and negotiated many diverse understandings of collective life.

Further readings

Hartog, Hendrik. 1989. *Public property and private power: the corporation of the City of New York in American Law, 1730–1870*. New York: Cornell University Press.

Kantorowicz, Ernst. 2016. *The king's two bodies: a study in medieval political theology*. Vol. 22. Princeton: Princeton University Press.

Maitland, Frederic William. 2003. *Maitland: state, trust and corporation*. Cambridge: Cambridge University Press.

Moumtaz, Nada. 2017. 'Is the Family Waqf a Religious Institution? Charity, Economy, and Religion in French Mandate Lebanon'. *Islamic Law and Society* 25 (1–2): 37–77. https://doi.org/10.1163/15685195-02512P03

Stern, Philip J. 2011. *The company-state: corporate sovereignty and the early modern foundations of the British empire in India*. New York: Oxford University Press.

Stern, Philip J. 2015. 'The English East India Company and the Modern Corporation: Legacies, Lessons, and Limitations'. *Seattle UL Rev.* 39: 423.

21
DATA

Jennifer Raso and Nofar Sheffi

Let's begin with a short exercise. Below write five sentences about 'data'.

- ____
- ____
- ____
- ____
- ____

Read what you wrote. What does it tell you about 'data'? Can you infer a definition of 'data' from your statements?

Data as inputting

'Data' often refers to a thing that is taken into account or that serves as the basis for a finding or decision. The word appears in sentences such as, 'After inputting all of the data, we'll be able to run an analysis'. Before we can input data, however, we must first classify what we enter using pre-established categories.

Imagine we must enter data about what we call 'data' into a database and, to do so, we must select one description of 'data' from the following list:

Data is

- inputting
- processing
- organization
- abstraction
- units
- aggregation
- resource

How frustrating! We must choose one of these options. If 'data' is many things at once, which box do we tick? If our observations about 'data' don't fit perfectly into the above categories, do we find the next best option? Do we reconceive 'data' and thus modify the options available to us? What if reconceptualizing is impossible?

Data as processing

Data is a 'known'. It is assumed as known and used to express or 'discover' an unknown.

This is what mathematicians do: they solve problems by expressing unknowns in terms of knowns. Let's say: 'X = 3(Y) + 5'. Mathematicians are asked to discover the value of X and are given input data, such as 'Y=5'.

X and Y are the 'variables' in the equation.

- A 'variable' is an abstract symbol that represents the unknown in an equation or function. Theoretically, it can represent different numerical values. In our equation, both X and Y can potentially assume different numerical values.
- Because, in this case, the value of Y is given as input, its value will be assumed and treated as a constant: as a known. The question posed to the mathematician effectively is: if the value of Y is 5, which value of X will make the equation statement true?

To solve the problem, the mathematician will substitute Y for '5' and express X in terms of Y. According to the rules of algebra, she'll tell us that, if the value of Y is 5, the value of X is 20.

Expressing the unknown in terms of the known is also what computer programmers do. They're given problems and write executable computer programs to help solve these problems. Perhaps a coder is told to write a program that helps producers of goods calculate their net profit. How would she approach the task? She'll write a function that considers defined variables – the number of units produced, the number of units sold, price per unit, production, operation, and marketing costs, etc. – to calculate profit.

- The predicted profit is the dependent 'variable' – the 'unknown'. It is what the producers want to make known through software.
- Production costs, marketing costs, the number of units produced, etc. are variables for which a value will be inputted each time the program runs. These variables will be assumed by processors as a given. They'll be treated as 'knowns' of fixed value.
- The program will instruct the processor how to calculate profit, taking into account the 'known' value of multiple variables. The 'knowns' will be manipulated to imbue the 'unknown' with value, making it 'known'.

The programs coders write are executed by our computers each time a process is initiated. Computers also express the unknown in terms of the known. They're given programs to execute and data to input.

- These programs instruct them how to generate knowledge.
- The input data infuses certain variables with value, rendering these unknowns temporarily known.

Positing the value of various variables, computer processors follow program instructions and discover unknown(s) value. They generate output data: knowledge that can be acted on.

Legislators also write executable programs, in a way. These programs (or codes) 'solve' social problems or 'implement' higher order programs. These actors process (as input) the data supplied to them, such as predicted responses to new statutes. They generate output: legislative programs that will be executed by a variety of actors in a variety of settings.

Tasked with executing these codes, administrative agents are (like) processors. They 'discover' and input facts, rendering them known. Following procedure, they apply 'the law' to 'the facts'. Fulfilling a function, they determine the legal meaning and effect of facts. This act generates recordable output data, which can be acted on.

Similarly, we all use the known to make sense of the unknown, to process events, and to decide on a course of action. We frame life as a series of problems, questions, dilemmas. We learn from experience (data, memory) to face the unknown.

Data as organization

Ok, so data is a known that's inputted and processed to make known an unknown. The knowledge generated in this execution process is called 'output'. Output is also data. This output data can be acted on at the moment it's generated. It can also be stored for future use.

Entities that act on data require memory to function. They must be able to save data in a memory, a database, an archive, a library, or a file. Saving data creates a recollectable past. It enables these entities to learn from experience, grasp the present, and plan for the future: to write programs for future execution and save these as data for future use.

Memory is distinct from a data dump. Imagine preparing for an interview. It's in an hour and for your dream job. YOU HAVE TO SUCCEED! You REALLY want to wear your lucky socks to the interview. You run into your room but it's a mess. EVERYTHING is in a huge pile, as tall as the tower of Babel, which makes things complicated. It would take forever to find exactly what you're looking for. If only you were organized. If only you had assembled the storage solution you bought at IKEA six months ago. Retrieving your lucky socks would then be much simpler, no?

Information systems, including our brains, don't just save data. They archive or keep records. To structure their memories and to render data locatable and retrievable, they use a knowledge organization system, which labels, classifies, and maps data. Information systems also introduce procedures that govern database access and maintenance. Without such systems and procedures, stored data is inaccessible and irretrievable. Imagine you're searching for a $10 bill that you know you had last week. You recall tucking it into the pocket of your jacket … or was it your jeans? You retrace your steps but you just can't find the $10 bill. See the issue?

To function, an organizational system requires regular maintenance. This work is performed by many actors. If you purchase a new pair of socks, you must arrange them in relation to your other clothes. Perhaps they're even luckier than your old lucky socks (maybe those old socks didn't get you the job after all). As you sort your socks, you find some have lost their partners and others are full of holes. They may not be great socks anymore, but could they become dust rags? For your storage system to remain functional, you may need to reclassify these items. Given the huge tower of things in your room, you may need a decluttering expert to make sense of it all.

Only that which is searchable can be taken up as data. If we fail to save data, that data becomes impossible to retrieve. If potentially relevant data is stored under a different label, it becomes undiscoverable and unaccountable. The structure of the database defines what can be 'searched' for: the terms of search. The search terms we input affect what can be 'found'. They influence the search process's design, the outcomes that process generates, and the construction and outcomes of the software that takes up inputted search data.

Data as abstraction

Imagine a pointillist painting. Look at it up-close. As close as you can! Come closer. Closer. Closer. Closer. Does your nose touch the painting? Great! Now, have a look. What do you see? Just dots, right? Can you make sense of what you see?

Now, move away from the painting. A bit further. Further still. Things make more sense the further you move from the painting, right? Suddenly you can connect the dots. You gain perspective. You spot windows, then a house. Some figures. A cloud. You identify a subject. You begin to interpret the scene.

Data is always an abstraction. It is a symbolic notation that signifies an object or a subject. Even its smallest bits are an abstraction. Think about the number '1'. IT's ONLY A SYMBOLIC REPRESENTATION. What does the number '0' represent? IT TOO IS A SYMBOLIC EXPRESSION. But of what? Good question. It depends.

In the decimal number system we use to count and calculate, each digit – from '0' to '9' – represents a different quantity. In mathematics, each decimal digit represents a numerical value.

Digital computers use the binary number system. They communicate with their operators using only two symbols: '0' and '1'. For these computers, '0' and '1' do not denote numerical values or quantities. They represent two states of a switch – 'on' and 'off' – that controls the flow of current through a specific electric circuit. '0' tells the computer to ensure an indicated gate is open, preventing the flow of electricity. '1' instructs it to ensure the gate is closed, allowing current to flow through the circuit.

Data is an abstraction also because it's meaningful only in perspective and in relation. To make sense of data, we must first move away and gain perspective. For example, to know what the digit '1' represents in a particular setting, we must identify it as a digit. However, we must also know which number system is used (decimal or binary) or if a different system is deployed. We also need to know how the system works and what each of the digits it uses represents.

Often data refers not to one data unit but to multiple units or a class, a category, an attribute, or a type. It is an abstraction because it is, more or less, a generalization. Think about it – what's the plural form of 'data'? That's a tricky question. The word data is already plural. Its singular form is 'datum'. We never (or rarely) use it. But why? Precisely because data is an abstraction.

Data units

Let's zoom in closer. Closer. Closer. Closer. Closer … THERE. Data is more than an abstraction. It's also these tiny units, or bits, of data. In a computer's language, the binary number system described above, the smallest units of data are 0 and 1. These Binary digITS are called 'bits'. Bits are units of (digital) data.

Programmers make things computable and aggregatable at the most minute level by transforming these 'things' into multiple units of data. Yet, to programmers, bits also represent algebraic operations and numerical values.

Alone, as we saw above, a single bit lacks meaning. To offer meaning, to communicate, data units are treated to what we might describe as a process of 'datafication':

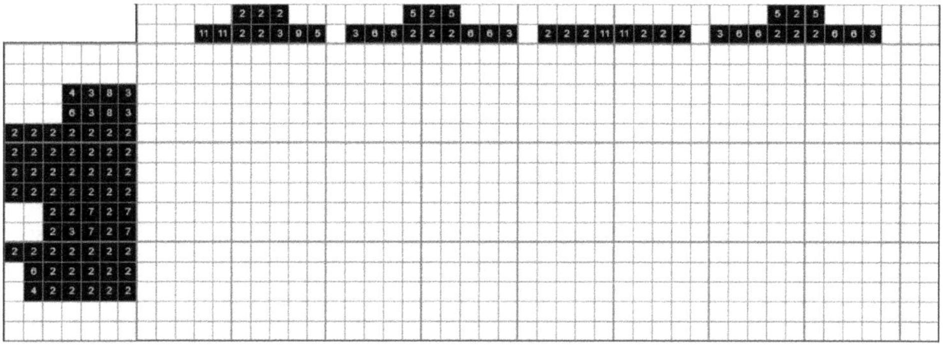

The process by which digital data units are created, counted, and (re)combined uses binary digits and switches. At the micro level of a data unit, ambiguity is purportedly intolerable. Something is either a 1 or a 0. A switch is either on or off. A punch card's surface either has openings or remains unperforated.

These units or morsels of information can be stringed, aggregated, pieced together, or broken down to generate new data.

Data as aggregation

Now let's zoom out. Oh my, that's (a) BIIIIIIIIG DAAAAATAAAAAAA (set)!

Data is generated according to rules for categorizing, archiving, retrieving, and treating inputs and outputs. This action makes data retrievable in the future for managing and processing. In this form, data becomes a (BIG or small) dataset.

Big data is collected and constructed. Commonly, big data is described as objective, accurate, and potentially predictive because we can use it to identify general features that recur together. These assumptions about big data, however, confuse data and process (if we distinguish between data inputs and data outputs). Data is process (among other things).

Counterintuitively, big data rarely describes the quality or size of a dataset. Most datasets *are* 'big' in that they are large, but they are never all-encompassing. No aggregation of data can ever be complete. What's more, there are many different big datas; some are bigger or smaller. What we do is select one big dataset over others. 'Big data', for example, often leaves out the LARGEST datasets, such as national censuses.

Big data refers to dynamic and interlinked datasets that constantly update. In other words, big datasets are not just large; they change rapidly, 'in real time'. We may imagine them as moving and having speed. These features contrast 'big data' from censuses. Unlike big data(sets), censuses are conducted periodically (typically, every five or ten years) and present themselves as snapshots of 'society' at that particular moment in time.

Some additional distinguishing features of big data include:

- When we use big data to perform actions – to classify, search, aggregate – our performance is contingent on the improvised decision that produced the (big) data(set).
- The aggregated incidences represented by a (big) dataset must have meaning. They cannot be random; they must represent a linguistic, classifiable, coherent phenomenon.
- Data is not simply extracted from big(ger) data. Big data users must look for something specific; the (big) data must 'say' something.
- Aggregation – data collection, dataset construction, curation, and analysis – requires massive amounts of (human-performed) labour.

Instructions for processing data are also data. Distinct from 'big data' they may nonetheless be aggregated into a 'code' or the 'codex'.

Data as resource

'Data is the new oil'. It is 'mined', extracted to produce value. It's traded on data markets. It's a fuel for data-driven endeavours. It's a valuable 'resource'.

But can data be owned? If data ownership is possible, who is a potential data owner? Can data be traded? If so, what is its value? Is it a resource with its own innate value?

Sometimes data is described as though it has its own intrinsic meaning, as in: 'With enough data, the numbers speak for themselves'. Yet, if data is a resource, what are the conditions and limits on its use? Is it renewable, exhaustible, something else? Do we burn through data, and if we do, what are the implications of this data combustion? Certainly 'data-driven' technologies, such as Google searches and predictive policing tools, require massive amounts of data to function. And this data dependence (including the need to store data somewhere) causes huge carbon emissions and contributes to climate change.

Is data more like water then? When too much data is collected or accessed, we dive into an ocean of data, we are flooded by a data deluge, or we are drowned in a data tsunami. We become immersed in data, like a child immersed in a pool. Droplets of data may coalesce in a cloud. Data flows across jurisdictional and geographic boundaries. Data streams are diverted, polluted, or conserved. To facilitate data transport, we're told, we must build new (or maintain existing) pipelines. But, of course, data isn't water. It relies on material infrastructures for its creation, storage, and transport. Perhaps ironically, existing data transport infrastructures (networked cables) run along coastlines and across the ocean floor. Their design and location fail to grasp the threat posed by climate change (extreme temperatures, rising oceans).

If data is a resource, what is its future?

Source: data tsunami image from gettyimages.com.

Further readings

Gitelman, Lisa, ed. 2013. *"Raw Data" Is an Oxymoron*. Cambridge, MA and London: MIT Press.

Gray, Mary L. and Suri, Siddharth. 2019. *Ghost Work: How to Stop Silicon Valley from Building a New Global Underclass*. Boston, MA: Houghton Mifflin Harcourt.

Vismann, Cornelia. 2008. *Files: Law and Media Technology*. Stanford, CA: Stanford University Press.

22
DOMESTIC WORK
Transnational regulation

Adelle Blackett

Situating domestic work

To the surprise of many observers, in March 2008, the governing body of the International Labour Organization (ILO) adopted a resolution requiring it to prepare to negotiate a new international treaty on decent work for domestic workers. A century-old institution that survived the beleaguered League of Nations to become the United Nations' first specialized agency, the ILO had repeatedly taken resolutions on domestic work since the 1920s and even considered it urgent to act in 1965. But nothing had happened. Now domestic workers had organized themselves, to force the issue onto the agenda. I was asked to serve as the ILO's lead expert. There was a ton of work to do.

The ILO ultimately defined domestic work in the resulting Decent Work Convention, 2011 (No. 189) broadly as 'work performed in or for a household or households'. In contrast, it limited 'domestic workers' to those undertaking domestic work in 'an employment relationship'. Even by that definition, there are at least 67 million women and men in paid domestic work; one in every 25 women workers worldwide is a domestic worker. The 11.5 million migrant domestic workers leave family behind in countries such as the Philippines and send back remittances. Worldwide remittances from migrant workers exceeded $575 billion in 2016. Migration to provide what can be labelled as social reproductive services (since domestic labour, paid or not, is the key to social as opposed to biological reproduction) in or for a household through an employment relationship, transnationally, has become central to a contested development strategy. Social reproduction, in this sense, according to Silvia Federicci, refers to '*all the activities that reproduce our daily life and at the same time, in a capitalist society, also reproduce labour-power*'. The work of social reproduction enables and sustains labour markets. It also sustains our humanity. Yet, paid domestic workers are required to labour under conditions that reinforce and reproduce patterns of structural inequality, transnationally.

But who are the specific women and men who do domestic work in other people's households? Asking that question brings to the fore unsettled histories, particularly punctuated by the global institution that is the centuries-long transatlantic slave trade. It encompasses the colonial encounter in Africa, Asia, and the Americas. In other words, it is part of the transnational history of racialization, whose legacies engender the racialized and Indigenous women (and some men) who were expected to do the work, on the basis of some supposedly innate predisposition to domestic servitude. Women migrant workers make up approximately 50 per cent of all migrants, and a significant majority of those women are or become domestic workers. This traditionally feminized labour remains economically and socially undervalued.

The common sense way in which the domestic work relationship is understood and regulated is part of a global legacy of subordination and servitude that operates in particular places and in particular ways on particular women's bodies. I refer to it as the asymmetrical, unequal, and largely invisible law of the household workplace. That is, nothing about the relationship was made to allow us to see just how unjust it is. We are meant to take it as a given that we are not supposed to see domestic workers' needs, desires, or aspirations.

The law of the household

Although live-in domestic work has been considered archaic and was predicted to all but disappear as societies became more egalitarian, there has been a stunning resurgence. Some people attribute this to the rise of the dual-income career family, whose employment is not easily mapped onto 9-to-5 arrangements. This is not the whole story, however. It is part of a story of expectations of perpetual availability: the domestic worker is expected to subordinate her life to those of the people for whom she works. In this sense, her work is boundariless.

Domestic workers receive substandard pay, which is often below minimum wage and for long, often indefinitely long, hours. This is the case whether the households are wealthy or of moderate income. Moreover, providing room and board is typically seen as the basis for a deduction from wages, despite the perpetual availability for employers that live-in work is understood to create. The tasks are intimate and subject to the risk of remarkably high degrees of supervision and the constant navigation of a one-on-one relationship. Hence, the general expectation of available labour power is accompanied by an expectation of servitude. A visible representation of this is the domestic workers' uniform, representing an unequivocal demarcation between the family and the domestic other.

Employers consistently use the frame of 'like one of the family', with the emphasis on the word 'like'. Most ethnographic analyses dismiss this terminology as opportunistic. Even if we were to take the family metaphor seriously, drawing on the history of the household economy, I would argue that rather than dislodging obligation, it asks: what kind of obligation? Even if we were to concede the subordination inherent in a patriarchal family's relationships, the domestic worker is not the spouse or the daughter. Her status in the family remains that of the other who is expected to deal with all of the physical and emotional labour and thus is visible when needed, but whose needs remain invisible. A critical part of the emotional work in domestic work is the obligation to be omnipresent but simultaneously invisible unless called, entirely loyal and trustworthy but knowing one's place.

It is precisely these expectations of behaviour acutely shaped along the lines of class, gender, race, age, and immigration status that mold and entrench the law of the household workplace. The boundarilessness and subservience attached to the work – couched as it may be in maternalism but also, unmistakeably, in the vestiges of the master-servant relationship to which servitude was central – is resolutely recast in light of the needs of the current era of global capitalism.

The presumed universal coverage of formal state law, within the framework of Westphalian sovereignty, may be the true fiction. Some labour codes in fact only apply to a very small percentage of workers in the formal economy – sometimes less than 10 per cent of the working population. In the global South, informal labour is on the rise, due to structural adjustment policies that touched groups like civil servants or employees in formerly nationalized firms who had moved into post-independence formal economy work. In Côte d'Ivoire, as in most nations of the global South at varying degrees of development, labour market informality – thus domestic workers' invisibility – is part of that norm. Some law and society scholarship emphasizes the fact that there is a gap between the law on the books and the law on the ground. My approach goes beyond that. Focusing on domestic work from a legal perspective helps us see what we might be missing when we move too swiftly away from the work relationship and the legal normativity that flows through it. In formalizing informality, we may jump too quickly to the question of social protection, without sufficiently contemplating legal relationships. Attention is needed to the other sources of law – beyond the state – in a manner that considers relative relational power and structural inequality

Domestic workers embody a long tradition of resistance that refuses to confuse laws that exact servitude with social justice. Domestic workers have been anything but passive onlookers in their own lives and history: they have repeatedly engaged in individual acts of resistance to exercise control over their working time. Even writing cookbooks – with their own vision of the law of the household workplace – can be seen as a form of transgression, and a source of law. Domestic workers have tended to reject live-in work and have opted to live out whenever they have been able to do so. Domestic workers have also acted collectively – for example, through media campaigns, lobbying efforts, hunger strikes, and calls for one-day work stoppages. Their goal has been clear: to transgress and in the process transform the unjust law of the household workplace. Similarly, domestic workers' dynamic participation was crucial at the ILO. Domestic workers' agency – their militancy – needed to be made clear, to prevent the myth of their passive acceptance of their lot from stalling standard setting on decent work for domestic workers any longer.

From invisibility to visibility

At the ILO, domestic workers embraced the preparatory work that articulated their claim to the human right to be included as workers. The social movement left an indelible imprint on the ILO's work, not only on the substance of the new convention and recommendation but also on the standard-setting process itself. Tripartite social dialogue (involving member state governments, employer's organizations and workers' organizations) played a central role and took on a new meaning. The ILO's process became much more contextualized and concrete. There was nothing abstract about the earnest domestic workers sitting in the room, forcing the delegates to see them. All those in the committee room recognized sooner or later the challenges and limits of their own representativeness on the subject. Several participants and observers could not stop themselves from noting that employers of domestic workers were in each of the delegations and indeed among the international officials who had brought the issues forward. However, in the official delegations, domestic workers were few and far between. Still, the committee was filled with actors new to international law, many of them racialized or Indigenous, determined, dynamic, and often charismatic women (and some men). They represented local associations from places as disparate as the Philippines, Colombia, Hong Kong, the United Kingdom, South Africa, Spain, Trinidad and Tobago, Canada, and the United States – and they came prepared.

A big part of the standard setting on decent work for domestic workers was working against invisibility and the unjust law of the household workplace. The standard-setting process deliberately set out to root out intersectional subordination faced by Black and other racialized women, a necessary precondition to make the law governing the household workplace equitable. Regulating decent work for domestic workers at its heart calls for labour to be regulated as if the most marginalized workers matter. Their marginalization has to be unsettled and they must be brought to the centre of transnational labour law's concern. Otherwise, the asymmetrical law of the household workplace will persist.

I wanted the ILO's standard-setting process to be part of a rethinking of labour law's boundaries. The biggest challenge was to ensure that domestic workers have the right to be included in labour law. This was significant because the Domestic Workers Convention and the Domestic Workers Recommendation were not intended to be symbolic instruments or even an abstract charter of rights. They were detailed and comprehensive. They were meant to give broad meaning to the notion of decent work that the ILO had been championing since 1999. Under the convention and recommendation, decent work would come to include decent working conditions and much more.

It would come to mean recognizing domestic workers' rights to equality and the freedom of association and providing protection against forced labour and child labour. It would come to include access to social protection such as maternity leave, occupational safety and health, and social security protections – even though there is some recognition of the need to improve some protections progressively. But there was still more in the documents. A premium was placed on making sure that mechanisms for inspection and dispute resolution were available. Decent work would also mean that special attention was needed to the nature and terms of contemporary labour migration, at the very least to rein in exploitative practices.

It was late, and victory seemed nigh for those seeking a new convention and recommendation. The employers' group threw its significant support behind the adoption of the historic convention and recommendation. There was singing and dancing in the committee room. Domestic workers were seen and heard.

Looking to the future: thinking transnationally

The international standard setting was quickly recognized around the world as historic success for the movement for domestic workers' rights and recognition. It offered a huge boost for ILO standard setting at a time when some people had lost faith in the process. Because of the daunting challenge of making decent work for domestic workers a reality, there was a call for international solidarity to promote and deepen technical cooperation, centred at the ILO.

Before the standard setting for decent work for domestic workers, it might have seemed improbable that domestic workers would be the ones to direct us to a place where the need to rethink the transnational would seem so obvious. But we have seen that domestic workers do a lot of border crossing: between production and reproduction, histories of subordination and servitude and their contemporary legacies, the national and the transnational, and state law and the law of the household workplace. Domestic workers embody the transnational that they have helped to build. They are also well placed to call for a transnational labour law that rethinks how people move and under what conditions. The limits of bilateral agreements coupled with a proliferation of international and regional texts affirming domestic workers' rights might signal that this is the time to make a different kind of case. The energy alone needed to identify and punish the proliferation of rogue recruitment agencies should reorient policy from the individualized solutions to the structural. I acknowledge that liberalizing migration within a framework of reasonable labour market access for the movement of people, with a premium on rights of permanent migration, may seem a particularly unlikely direction in the current moment of significant backlash against current migration flows.

The ILO Convention exemplified how a transnational legal order on decent work for domestic workers has been built, in a manner that addressed rather than sidestepped unequal power relations. The new transnational legal order was built to dislodge the asymmetry. It relied on multi-directional learning from ILO Members that had experimented significantly with regulating decent work for domestic workers. The standard-setting transgressed directional presumptions, to move in many cases from countries in the global South to those in the global North. The new transnational legal order unsettles the unequal law of the household workplace. It seeks to transform the domestic work relationship on an ongoing basis. As law remains central to globalization, thinking about transnational labour law should avoid overstating the disruptive power of plural sources of law. Decent work for domestic workers will simply not happen unless the transnational solidarity that brought the issue to the ILO's standard-setting table is sustained, and unless domestic workers themselves remain front and centre.

Further readings

Ally, Shireen. 2009. *From Servants to Workers: South African Domestic Workers and the Democratic State*. Ithaca, NY: Cornell University Press.

Anderson, Bridget. 2013. *Us and Them: The Dangerous Politics of Immigration Control*. Oxford: Oxford University Press.

Blackett, Adelle. 2019 *Everyday Transgressions: Domestic Workers' Transnational Challenge to International Labor Law*. Ithaca, NY: Cornell University Press.

Boris, Eileen. 2019. *Making the Woman Worker: Precarious Labor and the Fight for Global Standards, 1919–2019*. Oxford: Oxford University Press.

23
EXTRACTIVISM
Socio-legal approaches to relations with lands and resources

Dayna Nadine Scott

Resource extraction is an activity that humans can undertake in relation to land, resources, or elements of the earth. Mining, forestry, drilling for oil, and fracking all entail extraction. On its face, it describes the act of taking, from nature, resources for human use, typically non-renewable resources. But 'extractivism' connotes not an activity, but a relation: here, the concept does not turn on the type of resource taken, but on the underlying political economy. That is, extractivism is a particular *way* of relating to nature. The term is not reserved for fossil fuels and mineral extraction; neither would it apply to the extraction of those materials in all contexts and circumstances.

Extractivism is a mode of accumulation in which a high pace and scale of 'taking' generates benefits for distant capital without generating benefits for local people. It is a way of relating to lands and waters that is non-reciprocal and oriented to the short-term. Extractivism refers to a particular logic endemic to and intensifying under contemporary global capitalism, but also with a very long history. It is a system that was installed and entrenched as early as the 16th century, by many accounts, as colonialists chased timber, rubber, furs, petroleum and other bounty, and rendered them into global commodities.

Extractivism has been theorized in critical development studies, with leading contributions from Roger Merino Acuña and Eduardo Gudynas, as a specific type of developmental path in which the social and environmental costs of resource extraction exceed its benefits. Further, the benefits tend to be highly concentrated, while the costs are diffuse and disproportionately borne by poor and marginalized local residents. The meagre benefits that are generated (or promised) locally, such as employment, tend to be inequitably distributed, with gender being a crucial determinant. Jobs and training opportunities associated with extraction typically benefit men in the communities, while major extractive projects, through the introduction of work camps (or 'man camps') and the opening up of remote areas with new infrastructure, often secured by private security services, pose profound risks to local women and girls.

Although its adverse effects may be atmospheric (as is the case for the extraction and burning of fossil fuels), or may be dispersed across landscapes through polluted waterways (as is the case in mining and fracking) or by associated infrastructures such as pipelines (as is the case for oil and gas), the dynamics of extraction are shaped by the reality that the 'taking' by necessity must happen in the specific places where the resources are found. Where the resources are found is in the 'extractive zone', as Macarena Gómez-Barris terms it: resource-rich regions of high biodiversity targeted by colonial paradigms, and where complex social ecologies and lifeways are dismantled to achieve conversion of environmental elements into 'resources' for capital. In the process, affected communities are often dispossessed from their means of social reproduction, land-base, social and ecological wealth, and economies, and evaluated through extractive logics, for their capacity to provide wage labour for the extractive industry.

But extractivism meets fierce resistance from local communities worldwide, especially Indigenous peoples. Often this resistance exposes deep rifts between land-based peoples' cosmologies that rest on principles of deep relationality with lands, waters, and other beings, and extractive logics, in which elements of the earth are understood as resources presumed to exist for the sole purpose of providing sustenance to humans. The idea that humans may be in relationships with these elements, that these elements exist in relation to each other, and that those relations may include not just the right to 'take' but obligations to protect or to steward, is emerging as a direct challenge to extractivism. In fact, this challenge is now most forcefully expressed as one of a 'conflict of laws': with Indigenous peoples pointing to their own pre-existing, still operating, and always evolving, legal orders as the source of their obligations to protect and defend the land.

Across the extractive zone, struggles over territory, jurisdiction, and authority are transpiring as state law, international laws and norms, and new transnational governance mechanisms are coming up against local Indigenous or customary laws and practices, forging dynamic conditions of legal pluralism. Extraction is regulated on the global, transnational, and international levels by a dizzying array of overlapping and potentially conflicting norms and processes. These include measures in public international law, such as international trade and investment law, environmental law treaties such as those governing pollution and wastes, and the Paris Climate Agreement, and transnational governance tools led by non-state actors such as the Equator Principles (for socially and environmental responsible lending) or the Kimberley Process (for ethical diamonds). At the same time, there are international legal norms governing the use of 'resources' that do not adopt the aims of stewardship or protection of global commons, such as atmosphere or oceans, but rather the aim of enabling efficient exploitation and protection of states' 'permanent sovereignty over natural resources'.

In direct tension with states' claims to exclusive authority, of course, are the claims of Indigenous peoples also articulated in international law norms such as the United Nations Declaration on the Rights of Indigenous Peoples (the 'Declaration'). In struggles over extraction, the crucial contestation has centred on the right of Indigenous peoples to 'free, prior, and informed consent' ('FPIC') in respect of developments in their territories. The claim of FPIC attempts to intervene on the governance of extraction at the global scale and is having widespread impact. The most highly contested questions, then, for socio-legal studies have become: What constitutes 'consent' to extraction in the contemporary moment? How can we be confident that it is freely given? How much room for alternatives to the extraction, or for different conceptions of 'development', have to be on the table for people to be understood as actually 'choosing' an extractive project?

One marker often put forward as evidence of consent to extraction is a signed contract, or 'access and benefits agreement'. In fact, the dynamics of resource extraction across the globe are increasingly shaped by the negotiation of a complex web of contractual agreements. These include ostensibly 'private' contracts negotiated between extractive companies and local communities, now often called Impact-Benefit Agreements, Community-Benefit Agreements, or Mutual-Benefit Agreements. They also include resource-revenue sharing agreements between regional organizations or tribal councils and various state or sub-state authorities; early exploration agreements and MOUs between Indigenous communities and mining companies; framework agreements between communities and various governments and agencies over infrastructure or environmental assessment funding, among others. Over the past several years, companies in the extractive sector have come to accept that this 'deal-making' with local communities is a necessary part of the game – sometimes described as getting 'social license' to extract.

But across the extractive zone, many approved or state-authorized projects are not being built because of land defence actions by local peoples, or lengthy court challenges. Often, corporations' success in the courts cannot even guarantee that projects will proceed, because of the increasing capacity of Indigenous land defenders to disrupt extractive operations by making compelling claims of competing jurisdiction, or advancing arguments based on FPIC. In these contexts, the framework provided by state law is typically not capable of achieving the 'resource certainty' that industry demands. It is here that we can observe the distinct alignment of state and industry aims that characterizes the extractivist orientation. In these conditions, negotiating a deal becomes the first priority of companies interested in advancing a controversial extractive project, and the task of facilitating those deals falls to state actors invested in re-legitimizing those projects in order to safeguard 'investment'.

This regime of contractual negotiation, or 'consent by contract', is a mode of governance that attempts to define the social, political, ecological, and economic relations regarding the use of lands solely through confidential bargaining and agreement-making between private extraction companies and affected communities, but in fact affords the state a key role in setting the terms. Since their emergence in the 1990s, despite being formally 'voluntary', the negotiation of an impact-benefit agreement has become in practice almost compulsory. The basic deal is as follows: the affected community typically receives promises that the company will minimize or monitor environmental impacts, provide various economic benefits such as annual payments, and meet local employment targets. In exchange, the community promises to provide access to its lands, and offer varying degrees of support for, or acquiescence to, the project.

Many scholars have raised concerns about the way the deals are struck. These include complex questions of representation with respect to who has the authority to provide consent on behalf of a community. Confronting these questions means acknowledging the potential applicability of a plurality of applicable legal orders, including Indigenous legal orders, generating challenging questions about where authority lies, over which lands, and how community deliberation shapes its exercise. There are concerns about confidentiality clauses, which typically prohibit the communication of the terms of the agreement to anyone outside the negotiating process. There are also legitimate worries about the support and cooperation covenants, or what critics call 'no-protest clauses', that seek to prohibit community beneficiaries from taking any actions that could undermine the company's social license.

Thus, the general thrust of critical socio-legal scholarship on the contractual regime of today's extractivism is that it tends to undermine the practical ability of Indigenous peoples to envision priorities for their lands for themselves and determines for them a set path towards extractive capitalism and a wage economy. But it is also true that the opportunity to negotiate a deal represents a chance for a community to have some influence over the form, scale, and pace of development – an outcome that is hard to otherwise come by through state public law frameworks. Thus, as political ecologists have demonstrated, even new forms of 'progressive extractivism' that incorporate benefit sharing with affected communities – a function of the contractual turn – tend towards standard forms of economic development and a wage economy. This is because they depend on growth, an ongoing and linear process of material 'progress' that forces communities to accept a re-definition of relations with land into capitalist terms of revenues, assets and individual gain, and make affected communities more dependent on revenues generated through extractivism in order to meet their basic fiscal needs.

The fundamental problem is that the Indigenous communities whose lands are threatened by extractive projects are most often not recognized as holding the jurisdiction to decide whether permits should be granted in the first place. Notwithstanding the widespread adoption of the UN Declaration, the idea of FPIC – and conversely, the possibility that 'no' could mean 'no' – is not yet a feature of the public law regime in most jurisdictions. Thus, the contracts themselves cannot be taken as evidence of 'consent' without investigating the background context for the negotiations, including the underlying power relations. And here it is clear that the backdrop set by the formal state law supports and structures the engagement process, providing valuable resources, and bargaining entitlements to the parties. State law clearly allocates the legal rights and duties that come to shape the contractual order. The most crucial of these allocations allows the nation state the underlying title to, and jurisdiction over, all of the lands within its borders, thus securing its ability to provide access to Indigenous lands for extractive capital.

Further readings

Gómez-Barris, Macarena. 2017. *The Extractive Zone: Social Ecologies and Decolonial Perspectives*. Durham, NC: Duke University Press.

Seck, Sara L. and Simons, Penelope. 2019. 'Resource Extraction and the Human Rights of Women and Girls'. *Canadian Journal of Women and the Law*. 31(1): i–vii. https://doi.org/10.3138/cjwl.31.1.01a

Scott, Dayna Nadine. 2019. 'Extraction Contracting: The Struggle for Control of Indigenous Lands'. *South Atlantic Quarterly*. 119(2): 269–299. https://doi.org/10.1215/00382876-8177759

Szablowski, David. 2010. 'Operationalizing Free, Prior, and Informed Consent in the Extractive Industry Sector? Examining the Challenges of a Negotiated Model of Justice'. *Canadian Journal of Development Studies* 30(1): 111–130. https://doi.org/10.1080/02255189.2010.9669284

24
FINANCE, BANKING AND DEBT

Mariana Valverde

The world of finance is not accessible to everyday people. Business reporters tend to present Wall Street as a god-like force capable of solving all problems, while critical scholars see finance as an evil force, but still present it as unified and as able to control every detail of life. As recent work in the anthropology of finance (by scholars such as Karen Ho, Anna Tsing, and Annelise Riles) has shown, the financial industry – which is in fact highly fragmented – is often presented in both pro-business and critical accounts as a divine force that mysteriously appears in different parts of the globe simultaneously and possesses the ability to make both miracles and disasters happen.

In recent years, work in the area known as 'social studies of finance' has opened up the black box of international banking to reveal the organizational and human foibles that make up the reality of this seemingly divine entity. This research sheds some light on the legal and regulatory structures that put some limits – often ineffectually – on the greed of banks and bankers, but often only incidentally. Socio-legal research is very much needed to document in detail how laws and regulations meant to make the financial system more transparent and fair work, or don't work.

The main point of this essay is that there is no such thing as 'finance' in general, and that it is time to move beyond both praising and condemning finance as such. We need to understand the very different phenomena and actors that make up various sectors of the financial industry if we are going to suggest better regulations.

Here we will cover only three types or levels of financing: financing individual and household goals; financing corporations; and financing government activity. The striking differences amongst these three common types of financing mean that different types and methods of regulation need to be applied.

Household and individual finance

In the United States, as in many other countries, financing for individual and household needs and wants is generally provided by commercial banks, specialized mortgage lenders, and credit card companies. In countries where long-term renting is common, such as Germany and Spain, residential mortgages are not as central as they are in countries, such as the US and Canada, where long-term legal protection for tenants is non-existent and there is a powerful culture of home ownership.

The dangers of loosening up mortgage-lending regulations became apparent during the 2008 world-wide financial crisis, which as is well known was triggered by unregulated 'sub-prime' mortgages. What is less well known, however, is that the huge and growing size of the college student debt market also poses a major threat to the financial system as a whole. In 2012, for example, student debt in the US amounted to more than the sum of residential mortgages, car loans, and credit card debt, all put together.

Debt is the bread and butter of the financial industry. Often all that is known about a type of debt is the total amount: how much credit card debt exists in a particular country at a point in time, for example. But debt matters to both individuals and economies for other, more qualitative factors. A good example concerns what has been happening to the massive student loan debt market in the US. Given rates of graduate unemployment and given the fact that student loans, unlike mortgages, are often issued without any security, student loan debt has been seen as quite risky, and governments have often decided to backstop these debts so the banks don't suffer and don't stop lending to college students. But to boost the market in student debt, private lenders aggregated large numbers of individual student loans and reassembled them into new products known as SLABS – student loan-backed securities (paralleling the now discredited mortgage-backed securities). During these years, the US government passed laws making it much more difficult to defer student debt payments or avoid them altogether, e.g., by changing bankruptcy rules. Thus, in keeping with the general trend of law to further the interests of capital rather than of individuals (recently documented in Katharina Pistor's book *The Code of Capital*), a pincer movement was created whereby greater financial pressures were compounded by unfavourable legal changes.

This story makes the larger point that social studies of finance can help us understand the important **qualitative** differences amongst the forms that debt can take, and thus shed light on the old socio-legal question of which interest groups are able to influence government regulations and which groups are disadvantaged by changing regulations.

Corporate financing: stock exchanges vs investment banks

All efforts to seek financing are ways of incurring debt. But when large corporations seek financing for their projects, or just to survive, they do not visit the local commercial bank manager. Corporate finance can be very complex, but some general points can be made that help demystify the game. First, a corporation can 'go public', which means it can start issuing shares – little slices of ownership in the company – on stock exchanges. The issuing of shares, a method of financing that is not available to individuals, may seem like an easy way to get funds. However, issuing shares comes with many responsibilities. Since stock exchanges can only operate if potential purchasers have sufficient information enabling them to choose between the different companies listed on stock exchanges, the decision to 'go public' necessitates making public information about the company that owners and managers may have preferred to keep secret. Annual financial reports must be made available, and the corporate directors' names must be revealed. In addition, regulatory bodies, such as securities exchange commissions, have some powers to investigate the listed companies' claims. Securities exchange commissions are notoriously under-resourced and rarely impose major penalties. But, in general, the issuing of shares as a method of financing is subject to certain obligations, especially in regard to transparency.

By contrast, if a company chooses to seek financing outside of stock exchanges, little or no information needs to be made public. In part for that reason, there is today a far greater reliance on loans and investments that do not go through stock exchanges than in the past.

Another key difference between the issuing of shares and the seeking of loans from more secretive sources is that the sale of shares and the price paid must be made public, and anyone can check the ticker tape to see what shares in company X are now worth. By contrast, privately placed debt can be sold and re-sold without disclosure. I discovered this when investigating the way in which my own university finances its building projects. The university, like other public corporations, relies increasingly on issuing large-denomination 'debentures' or bonds, which do not require posting collateral. These are privately placed, not listed on stock exchanges, and are only available to institutional investors. The investors who buy them (pension funds and the like, who only deal in multiples of a hundred million dollars) can then sell University of Toronto debentures as they see fit. As far as I could determine, the university is not informed about subsequent sales of its large-denomination debts.

One area of corporate finance that has become prominent in recent years features investment funds classified as 'venture capital'. The term 'venture' indicates that the fund is willing to invest in untried or newer companies – particularly companies creating data-based inventions such as facial recognition software or robots.

However, the venture capital funds are usually connected to or are subsidiaries of larger investment funds that take a safer approach. Indeed, venture capital funds or their parent firms normally hedge the risks associated with 'start-ups' by also investing in safer securities – government bonds, say, especially US treasury bills, or infrastructure projects that have government backing.

The market in corporate debt used to be mainly driven by perceptions about a company's products. However, research has shown that lately the corporate debt market is more strongly driven by investors' concern for an overall 'balanced risk portfolio'. In his fascinating book *Portfolio Society: On the capitalist mode of prediction,* Ivan Ascher argues, with plentiful evidence, that large-scale finance is driven mainly by a desire for an optimum risk distribution in the total portfolio. So, a large investment fund might buy a particular currency or finance an airport or a bridge because a mathematical finance expert, or even just an algorithm, has decided that the overall portfolio of the fund needs a slight adjustment from the point of view of risk. Both firms and products have a diminished importance – which is what people mean when they say that financial markets have broken far too loose from 'the real economy'.

Public finance and government bonds

In order to pay for government services and activities such as waging wars, governments have for centuries sought financing, from private banks (the Rothschilds made their fortune by lending to European governments) but also from the general public, through government bonds. These bonds were often the only securities bought by ordinary people. Even today, when far more complex securities are available, many people will buy government bonds as an investment that does not pay a high interest rate but is quite secure.

But public finance does not only feature central governments. All manner of sub-national entities, from municipalities to transit authorities and urban development corporations, are players in the public finance game. These entities are in some cases allowed to issue their own bonds and thus obtain credit without posting collateral. In North America, the New York Port Authority is credited with having invented the tax-free non-government public bond in the period between the two world wars. Its bonds were tax-free because it was a public entity carrying out public works, but its financial doings were exactly those of for-profit corporations. The Port Authority had a legal monopoly on bridges and tunnels across the Hudson River, which reassured bond buyers that revenue would exist decades down the road when the bonds matured. Despite the tax-exempt status of its bonds, which made the Port Authority akin to municipalities, the general public did not have any input into road and bridge tolls, or indeed into decisions about which projects the Port Authority pursued.

The Port Authority has unusual legal and financial powers. But most other entities that borrow money, be they actual governments or agencies, live in fear of the ratings issued publicly by the big US agencies (Moody's, Standard and Poor, etc.). Often, the factors that go into credit ratings are quite outside the control of the entities being rated. For example, political tensions between Canada and China in the winter of 2019–2020 led the Moody's agency to issue a note warning that the credit rating of several major Canadian universities was at risk because China might at any moment stop the flow of Chinese students into Canada.

Similarly, in the global South, many countries have undergone major financial crises causing hardship whose causes were quite out of the country's control. If a transnational fund, for instance, makes a large-scale bet against a country's currency, this can result in severe economic difficulties not only for the government but for the people.

Of course governments, in the global South, but also in countries such as Greece and Italy, sometimes contribute to the problem by over-borrowing; but excessive government borrowing has often been actively promoted by investment banks that need some place to park their vast sums. Since governments can always raise taxes and lower social benefits, and they regularly do so to pay the national debt or even just the interest on it, governments can look like excellent investment sites even when the 'real economy' of the country is not promising.

And once they have borrowed money, often at high interest rates, what do governments do with the borrowed funds? Sometimes they build infrastructure. Critical research has shown conclusively that in the global north, issuing government bonds is a cheaper way to finance activity including infrastructure than resorting to private financing. But that is not a universal truth. In the recently bankrupt city of Detroit or in emerging countries such as Laos, there are few choices for public finance. In the global south, UN-derived entities such as the Asian Development Bank specialize in financing governments with poor or non-existent credit ratings. But these are hardly benevolent. They have often imposed harsh conditions (including reducing the size of the government itself) that have more force than that of law. International law, notoriously weak in general, does virtually nothing to regulate powerful transnational investors.

Conclusion

We have seen that although the regulation of the commercial banking sector serving individuals is weak, finance at higher levels is even less regulated. Indeed, public finance (that is, government borrowing) is governed more powerfully by the large-scale creditors/investors than by any law. This is not helped by the increased use of secretive arbitration tribunals instead of courts to settle disputes between investors and the countries they invest in. That lenders do not have the same interests as borrowers is obvious at the individual scale: a homeowner with a large mortgage does not see the bank as a friend. But at the transnational scale, the subjection of democratic governments and subnational public entities such as municipalities to the dictates of transnational financiers is not as easy to see.

Governments are supposed to be sovereign; but for many, perhaps most, countries in the world sovereignty – that is, the ability to make laws and make them stick – is non-existent when it comes to seeking financing.

Further readings

Ascher, Ivan. 2015. *Portfolio society: on the capitalist mode of prediction*. New York: Zone Books.

Karen Ho. 2009. *Liquidated: an ethnography of Wall Street*. Durham: Duke University Press.

Pistor, Katharina. 2019. *The Code of Capital: how law creates wealth and inequality*. Princeton: Princeton University Press.

25
FOOD SOVEREIGNTY AND FOOD JUSTICE

Carmen G. Gonzalez

Environmental justice is an important framework for understanding the North-South divide in many areas of international law and policy, including food policy. An environmental justice analysis makes visible the ways in which the global North benefits from unsustainable economic activity while imposing the environmental consequences on the global South and on the planet's most vulnerable human beings.

From the colonial era to the present, the North has exploited the fertile lands, forests, and vast mineral resources of Asia, Africa, and Latin America, and wreaked havoc on the livelihoods and ecosystems of the region's inhabitants. In recent decades, Northern countries have also used the global South as a dumping ground for hazardous wastes and a haven for polluting industry. And the global North has contributed disproportionately to climate change by utilizing more than its fair share of the atmosphere to deposit its greenhouse gases and by maintaining per capita emissions that dwarf those of the South.

The global food system is a paradigmatic example of environmental injustice. Decades of Northern aid, trade, finance, and investment policies have devastated the livelihoods and ecosystems of rural communities in the global South, while producing hefty profits for the Northern transnational corporations that dominate the global food system. Despite global agricultural yields sufficient to supply every person on the planet with approximately 2700 calories per day, nearly a billion people, most of whom reside in the global South, experience chronic undernourishment. This chapter applies an environmental justice analysis to the global food system, and identifies the ways in which this system perpetuates food injustice among and within nations. Food justice has three aspects: ecologically sustainable food production, equitable access to food and food-producing resources, and democratic local and national control over food and agricultural policy. The chapter analyses the underlying causes of food injustice, and outlines several strategies to create a more equitable and sustainable approach to global food governance.

From movements to secure access to clean water to popular mobilizations against what another chapter in this handbook calls 'extractivism', grassroots environmental movements have developed North-South and South-South transnational networks dedicated to specific issues, including water justice and food justice. Known collectively as 'the environmentalism of the poor', these grassroots environmental justice movements dispel the myth that environmental protection is a luxury that the South can ill afford.

The inequities of the global food system have sparked a variety of movements in both the North and the South, most notably the food justice movement and the international movement for food sovereignty. In the US, the food justice movement denounces the social and economic factors that prevent low-income American communities of colour from purchasing or producing healthy, nutritious, environmentally sustainable, and culturally appropriate food. It seeks to empower these communities to create local food systems that meet their needs as a transition to a more equitable and sustainable food system.

The international food sovereignty movement seeks to dismantle the corporate-dominated trade policies that have devastated rural livelihoods and environments in both the North and the South. It also promotes the redistribution of land and water rights to small-scale farmers, and advocates the right of peoples and nations to define their own food policies and control their food-producing resources. While international aid agencies strive to achieve food security defined as access to sufficient, safe, and nutritious food regardless of where, how, and by whom such food is produced, the food sovereignty and food justice movements demand a structural transformation of national and global food systems to promote democratized, localized, equitable, and sustainable food production. How food is grown or made is key here, not just how many calories per capita might be available.

Food justice is the right of communities to grow, sell, and consume healthy, nutritious, affordable, and culturally appropriate food produced through ecologically sustainable methods, but also their right to democratically determine their own food and agriculture policies. It is based on the principles of inter-generational equity, intra-generational equity, public participation in decision-making, and economic self-determination.

Achieving food justice requires careful attention to North-South power imbalances that determine where, how, and by whom food is grown and consumed. According to the United Nations Food and Agriculture Organization (FAO), over 820 million people suffered from food insecurity in 2018. Chronic undernourishment is a result of poverty and inequality rather than food scarcity. Even though the world's population is expected to reach 9.6 billion in 2050 and 10.9 million in 2100, the global food system currently produces enough food to feed a global population of 12-14 billion. Thus, technological innovation designed to increase food production is not necessarily what is needed most. We will not eliminate chronic undernourishment unless we tackle its underlying causes – poverty and inequality.

Chronic undernourishment is largely a rural problem. The vast majority of the world's undernourished people are small farmers or herders who are net food purchasers because they have been consigned to plots of land that are too small, too hilly, too arid, or inadequately irrigated due, in part, to competition for land and water from large-scale agricultural producers. The livelihoods of these rural dwellers have been and continue to be undercut by Northern aid, trade, finance, and investment policies that favour large-scale industrial agriculture, accelerate environmental degradation, enrich local elites or Northern transnational corporations, and increase the gap between the rich and the poor.

The global food system: from colonialism to the Green Revolution

A useful framework for understanding the underlying structural causes of chronic undernourishment is the food regime analysis introduced by Harriet Friedman and Philip McMichael. A food regime is a system of production and consumption of food on a global scale.

During the first global food regime (1870–1930s), cheap food and raw materials from the colonies and from the independent settler states fuelled the industrialization of Europe. European control over a significant part of the planet's natural resources enabled it to achieve a standard of living far beyond the constraints of its own resource base while relegating the South to the production of raw materials and the purchase of manufactured goods. This pattern of trade and production has persisted into contemporary times, and has impoverished the South by subjecting it to the volatility of agricultural commodity prices (including boom and bust cycles) and the declining terms of trade for agricultural products relative to manufactured goods. Nevertheless, the global South was largely food self-sufficient during the first food regime.

During the second food regime (1930s to 1970s), the United States played a pivotal role in the global transition to industrial agriculture, and adopted aid and trade policies that dispossessed small farmers in the global South, undermined Southern food self-sufficiency, and laid the groundwork for the dominance of Northern transnational corporations. In the decades following World War II, the United States and Western European nations provided generous subsidies to their agricultural producers, and imposed both tariff and non-tariff import barriers to protect them from foreign competition. By contrast, most Southern countries taxed the agricultural sector in order to finance industrialization.

When Northern agricultural subsidies (as well as agricultural mechanization) resulted in overproduction, the United States and Europe exported their surplus food to the global South as food aid, or provided subsidies and credits to private grain traders to facilitate the purchase of this food by Southern countries. The provision of surplus food to Southern countries free of charge or at reduced prices (a practice known as export dumping) actually exacerbated poverty and hunger in the global South by depressing local food prices and undermining the livelihoods of small local farmers. As small farmers lost their lands and swelled the ranks of landless rural workers, wages for agricultural labour declined – increasing rural inequality and generating widespread undernourishment. Prime agricultural lands became concentrated in the hands of affluent farmers, who produced coffee, cocoa, beef, vegetables, bananas, and feed grains for export. Thus, many of the world's poorest countries became dependent on food imports.

In the 1960s and 1970s, the United States sought to alleviate chronic malnourishment in the global South and forestall communist revolutions by exporting not just food, but the industrial agricultural model, including new high-yielding seeds, fossil fuel-based pesticides and fertilizers, machinery, irrigation, and mono-cropping. Known as the Green Revolution, this industrial agricultural model increased global food production, but displaced ecologically sustainable agricultural practices and fostered dependence on agricultural inputs manufactured by Northern transnational corporations. The Green Revolution's impact on undernourishment remains fiercely contested. While some observers contend that the Green Revolution enabled food production to outpace population growth, others point out that the Green Revolution increased rural inequality by benefitting the large farmers who could afford the agricultural machinery, irrigation systems, and other expensive inputs. An influential study analysing over 300 published reports on the Green Revolution concluded that the Green Revolution generally exacerbated rural inequality.

The primary beneficiaries of food aid and of the rapid industrialization of Southern agriculture were the Northern industrial farmers, grain traders, and input manufacturers that received generous government subsidies, access to new consumer markets in the global South, and the opportunity to supply Southern farmers with machinery, pesticides, fertilizers, and seeds.

The North's promotion of industrial agriculture also generated a variety of negative environmental consequences that currently threaten food production, including a dramatic world-wide decline in crop genetic diversity, dependence on fossil-fuel based inputs, massive soil erosion, depletion of aquifers, and rising greenhouse gas emissions. Approximately 75 per cent of the planet's food crop diversity was lost in the 20th century. This loss of genetic diversity increased the vulnerability of the global food system to pests, drought, floods, and other external shocks, including those associated with climate change.

The third food regime emerged in the aftermath of the economic shocks of the 1970s and 1980s, and is characterized by the unprecedented domination of global agricultural trade by transnational corporations. Prior to the debt crisis of the 1980s, Southern countries could insulate their farmers from unfair competition with highly subsidized Northern agricultural producers by imposing tariffs on imported food products. This policy flexibility was eroded by the requirements imposed on debtor nations by the International Monetary Fund (IMF) and the World Bank. Enticed into borrowing money at variable interest rates from Northern commercial banks to finance a variety of development projects, many Southern countries were unable to pay their debts when oil prices increased dramatically in 1973 and 1979–1980, causing interest rates to skyrocket. In exchange for loan repayment assistance from the IMF and the World Bank, three quarters of Latin-American countries and two-thirds of African countries were required to adopt a one-size-fits-all package of economic reforms known as structural adjustment.

The structural adjustment programs mandated by the World Bank and the IMF inaugurated the double standards that plague international agricultural trade to the present day: protectionism for the North and relatively open markets for the South. These structural adjustment programs required Southern countries to eliminate tariff and non-tariff barriers in order to open their markets to foreign competition and to slash government assistance to the agricultural sector (including marketing assistance, price guarantees, social safety nets, and agricultural research and education). However, Northern agricultural producers continued to receive lavish subsidies from their governments.

The reduction of Southern import barriers and the diminution of support to small farmers devastated rural livelihoods by placing small resource-poor farmers in direct competition with highly subsidized Northern agricultural producers. As cheap imported food flooded Southern markets, national food production declined, and waves of impoverished farmers migrated to urban areas. The structural adjustment programs of the IMF and the World Bank diminished food self-sufficiency in the global South by dispossessing small farmers and by requiring Southern countries to dedicate prime agricultural lands to the cultivation of export commodities (such as sugar, coffee, cocoa, flowers, fruits, and vegetables) in order to service the foreign debt.

The redirection of food trade from national to global markets reinforced the power of the transnational corporations that dominate the global food system. Supported by decades of overseas food aid programs, government subsidies, and public sector agricultural research, these transnational grain traders, seed and agrochemical corporations, and retail supermarkets chains wield unprecedented economic power. This market power enables these companies to pay farmers low prices for their agricultural output, charge high prices for agricultural inputs (such as seeds, pesticides, and fertilizers), and impose product quality standards that may be too onerous for many small farmers to satisfy.

The final threat to the livelihoods of small farmers in the global South is the proliferation of large-scale leases or purchases of Southern agricultural lands on terms that may deprive current users and occupiers of food-producing resources. These so-called land grabs have been driven by transnational corporations eager to exploit growing demand for biofuels, by Northern investors speculating on arable land in the global South, and by middle-income countries (such as Saudi Arabia, Qatar, China, India, and South Korea) seeking to engage in the offshore production of food to mitigate food price volatility on international markets and domestic shortages of arable land and irrigation water. These land grabs pose serious risks to small farmers in the targeted Southern countries, including eviction from their lands, depletion or pollution of water resources, and loss of access to forests, fisheries, and other resources. These large-scale land acquisitions are accelerating the South's transition to large-scale industrial agriculture at the precise moment when scientists and policy-makers are advocating a shift to small scale sustainable agriculture as a means of increasing agricultural yields, enhancing access to food, reducing dependence on external inputs, improving environmental quality, and protecting the traditional agroecological knowledge of small farmers and Indigenous communities.

In sum, the realization of food justice will turn on the global food system's ability to enhance the well-being of small farmers and promote environmentally friendly cultivation practices. Unfortunately, as explained above, the global food system has done precisely the opposite.

A justice-centred approach to Global Food Policy

A justice-based approach to global food policy must promote the human right to food, curtail the power of transnational corporations, mitigate North-South inequality, and ensure the full and effective participation of Southern nations and peoples in local and global food governance.

Food justice, like environmental justice more broadly, is grounded in human rights. The human right to food is recognized in the *Universal Declaration of Human Rights* (UDHR) and in the *International Covenant on Economic, Social, and Cultural Rights* (ICESCR). The right to food is also protected through Article 6(1) of the *International Covenant on Civil and Political Rights* (ICCPR), which guarantees the right to life, and has been interpreted to require the implementation of affirmative measures to eliminate chronic undernourishment. International human rights law must hold accountable the Northern states that are complicit in the widespread violation of the right to food. But the human right to mere food is not enough.

A human rights approach, but one focused on collective rather than individual rights, is essential to the achievement of food justice: it gives agency to communities experiencing chronic undernourishment, rather than treating them as objects of 'development'.

One of the most daunting obstacles to the realization of food justice is corporate impunity for abuses including agricultural export dumping and land grabbing. The unwillingness of Northern states to regulate the extraterritorial conduct of their transnational corporations enables these corporate entities to escape liability for their violations of the human right to food. Strategies to demand corporate accountability include strengthening the human rights enforcement capacity of Southern countries, holding Northern countries liable for failing to regulate the extraterritorial conduct of their corporations, enhancing the mechanisms available in a corporation's home state to adjudicate human rights violations abroad, and developing treaties that impose human rights obligations directly on corporations.

A key step toward a more just economic order is the implementation of differential treatment in international economic law, remedying past inequities by giving Southern countries more favourable treatment in international legal instruments. In order to address the structural causes of food injustice, international financial institutions (such as the IMF, the World Bank, and the regional development banks) and international trade and investment agreements must permit Southern countries to utilize a variety of tariffs, subsidies, and other protectionist measures to protect and enhance the livelihoods of small farmers, diversify and industrialise their economies, and end their crippling dependence on the export of primary commodities. Trade agreements must curtail Northern protectionism while giving Southern countries the flexibility to intervene strategically in the economy to foster long-term economic development.

In recent decades, national and transnational food movements have forged alliances across the North-South divide to demand a more equitable and sustainable food system premised on democratic community control over food-producing resources. The right to food sovereignty rejects the individual focus of Western human rights in favour of the collective rights of communities and peoples to freely choose their economic, political, and social system. In so doing, the call for food sovereignty echoes the rights of Indigenous peoples to self-determination.

In other words, the demand for food justice is ultimately a call for the vesting of the right to development and the right to permanent sovereignty over natural resources in *peoples* rather than states. The peoples would be regarded as the owners of natural resources, and the states would be viewed as trustees responsible for managing them for the collective benefit of the entire population. This approach would emphasize the duties of states to discharge their fiduciary obligations to their citizens in good faith, the democratization of control over productive resources, and the obligations of states and citizens to promote ecologically sustainable use of these resources.

This interpretation of food justice represents a paradigmatic break with the traditional notion of individual human rights. The collective right to food sovereignty politicises the struggle for food justice, promotes the right of peoples to democratically determine their food and agriculture policies, and facilitates the development of transnational alliances to challenge the corporate-dominated food system.

Further readings

Gonzalez, Carmen G. 2015. 'Food justice: an environmental justice critique of the global food system' in Alam, Shawkat; Attapatu, Sumudu; Gonzalez, Carmen G; and Razzque, Jona (eds.) *International environmental law and the global South*. Cambridge: Cambridge University Press.

Joseph, Sarah. 2013. *Blame it on the WTO? A human rights critique*. Oxford: Oxford University Press.

Martin Lopez, Miguel Angel. 2016. 'A study of the application of food sovereignty in international law'. *Groningen Journal of International Law* 4(2): 14–31 https://doi.org/10.21827/59db67f29999e

Wittman, Hannah; Desmarais, Annette; and Wiebe, Nettie (eds). 2011. *Food sovereignty: reconnecting food, nature and community*. Halifax; Winnipeg: *Fernwood* Publishing/Oakland: *Food First* Books/ Oxford: *Pambazuka Press*

26

GENDER AND LAW

Pallavi Banerjee and Pedrom Nasiri

Concepts of gender and their relationship to law vary considerably across space and time. In Europe and America, gender was long viewed as the outward expression of innate biological or psychological material. Scholarship has demonstrated how such perceptions were enfolded into law, producing legal systems where voting rights, custody rights, financial support, property ownership, employment opportunities, and bodily autonomy, to name but a few instances, all depended on gender. For instance, in many countries of the West, coverture was a legal doctrine whereby, once married, a woman's legal rights and obligations were subsumed by those of her husband. An unmarried woman or a widow, however, had the right to own property and make contracts in her own name.

By the latter part of the 20th century, new avenues of gender research emerged. For example, intersectional feminists argued that early examinations of gender treated the category as a monolithic group, at the expense of recognizing intra-group differences. By accounting for these intra-group differences, such as those based on race, class, religion, and so on, intersectional feminists unravelled how the law maintains structural inequalities by treating multiple and overlapping identities as if they were mutually exclusive. A now-classic example of the problem to which intersectional feminism seeks a solution is American discrimination litigation, which forces women-of-colour to fragment their lived experiences, advancing *either* a race-based discrimination claim *or* a gender-based discrimination claim. Intersectional legal scholars argued that such approaches fail to allow legal recourse for the distinctive forms of discrimination that women-of-colour experience as *both* women *and* of-colour.

Queer theory provided another avenue to the study of gender, highlighting the contingency of the concept's boundaries by foregrounding its socially constructed nature. For some queer theorists, inter-group and intra-group differences are important insofar as they elucidate ways people *perform* gender or make it *matter*. For example, theorists have demonstrated how ideological conceptions of 'sex as biology' and 'gender as social construct' make gender matter in legal systems by establishing a common-sense sex-gender distinction that suggests one's 'real' or 'objective' gender is the one they were medically assigned at-birth. The enforcement of this distinction has numerous repercussions from prison assignment for criminal offenders, access to appropriate medical care (especially for trans persons), medically unnecessary surgeries on inter-sex bodies, ability to compete in international competitive sports, marriage, and the more quotidian issues of access to washroom facilities.

There is a deep and varied corpus of knowledge on conceptions of gender and its relationship to law. Despite the existence of rich and consequential gender theories emanating from the global South, the majority of research on gender and law is developed in or uses theories from the global North. This reinforces the structural imbalance in the production of knowledge, continuing the disenfranchisement of 'southern theories' in favour of those produced in the northern hemisphere. We highlight this, not to discount scholarly works from the global North and their many significant contributions, but to emphasize that this knowledge inhabits and emerges from a particular space, and along with it, a particular history.

In the remainder of this chapter, we approach the concept of gender and its relationship to law from the perspective of 'southern theory'. Southern theory begins from the position that, to correct this imbalance, one must move beyond notions of diversity and acknowledge in its entirety the structural inequalities in the global production of knowledge and the reach and influence of Euro-American colonialism across time and space. Southern theorists recognize the global South as spaces filled with rich theories of critical importance. Working from this standpoint, we will introduce three themes: gendered contestation over land; the gendered politics of hunger; and the social analysis of terror and queer subjectivities.

Land

Usurped and unceded land as a tool for colonization and gender oppression is well documented in historical documents. Yet, the issue of land is starkly under-theorized in gender and legal studies produced in the global North. Dispossession by the forcible acquisition of land defines colonization, both in settler colonies like Canada, Australia, U.S, and colonies of rule in various parts of the global South. Relationship to land was central to how colonized societies formulated their legal gender relationships.

In settler-colonial contexts, land and land rights have been central to Indigenous people's politics, and the issue has always had a gendered dimension. In Canada, for instance, the creation of 'Indian' as a juridical category and Indian reserves as the signifier of land dispossession frames a historical crossroad that established settler-colonial governance and a legalized extraction economy as Canada's colonial project. In this context, the intersections of Indigenous subjectivity, gender, and land acquisition structures dispossession, and has spurred Indigenous women's movements for land rights. In an important collection of Aboriginal writings in Australia, called Our Land is Our Life, Marcia Langton argues that in the face of colonial violence, women's system of law and older women's ties to place were crucial to community survival.

Colonial imperialism was based on the forcible or legal take-over of land and entrenched in the ideology of economic exploitation of the colonies. For instance, the restructuring of the agricultural sector through the expansion of palm oil production in Nigeria by British colonialists in the 1920s created a strict division of gendered labour with men as farmers or plantation workers and women as their wives. Women were systematically kept out of farming, as the colonial financial support schemes were discriminatory toward women farmers. In the South Asian post-colonial context, Bina Agrawal shows how colonial practices continue the systemic dispossession of the colonized from their land. Even in governmental legal initiatives that foster people's participation in issues of development of agricultural land, women's participation and voices are systematically excluded.

Hunger

Like land, hunger and its legal ramification for gender politics have also been largely ignored by legal and gender scholars in the North. Hunger, in the colonial and post-colonial world, shapes the landscape of gendered access. Hunger drives displacement of peoples, polarizes the urban and the rural, the global North and South; and within these dispensations, hunger is gendered. Colonizers used famine and hunger as mechanisms to control Indigenous and colonized populations and, within these populations, women and children tended to die of starvation as the less productive labour force.

In Ulanowicz's and Basu's volume on the politics and aesthetics of hunger, contributors examine how colonial economic laws created the Bengal famine of India in 1943 that resulted in the starvation death of millions; the deliberate starvation of Native populations in the United States; the Egyptian poor involved in the bread revolts of 1977; and how hunger was produced in Soviet Ukraine as a weapon of control, torture, and dehumanization of minoritized populations.

In addition, scholarship has documented that hunger as protest is a gendered phenomenon. Men in colonized spaces prominently led 'hunger strikes' or self-starvation as a form of protest against the imperial state – such as Gandhi's self-starvation against violent policies of the British rule in India. In recent times, women activists have been using hunger strikes to resist state power in the global South, and also in the North. Some visible women in the recent years have been Aung San Suu Kyi in Myanmar protesting military dictatorship; political activist Irom Chanu Sharmila and environmental activist Medha Patkar in India; political prisoner Ethel Byrne in the US; and Theresa Spence, the former chief of the Attawapiskat First Nation, in Canada, among many others. The response of state-actors to many of these protests has been the medicalized force-feeding of protesters, thereby undermining not only the sovereignty of women and their bodies but also international laws prohibiting the act of force-feeding.

Terror

Terror induces affect and may be embodied as moral outrage, immobilizing anxiety, or even a desire for vengeance. Such embodied responses galvanize individual, social, or national bodies into action which may be experienced as law's negative forces, resulting in oppressive forms of violence, or law's positive forces, producing and re-producing the very fabric of our reality.

In her exploration of the emergence of queer national subjects in Europe and North America, Jasbir Puar highlights terror's confluence with gender and law. She argues that, in the post-9/11 era, some LGBT human rights campaigns have facilitated the demonization of Middle Eastern cultures by re-producing familiar Islamophobic tropes that position Muslim men simultaneously as homophobic patriarchs and effeminate figures of a failed and queer masculinity. Drawing on Achille Mbembe's notion of 'necropolitics', Puar demonstrates that such tropes work to devitalize the masculinity of Muslim men and legitimize their legal separation from the boundaries of everyday life, while also legally securing hegemonic masculinity.

Writing on similar themes, Sarah Carter examines the widespread anxiety 19th century settler-colonists in Western Canada embodied in response to Indigenous polygamous practices and its alleged degradation and exploitation of Indigenous women. Carter shows how legislation and court systems designed to 'save' Indigenous women from these marriages served as vehicles for the (re)entrenchment of normative European gender and kinship practices. This had profound consequences for Indigenous women who found their identity and cultural history newly subsumed and defined by their husbands. Carter argues that, as such, the duties of Indigenous women as kin-persons, wives, or mothers become incomprehensible under the settler-colonial system without reference to law and legal categories.

In their own work, Nasiri's research employs a post-colonial perspective to highlight how colonial anxieties and moral outrage surrounding polygamy continue to impose constraints on the formation of multiple-partner polyamorous families in Canada. Until recently, one such constraint had been the restriction of legal parenthood to two persons, which historically excluded additional parents from their children's legal kin structure, and invisibilized parental, erotic, and kin relations that do not fit the traditional bourgeois nuclear family norm.

There can be no doubt that law has deep ramifications for gender relationships, gendered materiality, and gendered access to resources. The combination of a 'southern theory' perspective with an intersectional gender analysis reveals how law shapes land rights; controls access to food for women; and configures terror, especial in the interactions of the Global South with the North. Further, while law does have the power to correct gender-based inequities, our analysis shows that left unquestioned, laws can be designed and passed to enforce and maintain the hegemonic gendered order and consolidate power for the dominant.

Acknowledgements

This text draws from the authors' previous work, particularly, Pallavi Banerjee and Raewyn Connell's paper Gender as a Southern Theory. in the 2018 *Handbook of the Sociology of Gender,* Springer.

Further readings

Moreton-Robinson, Aileen. 2015. *The White Possessive: Property, Power, and Indigenous Sovereignty*. Minneapolis: University of Minnesota Press.

Banerjee, Pallavi and Connell, Raewyn. 2018. 'Gender as a Southern Theory'. In Risman, Barbara J.; Froyum; Carissa M.; and Scarborough, William J. eds. *Handbook of the Sociology of Gender*. Manheim: Springer. pp. 57–68. doi.org/10.1007/978-3-319-76333-0_4

Puar, Jasbir. 2007. *Terrorist Assemblages: Homonationalism in Queer Times*. Durham: Duke University Press.

Ulanowicz, Anastasia and Basu, Manisha. 2017. *The Aesthetics and Politics of Global Hunger*. New York: Palgrave Macmillan.

27

GENOCIDE

Nicola Palmer

In international law, genocide consists of a series of designated acts committed 'with the intent to destroy, in whole or in part, a national, ethnical, racial, or religious group, as such'[1]. The illegal acts include killing, causing serious bodily or mental harm, inflicting conditions of life calculated to bring about a group's physical destruction, imposing measures to prevent births and forcibly transferring children. This legal definition, and its widespread acceptance, has enabled important legal processes and at the same time obscured certain aspects of the collective violence to which it responds. A discussion of genocide shows how fitting social facts into legal definitions can result in various forms of distortion. Perhaps particularly, the legal definition of genocide belies the longer time-frames over which the essential foundations of the life of groups, be they religious, national, ethnic, or racial, can be destroyed and the reach of the inter-generational effects of this violence.

The legal definition was agreed through the drafting of the 1948 Genocide Convention, credited in a large part, although not exclusively, to the writings and advocacy of Raphael Lemkin and his commitment to ensuring a legally designated response to the Holocaust. While the practices of the Nazi regime provided the basis for developing the concept, it is necessary to resist the temptation to elevate the Holocaust to the status of the singular archetype of genocide. Lemkin's writings were strongly informed by earlier instances of colonial conquest and occupation, and the Convention was designed to capture a range of different events, the common core being a pattern of acts and omissions oriented around the goal of destroying social groups.

It is this group dimension of the crime that animates it as a legal concept and has generated extensive interpretive debates. Within these debates, two recurrent issues valuably illustrate how the law on genocide has interacted with the social contexts it has been used to describe. First, debates on the boundaries of the protected group, and second, debates on the establishment of specific genocidal intent shed light on how the law has struggled to capture the longer temporal frames in which genocides have and continue to occur.

The process of drafting the Genocide Convention saw the narrowing of Lemkin's initial descriptive term in two domains: political and social groups were excluded from the protected collectives and the idea of cultural and artistic destruction was removed from the applicable

1 Convention on the Prevention and Punishment of the Crime of Genocide, Art.2, December 9, 1948, United Nations Treaty Series. 78 (1951).

conduct. The International Military Tribunal (IMT) in Nuremberg included this narrowed notion of genocide in its indictments and resorted to it in rhetorical flourish. However, none of the Nazis were convicted of the crime, with the IMT prosecutors focusing strongly on the notion of aggressive war. The legal definition of genocide would be given new momentum, use, and interpretation in the second wave of institutional building in international criminal law in the early 1990s. It was the United Nations International Criminal Tribunal for Rwanda (ICTR) that first applied the Genocide Convention's definition to a social phenomenon – the deliberate targeting and attempted extermination of Tutsi civilians in Rwanda from April to July 1994.

The collective dimensions of the crime raised immediate challenges. With a shared language and a complex history of settlement in the country, determining the Tutsi as an ethnic group, distinct from the Hutu and Twa, was more difficult for the Tribunal than comfortable journalistic accounts of the violence presume. In its final determinations, the ICTR resisted the complex sociological accounts of identity formation and relied heavily on the issuing of identity documents that included an ethnic designation to determine the existence of a legally protected group. The next challenge was to determine whether belonging to the protected group was subjectively or objectively determined. In the *Bagilishema* case in 2001, the ICTR Trial Chamber held that 'if a victim was perceived by a perpetrator as belonging to a protected group, the victim should be considered by the Chamber as a member of the protected group, for the purposes of genocide'. By the time the *Semanza* decision was handed down in 2003, the position had shifted towards including objective particulars of a given social or historical context alongside the subjective perceptions of the perpetrators. These early efforts to apply the legal definition of genocide to a particular instance of violence highlighted the difficulties of determining both the protected group and its membership. In doing so, they draw attention to how the law has struggled to account for the longer time frames over which membership of racial, religious, ethnic, and national groups are formed and socially recognised.

The decisions of the ICTR also fed into the wider discussion on which groups should be included in the legal definition of genocide. The ICTR's first case concerning Jean-Paul Akayesu was heralded by many feminists as a landmark decision in which rape was charged as genocide on the grounds that it 'caused serious bodily or mental harm' with the intent 'to destroy, in whole or in part, a national, ethnical, racial, or religious group, as such'. Leading feminists used the reasoning in the decision to argue for the inclusion of women as a protected group in their own right, where the coercive context of patriarchy would render, at the least, an initial presumption of non-consent in any sexual relations. Other feminists, tracking these debates, questioned where individual choices would sit if the cohesive structures of war, genocide, or indeed patriarchy, automatically designated sex as non-consensual.

At the time of these formative decisions from the ICTR, the United Nations International Criminal Tribunal for the former Yugoslavia (ICTY) was providing contemporaneous decisions on genocide and encountering different challenges regarding group designation and membership. In the *Krstić* case, the Tribunal faced the necessity of determining the boundaries of the group and the evidentiary implications that followed. The Trial Chamber found that the protected group was Bosnian Muslims, rendering the Bosnian Muslim population in Srebrenica part of this group. As a result, the killing of 7000–8000 military-aged men could not stand alone as an act of genocide and was combined with findings regarding the transfer of the women and children to support the designation that a genocide was enacted in Srebrenica. As these specific examples indicate, determining which groups should be legally protected, how factually to establish belonging to those groups, and the outer boundaries of these groups, highlight the historic and social contingencies that have given rise to the current concept of genocide and its specific legal contours.

The other side of the legal reckoning with the collective dimensions of genocide is encountered in determining the required subjective element or *mens rea* of the crime. While designed to respond to groups-based violence, in establishing individual criminal responsibility for genocide, the intent has been resolutely individual. Despite establishing the occurrence of genocide in Srebrenica, General Krstić himself was not found guilty of the crime as the prosecution failed to prove that he had the specific intent to destroy in whole or in part Bosnian Muslims. In international criminal law, the genocidal intent of the individual must be established. While the law has stayed strongly tied to the importance of the individual holding the specific intent to destroy the protected group, in whole or in part, social science research focusing on the micro-foundations of genocide illuminate the constraints and capacities of individuals to make choices as to whether or not to align with, propagate, and join collective violence. This work has shed light on how centralised policies of discrimination and ultimately extermination spread though societies as local authorities start to gain power and legitimacy from their alignment with these views. In line with this work, scholars like Alexander Greenawalt have argued for a knowledge-based approach to genocidal intent, in which there would need to be evidence of collective intent in an overall genocidal plan and, for the particular individual, knowledge of that plan and recklessness or foresight as to the occurrence of the planned destruction. Despite quite wide-spread academic support for this view, it remains only theoretical.

Yet, this move towards the notion of genocidal intent that looks more like *dolus eventualis* – similar to recklessness in a common law context – could enable the genocidal intent itself to be found in the systematic or structural policies of States, occupying forces, or effective authorities. Such a shift away from the dominant focus on individual liability can open up wider temporal analysis. It is this reasoning that underpinned findings in 2019 by Canada's National Inquiry into Missing and Murdered Indigenous Women and Girls, that the murder and disappearance of a large number of First Nations, Métis, and Inuit women, girls, and '2SLGBTQQIA' people in Canada constituted genocide. This follows the Australian Human Rights and Equal Opportunities Commission's finding that the forcible transfer of Aboriginal children to non-Indigenous families and institutions constituted genocide. Many orthodox international lawyers have been resolutely critical of both of these decisions, with critiques centring on both the boundaries of the protected groups and on the necessary intent. At the heart of these debates regarding genocide in the context of colonial conquest is the capacity of the existing legal definition to capture longer patterns of abuse. The socio-temporal politics that enables and explains the continuing aspects of the crime of genocide can be obscured by the choices around the interpretations given to the legal definition. These are not new challenges. Lemkin himself saw the strong connections among genocide, occupation, and prior episodes of violent examination under colonialism but at the same time was comfortable with what he saw to be the slower diffusion of cultures into one another and more 'benign' civilising assimilation. How the definition of genocide both enables and elides the temporal dimensions of collective violence is a recurring and on-going issue.

In Rwanda, these temporal challenges initially manifested through the Rwandan government's opposition to the establishment of the ICTR. The earlier patterns of discrimination, property crimes, and preparatory killings were understood by the government, under the leadership of the Rwandan Patriotic Front (RPF), as key to determining the intention to destroy in whole or in part the Tutsi group. These and other concerns about the structure of the ICTR, coupled with the political decision to incarcerate large numbers of suspected *génocidaires*, led to the most extensive set of domestic proceedings for the crime of genocide ever undertaken. Trials for genocide occurred before localised *gacaca* courts and the Rwandan national courts and most recently through a set of transnational proceedings, with 22 countries undertaking trials, deportations or the removal of the refugee status of individuals suspected of involvement in the Rwandan genocide. Running concurrently with the international trials at the ICTR, the understood roles of these legal processes have diverged. International judges and lawyers committed to building a system of global criminal justice articulate their work in a different register to the lay locally elected *gacaca* judges who have argued for addressing the very immediate questions of where the bodies of particular victims are located. At the same time, the Rwandan government's resolute focus on genocide has fuelled and enabled its failure to account for the wider instances of mass violence, including against Hutu civilians, that occurred before, during, and after the genocide. As the writings of current Rwandan scholars such as Richard Benda remind us, these temporal constraints are not only backwards looking. There are pressing concerns in Rwanda relating to the inter-generational repercussions of both the violence and the resultant legal processes on familiar and national relations. The capacity for genocide to operate as a continuing crime requires a reckoning with the historical foundations and ongoing and future impacts of collective group-based violence.

Once legal definitions are enacted, they take on dynamic social lives as they are used by different actors in a variety of forums, often underpinned by plural normative claims. Genocide is no exception. What socio-legal scholarship offers is the potential to map the law's divergent uses and its empirical and theoretical significance while being mindful of what is erased in the violence that legal definitions themselves necessarily inflict.

Further readings

Benda, Richard. 2018. 'Time to hear the other side: Transitional temporalities and transgenerational narratives in post-genocide Rwanda'. In Mueller-Hirth, Natascha and Rios, Sandra. *Time and Temporality in Transitional and Post-conflict Societies*. New York, Abingdon: Routledge. pp. 122–142

Greenawalt, Aka. 1999. 'Rethinking genocidal intent: The case for a knowledge-based interpretation'. *Columbia Law Review*. 99: 2259–2294.

Moses, Donald. 2010. 'Raphael Lemkin, culture, and the concept of genocide'. In Bloxham, Donald and Moses, A. Dirk, *The Oxford Handbook of Genocide Studies*. Oxford: Oxford University Press. pp. 19–41. doi.org/10.1093/oxfordhb/9780199232116.013.0002.

Özsu, Umut. 2019. 'Genocide as fact and form'. *Journal of Genocide Research* 22 (1): 62–71 https://doi.org/10.1080/14623528.2019.1682283

28

HUMAN RIGHTS

Challenging universality

Ben Golder

As most human rights textbooks teach us, human rights are the rights we (*all of us, everywhere*) are entitled to simply because we are human. Citizenship rights depend upon our formal membership of nation states, and private rights might accrue to us because of particular contractual arrangements that we have entered into or statutory regimes that may apply to our particular situation. But *human rights* apply universally to all humans irrespective not just of our nationality or our legal status but of our race, ethnicity, sex, gender, religion, or any other distinguishing features that might go to make us the particular, recognizable humans that each of us are.

The much-quoted *Universal Declaration of Human Rights (UDHR)*, agreed to by the United Nations on 10 December 1948 in the wake of the Second World War and after much debate, drafting and deliberation, says that human rights are owed to 'all members of the human family'. 'All human beings are born free and equal in dignity and rights,' its very first Article reminds us. 'Everyone is entitled to all the rights and freedoms set forth in this Declaration,' the second Article states. 'No one shall be subjected to torture or to cruel, inhuman or degrading treatment or punishment,' Article Five intones. The *UDHR* speaks everywhere in the majestic tones of the universal: all humans, everywhere, no one left behind.

This vaulting vision of human rights law finds institutional expression at the level of international law not just in the words of the *UDHR* itself (a resolution of the UN's General Assembly) but in those of the *International Covenant on Civil and Political Rights* (*ICCPR*) and the *International Covenant on Economic Social and Cultural Rights* (*ICESCR*), two multilateral covenants that entered into force in 1976. Compliance with these three founding documents (collectively referred to as the *International Bill of Human Rights*) is today the increasingly complex task of an array of UN agencies, inter-governmental bodies, independent experts, *special rapporteurs,* and working groups (not to mention global civil society groups, national and regional level human rights courts, and various human rights institutions).

Many observers take the phenomenon of this post-WWII legal and institutional commitment to human rights protection to be a sign of humankind's moral evolution. And yet for all the rhetorical and normative appeal of this maximally inclusive regime of international human rights that extends to all humans, the universal status of human rights has been much contested (not the least by some authors in this *Routledge Handbook of Law and Society*). Human rights activists themselves have queried the actual reach and applicability of international human rights law mechanisms and of the good faith and political commitment of national governments supposedly tasked with respecting and protecting human rights, asking just how far the writ of international human rights really runs. Who is included and who is excluded from human rights protection, and how effective in practice are those very protections? It is depressingly common to observe that the rise of what has been called the human rights revolution – the post-WWII human rights dispensation of international laws, norms, courts, and monitoring bodies – has itself been attended by ongoing war, political violence, genocide, global poverty, systematic exploitation, and heightening material inequality. What might it say about the much-vaunted universality of human rights that they are more honoured in the breach than the observance?

Of course, querying the empirical application of human rights is one avenue to contest the universal claims of human rights. But alongside, and perhaps even in advance of these kinds of questions, a different and more philosophically probing form of questioning is possible. This questioning assails not so much the practical application of human rights norms but the very normative aspiration to universality in the first place: to what extent can we say that the content of human rights norms themselves is truly universal?

This is a perennial and oft-revisited criticism of any form of political universalism, not just of the post-WWII international human rights regime. The famous criticisms made by the British thinkers, Edmund Burke and Jeremy Bentham, of the French Revolutionary *Declarations of the Rights of Man and Citizen* at the end of the 18th century rest to a great extent on their suspicion of philosophers (and, one suspects, in particular, philosophers hailing from across the Channel) purporting to decide, metaphysically and in the abstract, upon a set of timeless and unchanging principles of human behaviour. And an aspect of Karl Marx's well-known criticisms of the same document, in his text, 'On the Jewish Question,' turns upon his trenchant observation that the so-called universal 'Man' of the French Declaration was in fact a particular man functioning within a particular historical context according to a particular set of social relations and with a particular set of interests: the man of civil society, of Hegel's *bürgerliche gesellschaft*; that is to say, egoistic and *capitalistic* man.

These same questions have bedevilled the late 20th century project of international human rights law. In the original committee tasked with drafting the *UDHR* at mid-century, one of their number, the French Catholic philosopher, Jacques Maritain, reflecting on the impossibility of reaching philosophical agreement amongst such a diverse group, famously quipped, 'Yes, we agree about the rights, but on condition that no one asks us why'. The question of the universal and the particular has been at the centre of the politics of human rights for a long time, and is visible in the fraught geo-political debates of the end of the 20th century (particularly in East Asia) over the balance of state-led developmentalism versus individual rights protection.

It was this perennial question of the particularities of supposedly universal norms that motivated the Executive Board of the American Anthropological Association (AAA) to issue its famous (in some quarters, subsequently *in*famous) 1947 Statement on Human Rights, addressed to the UN's Commission on Human Rights then engaged in drafting the *UDHR*. There we encounter a frank acknowledgment that the world's diverse peoples all seek to 'resolve[...] the problem of subsistence, of social living, of political regulation of group life, of reaching accord with the Universe, and satisfying [their] aesthetic drives' in vastly discrepant ways that 'differ, often strikingly, from one another' and which appear, to the proponent of one culture, 'less desirable than those to which he [*sic*] is accustomed'. Not only, moreover, does this situation of radical cultural incommensurability (and the 'scientific fact that no technique of qualitatively evaluating cultures has been discovered') mean that drafting a single document on human rights will most likely fail to be 'convincing to the Indonesian, the African, the Indian, the Chinese,' but will smack of hypocrisy and colonialism. 'How can the proposed Declaration be applicable to all human beings,' asks the AAA, 'and not be a statement of rights conceived only in terms of the values prevalent in the countries of Western Europe and America'? This is not simply a problem of conceiving and understanding foreign cultures: it is in fact a material, historical, and political one as well, as attested to by the AAA's express mention of the 'economic expansion, control of armaments, and … evangelical religious traditions' of the Western colonial powers whose global rise 'has been marked by [the] demoralization of human personality and the disintegration of human rights among the peoples over whom hegemony has been established'.

A long, productive, and ongoing tradition of human rights critique has subsequently emerged which takes issue with the universal claims of human rights and their applicability in a world marked by the injustice of colonial histories (and presents) and characterised by the cultural syncretism of the globalized era. These philosophical and political debates are often, unhelpfully, characterised as a struggle between universalists and cultural relativists, with proponents of some form of international human rights protection in its current form occupying the argumentative mantle of 'universalists' whilst those problematising or disputing the normative reach and universal legitimacy of human rights from the perspective of history or politics (as did the AAA in 1947, and as do any number of post-colonial, de-colonial, and anti-racist scholars today) referred to as 'cultural relativists'.

This framing is not helpful because it suggests a simple conceptual opposition between the universal and the cultural, rather than an ongoing dialectic or a mutual contamination between them. Seeing the universal and the cultural as opposed would be to suggest that the universal is itself not cultural, when it is more plausible to suggest that there is not one universal but in fact many competing versions of universalism, each culturally located. But the framing is unhelpful not only for how it positions us to understand who or what is universal (emanating ineffably from Geneva, or Strasbourg, or New York, but somehow not *of* these places, either) and who or what is the subject of culture (often, traditionally, spatially bound to places of the global South). It is unhelpful, politically, for what it leaves out of the frame, and hence beyond our critical gaze. If we take some of the emblematic examples of universalism versus cultural relativism debates in the human rights literature (questions of Muslim women's veiling, for example, in the jurisprudence of the European Court of Human Rights, or legislative campaigns against Female Genital Circumcision) what routinely gets elided in these discussions of whether particular cultural practices fall afoul of supposedly universal international human rights standards is the question of gender and sexuality. Why are contestations over the universal figuratively and literally fought on and over women's bodies, and how does our understanding of this as a question of culture versus universal values work to displace queer and feminist contestation of both culture and universalism?

Those questions about who gets to speak for and against culture recur also in a different context, namely the East Asian so-called values debate of the 1990s, in which proponents of a certain form of authoritarian state developmentalism (famously, Lee Kwuan Yew of Singapore and Mahathir Mohammed of Malaysia) mobilised cultural arguments (framed in terms of Confucianism and communitarianism) to refute the application to their societies of supposed Western-style democracy and individual rights protection as found in international human rights law. Seeing these debates as arguments about the correct institutional balance between universal values and local cultural practices (as, for example, in human rights doctrines such as the 'margin of appreciation') is, of course, also to not see things. What we might not see is the political contestation over the content of 'the cultural,' as well as the fact that these debates are equally debates about varieties of capitalism and forms of exploitation. And it might well be harder for us to see that in fact the positions of both self-avowed universalists and self-avowed proponents and custodians of culture share much in common: both arrogate to themselves the right to determine what is and is not truly universal and what is truly cultural, delimiting the boundaries and contours of each.

But if things are evidently more complicated than the simple opposition between universalism and cultural particularism, then how might we better think of the universal in human rights law? Let me conclude briefly with some comments about how scholars and practitioners of human rights have recently begun to think about a more supple, mobile, and politically contested version of universalism in human rights law. One way in which they have done so is to think in terms of *translation* and *vernacularisation*. Here the idea (best expressed in the work of the legal anthropologist, Sally Engle Merry) is that the values of international human rights law are put into practice locally, but when they are adapted by local activists, they undergo a necessarily cultural process of change. Human rights politics, on this account, is not about the top-down legislation of the universal but rather about its fraught (yet productive) diffusion, modification, and translation across borders. Another promising way in which to conceive of human rights universalism is to think about those universal values encapsulated in the *International Bill of Rights* with which we started not as the timeless encapsulation of what is truly human but rather as a kind of placeholder, a universal-for-the-time-being that demands that different cultural groups respond by articulating and arguing for their divergent understandings of being human. Again, this version of human rights politics is one that sees the universal claims of human rights as themselves, openly, political and the political contest one over what it means and might mean to be human. Clearly both these forms of politics start with human rights (and that itself might be one way of not asking the prior question of whether there are other and better forms of politics to address global suffering and make global justice claims today). But it is equally clear that these contemporary political trends are attempting to address injustices by, as it were, updating and modifying what has been the foundational premise of the human rights system to date; namely, its universality.

Further readings

Executive Board, American Anthropological Association. 1947. 'Statement on Human Rights'. *American Anthropologist* 49:4 [New Series]. 539.

Merry, Sally Engle. 2006. *Human Rights and Gender Violence: Translating International Law into Local Justice.* Chicago: Chicago University Press.

Mutua, Makau. 2001. 'Savages, Victims and Saviors: The Metaphor of Human Rights'. *Harvard International Law Journal* 42:1. 201.

Waldron, Jeremy (ed.). 1987. *Nonsense Upon Stilts: Bentham, Burke, and Marx on the Rights of Man.* London and New York: Methuen.

29

IMMIGRATION, LAW, AND RESISTANCE

Susan Bibler Coutin

In recent years, immigrants in many parts of the world have increasingly come under attack, a situation that has intensified in the COVID-19 pandemic. Rising nationalism in many countries around the globe has fuelled xenophobia, with the result that those who have crossed national borders – or who are perceived as foreign even if they have not – have been denounced as outsiders who allegedly threaten economies and national values. Stiffened enforcement initiatives, whether located at borders or in countries' interiors, and draconian detention and deportation practices, have, paradoxically, been coupled with claims that the laws are 'lax' and need to be strengthened still further to bar migrants' presence altogether. Meanwhile, members of immigrant communities have had to contend with a host of indignities, including racial slurs, confinement, lack of access to legal counsel, denial of work authorization, exploitative labour practices, and precarious living conditions. Migrants' resilience in the face of such challenges is truly extraordinary.

This short article will introduce the concept of 'illegalization' to explain how the condition of 'illegality' results from state practices that situate members of immigrant communities outside of national polities, rather than from a particular act on the part of immigrants themselves. The article will then present a case study of how immigrant communities in Southern California in the U.S. have resisted illegalization in an effort to live with dignity.

Nationalist narratives that are critical of immigration frequently depict unauthorized immigrants as law-breakers who are intrinsically 'illegal' because of crossing a national border without authorization. This nomenclature is misguided, for several reasons: some individuals (such as tourists, students, or business people) enter a country with authorization and then stay beyond the expiration date of their visas, individuals sometimes gain and lose status as the conditions of their lives change so boundaries between authorized and unauthorized groups are porous, and this terminology treats legal status as an essential quality of personhood rather than as the product of state policy.

In contrast, the term 'illegalization' highlights the process of categorization to which individuals whose presence is unauthorized are subject. Individuals are constituted as unauthorized through state policies that bar their presence, so, paradoxically it is *law* that creates the condition of *illegality*. Moreover, practices that produce illegalization are diffused throughout societies, through everyday practices that deny rights and services to those who cannot present the correct papers. Examples include employers who deny jobs to those who lack work authorization, schools that do not allow individuals who lack citizenship or proper visas to enrol, officials who prevent unauthorized immigrants from passing through checkpoints, clerks who deny licenses to drivers or those who wish to practice a trade. These everyday interactions are infused with power in that they formally constitute individuals as 'unauthorized' or as 'illegal', even without a formal court ruling or judgment to that effect. Moreover, justifications for illegalization are put forward through discourses of belonging and deservingness that link legal status to moral qualities, such as patriotism, individualism, and modernity, and illegality to foreignness, dependency, and disruption. Such dichotomous depictions of moral worth are deeply dehumanizing, so much so that the slogan 'No human being is illegal', originally coined by holocaust survivor Eli Wiesel, became an international campaign.

Illegalization is closely linked to two other processes: racialization, that is, constituting individuals as members of particular ethnic and racial groups, and criminalization, the over-policing of non-citizens coupled with criminally prosecuting individuals for immigration violations. Racialization is widespread and takes different forms in different countries. In Costa Rica, for example, immigrants from other Central American countries, especially Nicaragua, are regarded as racially and ethnically 'different' even though they are from the same region. Similarly, in Europe, racialization has religious overtones, as Muslim immigrants are often seen as the quintessential 'other'. The U.S. is a particularly stark example of racialization, as individuals of Latin-American descent are often presumed to be immigrants even if they are native born U.S. citizens. Public anti-immigrant rhetoric, such as U.S. President Donald Trump's allegation that immigrants from Mexico are rapists, often targets particular ethnic, racial, or national groups. Furthermore U.S. immigration policy enshrines racial distinctions through policies that impose more penalties on those – typically from Latin America – who cross a border without authorization than on those who enter the country as tourists – for instance, from Asia – and then stay beyond the expiration date of their visas.

The criminalization of immigrants is also a widespread and international phenomenon. Individuals who are deported are often presumed to have committed crimes, even if they may simply have been caught crossing a border or apprehended for working without authorization. In the case of those who have been convicted of crimes, deportation simply disperses individuals without addressing the underlying causes of offending. Furthermore, immigrants themselves are often victims of crimes due to their vulnerability to thieves, traffickers, organized criminal networks, and employers who violate labour law. The association between immigration and criminality is therefore belied by the many factors that make immigrants themselves vulnerable to victimization by others.

Illegalization, racialization, and criminalization combine to form a powerful nexus that ensnares immigrants. These processes underlie spatialized enforcement tactics that have been used in Australia, Canada, Italy, and elsewhere to 'remove' the spaces that migrants occupy from national territories for the purposes of restricting migrants' access to legal rights. For instance, migrants who were encountered at sea have been detained on islands, in an effort to define them as 'outside' of the nations where they seek refuge. Borders have also been externalized, as when U.S. authorities enforce U.S. immigration laws in Southern Mexico in an effort to keep Central Americans from moving northward. Such immobilizing tactics keep migrants in a state of uncertainty, preventing them from 'arriving' at their destination and suspending time as they put their lives on hold. These harsh practices have embodied effects on immigrants who may experience extreme deprivation, even death, and who may also be harmed emotionally and psychologically by discourses that define non-citizens as potential security risks.

Members of immigrant communities have not accepted these measures passively. The remainder of this article draws on fieldwork within immigrant-serving organizations in Southern California and interviews with immigrants themselves to identify three strategies through which immigrants – particularly those from Latin-American communities – have contested illegalization, racialization, and criminalization. Related strategies can be located in other parts of the world as well.

The first strategy used by immigrant community members is to construct *counter-narratives* that challenge illegalization. Counter-narratives can be articulated in private conversations or can be publicly proclaimed, for instance, through marches, through news media, or through organizations' publicity and organizing campaigns. Counter-narratives typically explicitly challenge the justifications for illegalization. For instance, immigrant community members can argue that they are in fact law-abiding, hard-working individuals who contribute to society. Such arguments put forward relatively expansive grounds for deservingness, basing deservingness in behaviour rather than on arbitrary factors such as how an individual entered or remained in national territories. A limitation of such approaches is that they allow anti-immigrant sentiment to set the terms of the debate, thus engaging in what has sometimes been termed a 'politics of respectability', according to which some individuals (for instance, those with criminal convictions or those who are unemployed) are deemed unworthy. One example of a counter-narrative used at an immigrant rights march in Los Angeles in 2017 was 'Immigrants Make America Great'. This slogan evoked U.S. President Trump's slogan, 'Make America Great Again', but inverted it, suggesting that immigrants are a source of rather than threat to 'greatness'. Another marcher's sign read, 'Illegal President Not Elected by the People, Very, Very Bad'. This sign turned language of illegalization back against Trump by highlighting his loss of the popular vote in the 2016 election and also mocked his speech style through the words, 'very, very bad'. Such signs and slogans assert immigrants' value and self-worth, accusing the state, rather than immigrants, of illegality.

The second resistance strategy identified in this study was to *practice legal craft* by submitting claims for legal status. Doing so allowed immigrant community members and their allies to re-interpret law in ways that defined immigrants as deserving of legal recognition. In the hyper-legalized context created by immigration enforcement activities, many unauthorized immigrants interpret the documents – check stubs, rent receipts – produced through daily life as potential evidence in a future immigration case. By saving such documents, immigrants – who are accused of being undocumented – may experience hyper-documentation. By preparing for the day when they can submit these as part of a legal claim, immigrant community members help to bring that day into being. Working with attorneys and legal advocates, community members sought to demonstrate the *normalcy* of their lives, demonstrating through their saved documentation that they had volunteered in schools, been employed, supported family members, and engaged in social activities. Having to prove what citizens could take for granted was a burden, but also enabled community members to push law to recognize the value of their family relationships and life goals. Of course, like counter-narratives, the strategy of filing for legal status in some ways reinforced the very system that underpinned illegalization. Nonetheless, taken collectively, these many individual cases staked a broader claim to presence, meaning, and the opportunity to have futures.

The third strategy noted by researchers was *community resistance*, such as advocating for localities to declare themselves sanctuaries in support of immigrant communities, or adopting local measures designed to promote social inclusion. The term 'sanctuary' has a long history, drawing on medieval traditions of creating sacred spaces that shelter those who are targets of state action. During the 1980s, sanctuaries were established in the U.S. for refugees who were fleeing civil wars in Central America and who being denied political asylum. In the 2000s, sanctuary has once again become a common resistance practice, as some U.S. cities and states have resisted federal pressures to instruct police to collaborate in immigration enforcement practices. 'Sanctuary' cities and states, in contrast, typically limit such information-sharing and even go further to actively include immigrants in ways that produce localized understandings of citizenship. For example, the state of California issues driver's licenses and business licenses to individuals regardless of immigration status, as well as making educational opportunities more broadly available. Some Southern California immigrant community members were able to take advantage of these measures to mitigate impacts of illegalization. For instance, community members who secured local licenses could practice professions without having to be employees, and therefore did not need work authorization. Through community resistance, immigrants refused to allow legal uncertainty to define their lives.

Counter-narratives, legal craft, and community resistance engage law in complex ways. These strategies take up elements of law – such as the notion that presence in the U.S. confers at least minimal due process rights or that good moral character is part of a status claim – and interpret them in ways that challenge illegalization. Claims are staked through each of these strategies, whether through public media, legal processes, or city and state initiatives. Of course, such strategies can be challenged through media coverage that promotes narratives of illegality, denying or restricting legal opportunities, or adopting local measures that cooperate with federal immigration enforcement, as occurred when the state of Arizona passed legislation requiring police to check individuals' immigration status as part of routine interactions.

The nexus between illegalization, criminalization, and racialization is powerful and pervasive, infusing a myriad of government and private actions with immigration enforcement dimensions. Thus, when employers check job applicants' work authorization or when colleges and universities determine whether individuals have the right 'papers' to qualify for educational opportunities, immigration law is being enforced. Yet, these interactions also create opportunities to contest enforcement practices, and to redefine immigrant community members as deserving residents. These dynamics demonstrate law's complexity: law can be a powerful force that 'others' non-citizens, but it also sets out grounds for social inclusion. And sometimes, through the actions of immigrant community members and their allies, law's potential to promote social justice can be realized.

Further readings

Álvarez Velasco, Soledad. 2020. 'From Ecuador to Elsewhere: The (Re)configuration of a Transit Country'. *Migration and Society* 3(1): 34–49. https://doi.org/10.3167/arms.2020.111403

Chacón, Jennifer M. 2009. 'Managing Migration through Crime'. *Columbia Law Review: Sidebar*. 109: 135–148.

Constantini, Peter. 2019. 'Manufacturing Illegality: An Interview with Mae Ngai'. *Foreign Policy in Focus*. January 16. Available at https://fpif.org/manufacturing-illegality-an-interview-with-mae-ngai/.

Fassin, Didier. 2016. 'From Right to Favor: The Refugee Question as Moral Crisis'. *The Nation April 16*. Available at https://www.thenation.com/article/archive/from-right-to-favor/.

30

IMPERIALISM AND LAW

Jothie Rajah

Imperialism is, at its heart, an exploitative relationship in which the interests of a dominant state or states are furthered at the expense of a subordinated state or states. Conventionally, imperialism has been understood as the kind of control and power exemplified by, for example, the consolidations of 19th century British empire, in which political power, trade, governance, armed forces, and ideologies of 'White' racial superiority, worked together to enrich and empower empire's metropolitan centre while extracting resources, labour, and territory from colonies. However, beyond this conventional understanding, imperialism's power and influence can take multiple forms, spanning a continuum from obvious and more coercive expressions of imperialism, to more subtle imperialisms. These forms include political, military, social, economic, and cultural imperialism.

Although usually understood as a powerful coercive force, imperialism nevertheless exposes the fragility of sovereignty. A contemporary example of sovereignty's fragility lies in the fact that Iraq and Afghanistan are both member states of the United Nations despite the fact their sovereignty has been dismantled by the militarized imperialism of the United States. These examples demonstrate that while states, in formal legal-political terms, bear sovereign autonomy and are in (notional) parity with other sovereign states, in *realpolitik* terms, sovereignty is not an all-or-nothing condition. Imperialism can diminish the sovereignty of some states without actually taking them over, while augmenting the expansive reach and power of other states.

Both historically and in our present, law has closely partnered with imperialism's dynamic of domination and subordination. Historically, while some of the earlier known empires were centred outside Europe, the yoking of imperialism to law has been a strong feature of modern European imperialism. Modern European imperialism stretches in time from the late 15th century into the formal decolonization that occurred after World War II. Terms such as 'Western imperialism' and 'Euro-American hegemony' reflect ideological convergences between European and US imperialisms, even though the US has had few formal colonies.

The co-constitution of imperialism and law is traceable to the legal treatises of Francisco de Vitoria (1485-1546). Employed by the Spanish government, Vitoria justified the Spanish colonization of what was then called the Indies. Vitoria's legal authorizations of Spanish imperialism inform and shape justifications of empire reaching into our present. For example, Anthony Anghie traces continuities from a treatise scripted by Vitoria in 1530, to the post-9/11 Bush doctrine of preemptive self-defence. For the U.S., preemptive self-defence justifies the contemporary invasions and occupations of Iraq and Afghanistan, as well as the secretive, arguably illegal, conduct shown in its counter-terrorism operations.

In addition to the obvious imperialism of militarised invasions and occupations, imperialism today is often expressed in less obvious yet institutionalized ways. And because socio-legal studies are committed to understanding how law works in specific contexts, I detail below a specific instance of (partially masked) imperialism that illustrates the manner in which contemporary law and imperialism are articulated at the enmeshed levels of global, national, and local.

In September 2011, major global news media, including the *New York Times* and the *Guardian*, reported on a conflict over arable land in Mubende district, Uganda. According to people who had been living on the land, they were violently evicted after a British commercial entity, New Forests Company, acquired a 50-year license to grow trees on Ugandan land in order to trade in carbon credits under the Kyoto Protocol.

In a competing account, the Ugandan government and New Forests Company asserted that the residents were illegal encroachers who had left the land peacefully and voluntarily. The affected communities, supported by Oxfam, and the Uganda Land Alliance, filed a complaint with the World Bank's Compliance Advisor Ombudsman. Perhaps because the World Bank's private investment arm, the International Finance Corporation was a major investor in New Forests Company, the World Bank's Compliance Advisor, Ombudsman, brokered an (apparent) settlement to the conflict. In the four years between the evictions and the settlement, the displaced people struggled with poverty and despair. The terms of the eventual settlement are only partially disclosed by the documents on the Compliance Advisor Ombudsman's website. Reading between the lines, the settlement appears to situate the dispossessed people in permanent relations of dependency upon New Forests Company.

As this brief description illustrates, issues, actors, arenas, texts, and legalities at plural scales – local, national, intra-national, global – have played interlacing roles in these events. Alert to the complexities of grappling with multi-scalar issues, I adopt a global socio-legal approach in order to excavate the workings of imperialism in these events. A global socio-legal approach is Eve Darian-Smith's analytic toolkit for socio-legal scholars seeking meaningful ways to address pressing, border-transcending concerns. The key pillars of a global socio-legal approach are: first, a normative orientation towards the counter-hegemonic; second, an interrogation of the gap between law on the books and law in action; and third, a move away from narrowly state-centric law. This is clear in the Mubende land grab just mentioned: the global economic, political, and cultural power attaching to the UN-led Kyoto Protocol, and to the World Bank as an international institution, have shaped the domestic law as it played out within Uganda. It is thus impossible to neatly separate domestic law and domestic politics from the international.

In addition to facilitating analysis of multi-scalar situations, two of Darian-Smith's orienting axes for analysis are especially pertinent for understanding contemporary dynamics of imperialism. First, instead of framing law and society through the jurisdictional and territorial space of 'nation-state', Darian-Smith urges us to think in terms of 'communities' and 'societies'. A global socio-legal approach usefully excavates the manner in which powerful global and national elites have disregarded the potency of the categories – 'communities' and 'societies' – by imposing Euro-American legal knowledge and re-inscribing imperialism's silencing of alternative, non-Western, forms of understanding law, legal power, and political authority.

Some further detail about the 2011 Uganda land grab will usefully illustrate these points. The UK-based New Forests Company ('the Company') was incorporated in 2005, the same year that the UN-led international treaty that opened the door to carbon trading, the Kyoto Protocol, became effective. As a UK-based company, acquiring its license to grow forests from the Ugandan government, directing its activities to carbon trading, and securing investments from global private funds, (including the Hong Kong and Shanghai Banking Corporation, the World Bank's Agri-Vie Agribusiness Fund, and the European Investment Bank), the Company is, unambiguously, a global actor aligned with institutions in the global North, and engaged in capitalism through financialization.

Additionally, the role of the Kyoto Protocol prompts attention to how two major international institutions, the United Nations and the World Bank, are complicit with the Company, and with Ugandan national elites, in perpetuating a version of law which privileges the doctrinal legal category 'nation-state' while disregarding questions of social justice as seen by those within the nation-state whose experiences are more accurately captured by the category 'community'.

This context and set of actors emanating from the Kyoto Protocol speak to contemporary imperialism: global warming, the phenomenon precipitating the Kyoto Protocol, evidences the disproportionate harm suffered by the global South. In James Goodman's assessment, climate change bears an impact akin to a world war in which the global South suffers the most direct harm, but is also less overtly impacted by the kind of measures that have been put in place in the name of environmentalism, specifically carbon emissions trading and carbon credits.

It was after the Company acquired its license to the land, and received financial backing to engage in the speculative activity of carbon trading, that the Mubende people's troubles began. Brute violence, a familiar technique of imperialism, was very much at the forefront in that, as Oxfam reports, from 2009, New Forests Company's employees, were 'evicting, harassing, erasing … plantations, demolishing … houses, intimidating, [and] mistreating' the Mubende people. Agents or employees of the Ugandan government's National Forests Authority also appear to have participated in these acts of violence and intimidation.

In response, the Mubende community turned to the courts. The Nakawa High Court granted interim orders restraining evictions pending full hearings, but these court orders were violated when, in February 2010, army and police were deployed in the area to enforce the evictions (again as documented by Oxfam). With the army and police joined by casual labourers that the community believes to be the Company's employees, people were beaten, homes were burnt, crops were destroyed, and livestock was butchered. In the tragic unfolding of events, an eight year old child, unwell and left to sleep at home while his mother went in search of medicine, was burnt to death.

Against the community's accounts that Company employees played a role in the violent evictions, the Company asserts that authority lay solely with Uganda's National Forestry Authority to peacefully vacate 'illegal encroachers'. As a category, 'illegal' posits a sharp, binary distinction between legal/illegal, leaving no room for a consideration of histories and present moments through which the legal/illegal distinction in Mubende is underpinned by histories, and ongoing expressions, of imperialism and violent dispossession. In addition, the negative term 'encroacher' reproduces a central premise of 'Western' legal thinking regarding property – property law and property norms being a crucial element in imperialist politics and law.

By characterizing the Mubende people as 'illegal encroachers', the Company, in effect, designates the Mubende land as *terra nullius,* the supposedly empty land that imperial powers thought they saw in places such as Australia or Western North America. As Anthony Anghie and other critics of legal and cultural imperialism have shown, it was in part by accusing Indigenous people of not having properly occupied land and turned it into property that non-European societies were expelled from the realm of international law. These societies, whose own cultural and legal norms were regarded as mere 'savage customs', were seen as incapable of advancing any legally cognizable objection to their dispossession, and thus reduced to objects of conquest and exploitation.

Western legal thinking legitimised conquest as legal, and decreed that lands inhabited by people regarded as inferior and backward were *terra nullius* even if they were clearly inhabited. In other cases, imperial powers claimed that native chiefs had entered into treaties which gave those powers sovereignty over non-European territories and peoples, even though for other purposes, Indigenous leaders were not seen as having the legal power to sign anything. Further, the purpose and content of the treaties that did exist, in places such as New Zealand and Canada, was not necessarily clear and understood in the same way by both parties.

The Mubende conflict shows how, in a 21st century variation on the theme of *terra nullius*, the Company dealt directly with the Ugandan state at the stage of obtaining its license to the land. By acting in concert with the Company, the Ugandan state furthered imperial aims, even though in other contexts, Uganda might claim to be a sovereign nation that has emerged from a previous imperial past. In keeping with the terra nullius approach, in 2005, the Company secured from the Ugandan state a license to land that was already occupied, and that already sustained more than 20,000 people. The Company limited its recognition of 'law' to state law, thereby erasing legalities evident at the local levels of 'community' and 'society'.

Throughout these events, the people who lived on the land understood themselves to be doing so in a lawful manner. In Oxfam's 2011 assessment, the Mubende people 'had functioning villages and government structures, such as local council systems, schools, health centres, churches, permanent homes, and farms on which they grew crops to feed themselves and surpluses to sell at market. They paid taxes. Theirs were strong and thriving permanent communities'.

The legal relationship to the land of the Mubende people may not have conformed to Euro-American, state, or legal positivist categories such as owner, lessor, or tenant, but in the legal knowledge and consciousness that these people expressed, (through their words, through their court action, and in their conduct of themselves in relation to the land and to each other), they expressed the belonging, commitment, and rootedness in place characteristic of the personal life-world of a community. By disregarding community, and the forms of legality, legal knowledge, and legal consciousness expressed by the Mubende people, the Company, the Ugandan government, and the World Bank have allied to perpetuate imperial dynamics.

Consistent with the imperial dynamics of domination and subordination through which profit for the imperial centre is fostered at the expense of colonies, it is globalised, financialised, (and arguably, predatory) commercial power situated in the global North, aided by the international institutions widely perceived to be susceptible to the influence of the U.S. (the World Bank, the United Nations), that has prevailed. On the other side of the coin, it is marginalised people in the global South who have suffered poverty, dispossession, and diminished future prospects.

This example helps illustrate how entrenched categories of law (such as a state license to use land) and global structures of power (such as the Kyoto Protocol, and globalised corporations), build on the inequities of past imperialisms to engender imperialism today. A global socio-legal approach to events like the Mubende land grab helps to excavate the close links between globalization, capitalism, and imperialism, past and present.

Further readings

Anghie, Anthony. 2006. 'The evolution of International Law: Colonial and postcolonial realities'. *Third World Quarterly* 27(5):739–753. https://doi.org/10.1080/01436590600780011

Bush, Barbara. 2006. *Imperialism and Postcolonialism*. Harlow, England; New York, N.Y.: Pearson Longman.

Darian-Smith, Eve. 2013. *Laws and Societies in Global Contexts: Contemporary Approaches*. Cambridge: Cambridge University Press.

Grainger, Matt and Geary, Kate. 2011. *The New Forests Company and its Uganda plantations*. Oxfam International.

31
INCARCERATION
How to understand imprisonment trends

Maximo Sozzo

There have never been as many prisoners, in the world, as there are now. The International Centre for Prison Studies, which collects official data provided by each country, informs us that in September 2018, the total for prisoners around the world was 10,740,000. In both absolute and relative terms, the United States leads the way: its prison population of 2,245,000 (in 2017) makes up one-fifth of the global total, and gives the United States an extremely high rate of 669 prisoners for every 100,000 people. It is well known that prison populations have increased very rapidly in the U.S. over the past decades: in 1970, the rate of imprisonment was one-quarter of the current rate, that is, 162 people per 100,000.

Other countries, however, also exhibit alarming numbers and trends. For example, Brazil currently (2019) imprisons 746,000 people, which gives it a rate of 348 per 100,000. Even though that is only half of the current U.S. rate, Brazil's rate in 1990 was only 70 per 100,000, so that the increase in imprisonment in Brazil over the past three decades has been greater than that seen in the United States over a longer period. Turkey is another dramatic case. In 2000, the imprisonment rate was 73 per 100,000, but by 2018 this had increased to 326 per 100,000, resulting in a fourfold increase in prisoners over a very short time period.

Over the past 30 years, social scientists have made very significant efforts to try to explain the factors underlying these increases in imprisonment rates, using a variety of theories. Three approaches can be mentioned here that have been widely discussed in the 'punishment and society' literature.

One approach has been to emphasize the connection of imprisonment trends with what is known as 'late modernity': a complex economic, social, political, and cultural process that has produced a new experience of crime. The new experience, scholars such as David Garland argue, is in part a consequence of substantial crime growth but also includes increased fear of crime, personal worries about victimization and related self-protection and home security measures, and punitive public opinion. These processes tend to lead to state crime control responses, including, as a key strategy, increased use of imprisonment.

The second approach used by scholars to explain imprisonment trends has focused on the connection between punishment and the rise of neo-liberalism, understood as a transnational political program that seeks to reshape and restructure the state. Loic Wacquant and Alessandro Di Giorgio are two prominent authors here. From the 1970s onward, neo-liberal policies have involved withdrawing resources from social welfare measures and relying more on punitive measures. The gradual rise of a so-called 'penal state' has increased reliance on imprisonment as a key aspect.

Thirdly, other analysts emphasize changes in contemporary capitalism resulting in what some call 'post-Fordist economies'. While older 'Fordist' models (here referring to Henry Ford's notion that his employees ought to be paid well enough to afford Ford cars) resulted in at least some sectors of the working class being economically secure, post-Fordism is more exclusionary: large parts of the population are simply excluded from the formal, long-term labour market, and have to rely on 'flexible' and precarious work. The increase in imprisonment is here seen as a social control response that neutralizes those who are permanently excluded from secure well-paid jobs.

Clearly, these three narratives have their strong differences, but there are commonalities as well. They all assume that the causes of imprisonment trends need to be sought in deep structural causes, macro processes that in the less circumspect accounts amount to epochal changes. A further common point is that all of these accounts tend to rely heavily on data from the U.S. But at the same time, they have explicitly claimed, more or less cautiously, the possibility of its application to other scenarios – though mainly of the global North. This has been so because, to a large extent, the background causes that they identify have in their same definition the propensity to cross borders – late modernity, neo-liberalism, 'post-Fordism'.

These two features that tend to recur across all three approaches mentioned - a quest for 'deep' underlying social causes and a desire to generalize across borders – produce strong limitations in our capacity of making sense of incarceration trends. I will suggest other ways of understanding and explaining imprisonment trends after presenting some facts that do not fit the thesis of a continuous and inevitable growth of imprisonment.

Many national contexts simply did not experience a growth of incarceration rates in recent decades. Some show a basic stability over time. Sweden, for example, has an imprisonment rate of 59 per 100,000; this was 58 in 1990, and 65 in 1970. Denmark had a rate of 70 in 1970, 67 in 1990, and 63 today. Germany's rate was 74 in 1993, just after reunification, and today's rate is 78. Finland is an interesting case because its imprisonment rate has even decreased significantly, from 111 in 1970 to 51 today. Outside of Europe, Canada's rate was 104 per 100,000 in 1970, and was 107 by 2017. On its part, India's imprisonment rate was 34 per 100,000 in 1970 and today it is virtually the same, 33.

Elsewhere, a different pattern is visible: sharp increases in imprisonment followed by significant decreases. This is the case in the Netherlands, where the imprisonment rate was a low 30 in 1970 but increased significantly from the 1980s on, reaching 125 in 2006: however, it has since decreased significantly, to 61 in 2017. While higher than in the post-war period, the current Dutch rate is half of what it was only a decade ago. Russia also has an interesting pattern. In 1990, just as the Soviet Union was falling apart, the imprisonment rate was 473. This increased in subsequent years, achieving an extremely high rate of 729 in 2000, but since then imprisonment has decreased, and the Russian rate is currently about half of what it was in 2000 and 23% less than in 1990. Similar ups and downs have been recorded recently in many Baltic and Eastern European countries like Ukraine, Belarus, Lithuania, Latvia, Estonia, or Moldova, as well as contexts of the global South like South Africa.

These strikingly different national trends present obvious challenges to this kind of macro-level explanations. Is it the case that countries that don't fit the incarceration growth trend did not experience the underlying socio-economic changes that were mentioned earlier? And if this is the case, why? Or was it the case that they did experience the macro-level social and economic changes, but countervailing factors worked to minimize the use of imprisonment as a means of social control? And if this is the case, which are those factors? Are they same everywhere? Why they exist? and How they work?

Certainly, there are other contexts that today have imprisonment rates that are much higher than 3 or 5 decades ago. Almost all of Latin America, not just Brazil, has seen such increases. For example, the incarceration rate in Peru in 2018 was 279 and was 82 in 1990 or in Colombia, it is now 246 and it was 96 three decades ago. The same happened if we compare the contemporary incarceration levels with those of 1970 in New Zealand and Australia or Italy and France. Some recent studies of punishment and society suggest that understanding of the increase and high imprisonment levels – but also, conversely, its descent and low levels – requires moving beyond the two common features that we pointed out in the three interpretative frameworks that have been crucial in the international debate over the past 30 years. These displacements are already taking place in the literature even when there is much to do in this regard.

First, it is not necessary to abandon the social science aim of arriving at generalizations, but it is necessary to modify how we go about arriving at generalizations. The fact that punishment and society research has studied the United States much more than other countries has led many to treat the U.S. as the leading edge, foreshadowing what happen elsewhere. But that is problematic. The very valuable studies about the U.S. that we have must be 'provincialized', put in international context. One cannot assume that American penal trends will inevitably spread to the rest of the world, although in some cases discourses and practices 'made in U.S.' are adopted and adapted in other contexts, travels that must be carefully studied in its own right.

In this direction, it is crucial to generate in-depth, thick studies of penal histories of other countries, including both those that exhibit increased imprisonment rates and those that have either stable rates or ups and downs that do not result in an overall significant increase. Progress has been made on some contexts that are very significant, especially in Northern and Western Europe. However, developments that refer to other regions, particularly of the global South, are more incipient, and need to be encouraged. Only from in-depth and embedded reconstruction of the historical processes of the penal field in each national scenario is it possible to build a comparative analysis that allows establishing similarities and differences and producing generalizations on a more solid basis.

Comparative studies of penality do already exist. However, in some cases, the comparisons use only a small number of quantitative indicators, tracing connections between factors of different type. Their results may be a starting point, but cannot be considered a point of arrival. Comparative studies that cover fewer countries but have deeper analyses of each case seem more promising if we are going to improve our understanding of trends of imprisonment.

Second, attention must be paid to the interrelated questions of the proximate causes and the 'how' of imprisonment trends. This means, from my perspective, building a more political – in a broad sense – interpretation of contemporary penal mutations, emphasizing the core role of strategies, rationalities, programs, and struggles of state and non-state actors around the exercise of the power to punish.

Some recent studies document the wide variety of institutional arrangements that shape penal processes. However, institutional arrangements cannot be studied statically, but must be placed in relation to the multiple struggles around penal decisions and practices. For one thing, not all actors have the same power. Economic and political elites have more power to shape penal policy, obviously; but elites are not a homogeneous block.

In addition, other actors, from prisoners themselves to human rights organizations, also affect policy outcomes in some cases (though not in others). Those who have dominant positions today in the struggles around the power to punish may cease to have them tomorrow, even if this does not seem plausible at present. Historical research on the penal field provides plenty of examples of unexpected shifts and outcomes. Political and penal rationalities and programs play an important role in structuring these constant struggles around the power to punish.

However, recognizing this does not imply to assume that the deployment of the power to punish is the direct and linear translation of political rationalities. There is always a certain degree of contestation between various rationalities and programs, and this results in perpetual and complex transactions. This does not prevent pointing out the possible existence in specific times and places of dominant directions, but on the condition of recognizing the persistence of other orientations, even those playing marginal roles. But also, crucially, penal decisions and actions are also the result of interests, which often may not be explicitly expressed in terms of ideals by the actors – to use the famous words of historian, David Rothman, the dynamic of 'convenience' goes beyond 'conscience'.

These struggles about penal policies and practices do not happen in a vacuum. They are conditioned – although not determined – by various types of broader economic, social, and cultural processes and conditions. Discerning the relative weight of these elements is certainly a central component for this type of exploration. An actor-centred approach should not fall into a kind of voluntarism in which limits of the possibilities of action have vanished. In this way, even deep causes, such as those pointed out in the explanatory perspectives that we refer to above – or others – could be considered as having causal effects, according to the evidence that can be built in this regard on specific scenarios.

A perspective like this, as is evident, cannot be constructed in a general level, but must be rooted in certain spatial and temporal coordinates. Hence, both of the displacements proposed here with respect to the grand narratives that constituted the benchmarks of the international debate in the last 30 years on the increase and high levels of imprisonment, are interlaced.

Further readings

Garland, David. 2013. 'Penality and the penal state'. *Criminology* 51(3): 475–517. https://doi.org/10.1111/1745-9125.12015

Goodman, Philip; Page, Joshua and Phelps, Michelle. 2017. *Breaking the Pendulum: The Long Struggle over Criminal Justice*. New York: OUP.

Lacey, Nicola. 2008. *The Prisoner's Dilema. Political Economy and Punishment in Contemporary Democracies*. Cambridge: Cambridge University Press.

Sozzo, Maximo. 2018. 'Beyond "the neoliberal penality thesis"? Punitive turn and political change in South America'. Carrington, Kerry; Hogg, Russell; Scott, John; and Sozzo, Maximo, eds, *The Palgrave Handbook of Criminology and the Global South*. London, Palgrave Macmillan. https://doi.org/10.1007/978-3-319-65021-0_32

32

INDICATORS

Socio-legal dimensions of quantification

Sally Engle Merry

We are used to using numbers to describe economic processes and viewing them as authoritative and reliable descriptions of economic activity. They are routinely accepted as a reasonable basis for policy decisions. Statistics such as gross domestic product, the unemployment rate, and credit ratings for individuals and for securities are typically assumed to be objective and reasonable measures of economic activity. Economic development is increasingly governed by a set of numerical goals such as the Millennium Development Goals, replaced in 2015 by a new set of Sustainable Development Goals. These numbers are widely used to measure national progress globally. But this is only the tip of the iceberg. Quantification now plays a major role in shaping the world we live in. Indeed, our contemporary reliance on numerical data for life decisions is growing every day as an ever-widening range of experiences and dimensions of social life is converted into numbers. Yet, we know relatively little about how these numbers are created, how we decide what gets counted and what does not, and what these numbers mean.

An important technology of quantification is the development of indicators: collections of statistical measures that have been joined together to produce a single number or rank for a complex idea, such as rule of law or access to justice. The single number allows ranking of all the measured countries, schools, or organizations that have been measured. They are widely used as the basis for policy-making, governance, and myriad individual decisions such as where to invest funds, which college to attend, which teachers deserve raises, and which city is most liveable. Through simplifying complex phenomena, they make ranking possible. Indicators can take the form of indexes, rankings, and composites which aggregate different indicators. Many of the best-known indicators are aggregations, such as the Human Development Index, the Consumer Price Index, and World Governance Indicators. They are typically used for judgements and decision-making.

Studying the process by which things are counted reveals a series of pragmatic decisions about what categories to include and which to leave out. Do those who count think it is important to include race? Gender? Disability? Do measures of poverty only count the poor, or do they also consider the gap between rich and poor? Someone makes these decisions. Committees of experts decide what to count, how to categorize, and how to deal with missing data. Experts typically come from the global North and rely on previous collections of statistical data to decide what to count. They tend to focus on things that are seen as countable while ignoring those that are not. For example, in the past, the unpaid labour of women in the home caring for children and the sick was viewed as uncountable. Thus, it remained uncounted and unrecognized. However, the range of phenomena open to quantification is expanding daily, as we now measure happiness, the number of steps we take, the hours we sleep every day, and many other features of everyday life.

Numbers are essential to making problems known and building social movements. Violence against women was long ignored and when activists began to focus on the problem in the 1970s and 1980s, a first step was counting how often it took place. The movement against trafficking has relied on trying to find data on how many people are trafficked, and reform projects such as the World Justice Project have developed a range of indicators of the rule of law as a way to pressure governments to reform their legal systems.

Yet, despite their value for making problems known and pressing for reform, numbers often provide a distorted picture of the situation they endeavour to describe. They suffer from limited or inaccurate data such as gaps where data is missing, of poor quality, or based on proxies that were designed to measure something else. Particularly, in poor countries, the cost of data collection means that data quality is often poor or missing. They also face the challenge of making diverse entities comparable. Any system of counting requires substantial interpretive work in deciding what to count, how to categorize it, and what to call what is measured. A complex phenomenon such as gender-based violence can be counted in many ways, such as the trajectory of the relationship, the victim's freedom to leave, or dominant gender ideologies. However, the most common approach is to focus on physical blows and injuries. This measure excludes the critically important wider context such as marriage and kinship practices, economic factors, and issues such as racism and migration status. Yet, to count something, it must be made similar to other entities so that they can be added together. Measuring something like gender-based violence is fundamental to raising its visibility as a social issue. Yet, it is also hard to incorporate details of kinship, gender ideology, marriage rules, and economic inequality in a quantified measure of gender-based violence.

Since systems of quantification have important knowledge effects, in that they create a picture of the world, the way they sort and tabulate information matters. They also have governance effects, in that they determine policy, so they are a significant site of power. Yet, these effects are relatively unrecognized and untheorized. Quantification can have both constructive and harmful effects: it can reveal patterns of discrimination and inequality, yet it can also conceal or obscure inequalities. It is important to peer into the black box of measurements such as social indicators to see how they are made and to what extent they distort and misrepresent reality. As the demand for quantification spreads to new areas, such as corruption, the rule of law, the quality of life, and even happiness, the importance of quantification as a form of knowledge and a basis for policy decisions becomes ever more significant, and the consequences of its distortions more serious. Some categories are not counted at all, such as women's unpaid labour in the home. Context is rarely measured, so that the incidence of events such as domestic violence maybe counted, but not the surrounding social, cultural, and emotional dimensions of an incident or its trajectory.

The turn to numbers as a mode of evaluating social and economic life is an historical development initiated during the emergence of nation-states in the 19th century. With the growth of the modern state, whose wealth depends largely on its population rather than on its territory or control over trade routes, it became increasingly important to know about this population: its health, education, capacities, and birth and death rates. States began to collect statistics about these issues through vital registration, censuses, and surveys. In the early 20th century in the United States, surveys expanded into new domains, such as political preferences and sexual behaviour. Americans became accustomed to being surveyed and counted. With the encouragement of the United Nations and global development agencies, the global production of economic data and population surveys has expanded, with poor as well as rich countries investing in counting and measuring their people. The turn to measurement was based both on a new concern with governing populations and with reform agendas such as improving public health and diminishing discrimination in school admissions. For example, the use of standardized tests for college admissions revealed the existence of quotas for religious minorities in the United States.

The need for numbers and the authority they exercise as a form of knowledge and an aid to governance is now largely taken for granted. Although there are some contemporary debates about privacy with the expansion of attention to big data, to a large extent the creation of quantitative data about populations and the use of this information in both private and public modes of governance is widely accepted. Yet, looking at how such numbers are constructed, the difficulty of challenging them, and the extent to which they are fundamental to knowledge and governance suggests that they are a largely unrecognized field of power. Some key economic numbers such as gross domestic product and unemployment rates are well established, but this mechanism is now being developed in newer domains, such as education, the rule of law, and anti-trafficking campaigns. As this technology of knowledge expands into less readily quantifiable fields, it becomes increasingly problematic. This does not mean that numbers are of no value as a form of knowledge and as guides to decision-making. It does suggest the importance of a healthy scepticism about what they measure and what they fail to measure and the need to question where they come from. Counting things is an invaluable strategy for producing knowledge, but its value depends on being clear about what is and is not being counted and how the numbers have been collected and interpreted.

What are the implications of relying on indicators for governance? First, indexes and ranks create knowledge that appears objective and scientific, in part because the precision of numbers themselves lends scientific authority to their claims. Yet, these ranks and indexes are made up of categories for counting and theories of how things should work that are typically unstated: they embody assumptions about what constitutes a good university, what should count as economic growth, and who is a slave who needs to be liberated rather than simply a person in a miserable job. There is power in the hands of those who create the categories, decide what to measure, evaluate vague data, and convert it into unambiguous numbers. These efforts reinforce ideas about what matters in economic productivity, good universities, and in creating a rule-of-law state. They influence which school a student attends, which countries appear to be doing well economically, and which ones can claim the status of good governance or an absence of slavery. Yet, the power to construct categories and populate them with data is rarely recognized, nor are the theories of social life underlying statistics generally made explicit. They are not uniformly distributed but tend to reside among experts and people from the global North.

A second feature of the power of quantification is the way it highlights some kinds of information and excludes others. The focus on what can be counted and what has already been counted leaves many important issues out, such as the social climate of a school, the contribution of natural resources to GDP, or the array of social conditions within which a person becomes caught in a slave-like condition.

A third feature is that the demands of making data commensurable so that comparisons can be drawn among schools or countries means that quantification can only take place on the basis of shared standards. Recognition of a diversity of standards for schools, a variety of forms of unfree labour, or variations in how a society should be governed cannot be included in the calculations. Indeed, quantified representations cannot consider the context of the things counted even though such contexts are deeply significant for understanding events as diverse as test-taking or becoming enslaved.

A fourth feature of the turn to quantification is its compatibility with a marketized, neo-liberal conception of society. Wealthy private actors often play a central role in developing indexes which express their view of the world. The rule-of-law indicator embodies a particular notion of good governance, based in the global North. The educational ranking system rewards a particular vision of education, also located in the global North, and promotes the branding of educational institutions within a market for educational services. Thus, these systems of quantification subtly endow Northern institutions with superior status and pressure Southern ones to follow suit, not explicitly but simply through establishing standards and evaluating countries according to these standards.

Finally, the growing reliance on data for governance raises enormous questions of cost and the value of investing in counting things in comparison with other activities. Clearly, this is a greater problem for poor countries than rich ones, but even in the school ranking system in the USA, schools are compelled to devote resources to tending their rank that they would rather use for other activities. The act of counting itself absorbs scarce resources in the name of reform, such as efforts to eliminate inequality. At the same time, this technology serves to foster an economistic, market-oriented vision of society. Given the goal of uncovering and exposing relations of power where they remain veiled – here as they are hidden by claims of science, objectivity, and rationality – it is critical to interrogate this burgeoning turn to measurement both as a mode of knowledge production and a dimension of governance. We may not like where it is taking us.

Further readings

Davis, Kevin E., Kingsbury, Benedict and Merry, Sally Engle. 2012. 'Indicators as a Technology of Global Governance'. *Law and Society Review* 46 (1): 71–104. https://doi.org/10.1111/j.1540-5893.2012.00473.x

Espeland, Wendy Nelson and Stevens, Mitchell L. 1998. 'Commensuration as a Social Process'. *Annual Review of Sociology* 24: 313–43. https://doi.org/10.1146/annurev.soc.24.1.313

Espeland, Wendy Nelson and Stevens, Mitchell L. 2008. 'A Sociology of Quantification'. *European Journal of Sociology* XLIX, 3: 401–436. https://doi.org/10.1017/S0003975609000150

Merry, Sally Engle. 2016. *The Seductions of Quantification: Measuring Human Rights, Gender Violence, and Sex Trafficking*. Chicago: University of Chicago Press.

Rottenburg, Richard; Merry, Sally Engle; Park, Sung-Joon; and Mugler, Johanna. 2015. *A World of Indicators: The making of governmental knowledge through quantification*. Cambridge: Cambridge University Press.

33

INDIGENEITY

Making and contesting the concept

Miranda Johnson

'Indigeneity' refers to the quality of being Indigenous, which literally means someone who has been born in the land in which they live. In richer, symbolic terms, an Indigenous person is *of* that land and belongs to it. As an expression of identity, Indigeneity is regularly associated with an ancient and spiritual relationship to land and water and to ancestors and descendants, including non-human beings, over generations.

Yet, the term is not simply descriptive of a long-standing identity. A modern discourse of indigeneity carries moral, political, and legal connotations. For instance, indigeneity differentiates prior occupiers from later usurpers, or the colonized-Indigenous from the colonizer-settler. In some settler state jurisdictions, proof of indigeneity is key to winning distinct rights to land in the form of 'native' or 'aboriginal' title.

While the discourse of indigeneity that has emerged in settler majority states in the last 50 years emphasizes rights and reforms for the dispossessed, scholars point out that autochthony was a key idea for ancient Greeks and earlier European struggles against and within various empires. The identities and identification of Indigenous peoples can encompass racial ideas and notions of biological inheritance or difference – and therefore be grounds for racist discourse – though indigeneity is not itself an exclusively racial discourse. However, while a modern discourse of indigeneity is often used to protest state and corporate incursions onto Indigenous soil and build transnational solidarity, it can also be used by states and political parties to exclude others and outsiders.

Making indigeneity

The term 'indigeneity' is a recent coinage and is primarily employed in academic rather than broader public discourse. Its increasing use demonstrates the scholarly need for a term that addresses contested and politicized claims to being Indigenous, and the rights associated with those claims both in local and global contexts.

In western settler states such as the United States, Canada, Australia, New Zealand, and parts of Scandinavia, the term marks a historical change in the symbolic value of Indigenous identity. This is a consequence of the resurgence of Indigenous peoples' political struggles that brought claims to (non-Indigenous) national and international consciousness. In particular, the term indigeneity can usefully help scholars in these countries distinguish contemporary assertions of identity from historical use of the term 'native' (although this term is still widely used especially in

North America). Native-ness was associated with peoples of the past who may not survive into the present. Natives were often regarded in western scholarship as 'people without history' and presumed to be incapable of adaptation (or adaptation was assumed to mean the loss of native culture and identity). Where natives survived the genocidal threats against them and ongoing land dispossession in the 19th century, governments aimed to assimilate surviving populations in the 20th century. But today, the discourse, of indigeneity emphasizes the survival and continuity of distinct identities despite colonial attempts to eliminate and dispossess Indigenous peoples.

Claims to indigeneity are now global and they are made by diverse peoples in many parts of the world from Latin America, to South Asia, and Africa. These claims for rights and justice that are now associated with a modern discourse of indigeneity, however, emerged first in western settler states of North America, the South Pacific, and Scandinavia. In the 1970s, Indigenous peoples' claims in these regions – many of which were long-standing and had been protested for generations following land and resource dispossession by settler colonizers – forced wider political changes. Indigenous activism was the consequence of various longer- and shorter-term forces, including historic land dispossession and the more recent mining boom in remote areas that once more threatened Indigenous land. Indigenous activists were influenced by global discourses of anti-colonialism and human rights, though Indigenous claims were distinct from both because activists were not demanding separate statehood but rather collective rights.

Repudiating official assimilation policies of the past and insisting on their rights, activists sought to undo previous policies and ideologies of assimilation. Indigenous legal and activist successes pushed governments to furnish new policies of Indigenous self-determination. By winning judicial attention in major court cases concerning their aboriginal title rights – a distinctive form of property title that recognizes Indigenous peoples' collective and inalienable rights to land – activists opened up new possibilities for land rights recognition in the late 20th century. They also drew attention to treaties and formal promises made to them in the past by imperial authorities and later broken by settler governments.

At the international level, Indigenous activists from the settler states created transnational organizations and took their claims for self-determination to the United Nations. In the 1980s, a special UN working group was created to draft a declaration on the rights of Indigenous peoples in the framework of human rights. This resulted eventually in the UN Declaration on the Rights of Indigenous Peoples passed by the General Assembly in 2007.

Indigenous peoples' claims made a distinctive argument about their identity as inextricable from the places and landscapes to which they belonged. Paradoxically, as Ronald Niezen has pointed out, an identity that emphasizes some peoples' claim to being of and belonging to particular landscapes has found global resonance. Today, transnational Indigenous organizations support and promote local struggles; they press for the recognition and implementation of Indigenous rights around the world; and they inform international professional and scholarly associations. However, contestations of the value of indigeneity spark new expressions of racism against Indigenous peoples; and policy shifts and legal requirements impose new constraints on Indigenous actors as well as offering them possibilities for solidarity and mobilization.

Politics of indigeneity

The relative success of Indigenous rights campaigns in the 1970s and 1980s met with new challenges and constraints. As courts and governments were forced to recognize Indigenous peoples' rights, they began to develop evidentiary paradigms for how land and treaty claims could be demonstrated; and they established mandating requirements for the representation of constituted claimant groups. More broadly, activist and policy success generated a wide-spread politics of indigeneity.

For Indigenous peoples, one aspect of this politics concerns the considerable tension between self-definition, a critically important principle that aims to ameliorate past policies of misrecognition and attempts by settler states to assimilate Indigenous peoples, and the constitution of tribal membership. In Indigenous political communities such as tribes, iwi, First Nations, and so on, membership is usually based on an individual's ability to prove descent from certain ancestors or those named on historic tribal rolls. Such proof can be hard for some to demonstrate, particularly in a country like Australia where large numbers of children were removed from their parents and communities – those today known as the 'stolen generations'. Indigenous nations must navigate the diasporic effects of dispossession and assimilation-era policies while being mindful of retaining the benefits and identifying the obligations of membership for their own people.

In broader public discourse in settler states, a sometimes sharp debate plays out concerning the authenticity of Indigenous claimants. The double binds of indigeneity – that the more modern an Indigenous person or people appears, the less Indigenous they are assumed to be and vice versa – have been widely critiqued in historical and cultural studies of indigeneity in the North American as well as the Pacific context, but they persist. Further, as indigeneity has become a central symbol in some settler states, settlers often raise the issue of authenticity and question how 'Indigenous' certain public figures are. They also question the moral grounds of indigeneity as a claim to being here first. In other words, they question how inherent such an identity is and whether asserting such an identity is the right basis for rights and the redistribution of resources. Ideas of Indigenous 'privilege' serve as a cover for a range of anti-Indigenous sentiments articulated through racist terms and long-standing stereotypes of 'lazy natives' and so on.

At the same time, settlers also make their own claims to indigeneity – another long-standing trope of what Jean O'Brien has called 'firsting and lasting' among inter-generational settler communities in North America. In New Zealand, 'Pākehā' – a Māori language term for white settlers and their descendants – claim to share in an inter-generational relationship to land and environment, sometimes in order to assert their own rights to land as in the case of high-county farmers in the South Island.

In the context of UN working groups drafting the declaration, another politics of indigeneity erupted concerning who was entitled to use the term. Representatives of newly post-colonial African and some Asian states said their national populations were all Indigenous, thus denying the particular claims of tribal and other groups within their territories who claimed long-standing and culturally particular rights to lands. Chinese delegates claimed that the term 'Indigenous' only applied to the settler colonial contexts where Indigenous peoples had been dispossessed and had not benefited from decolonization. These state delegates therefore argued that the term and associated rights did not apply to their countries. Indigenous activists countered by emphasizing the ongoing experiences of oppression, dispossession, and pressures of assimilation faced by minority 'Fourth World' peoples (the term popularized by Secwepemc leader and founder of the World Council of Indigenous Peoples, George Manuel) that they shared, in settler colonial as well as newly post-colonial states.

What happens in countries where an Indigenous party holds power, or Indigenous groups are in the majority? In both Fiji and Bolivia, Indigenous political leaders pre-dominate and deploy a modern discourse of indigeneity to various ends. Andrew Canessa argues that in the Bolivian case, many Indigenous peoples feel betrayed by Evo Morales, the 'world's first Indigenous president', whose progressive agenda nonetheless incited intra-Indigenous tension and even conflict (which resulted in his being eventually deposed). Like Kēhaulani Kauanui in the different case of Hawai'i, Canessa argues that discourses of indigeneity are not only matters of identity but very much of politics too. In his comparative study of Cameroon and the Netherlands, Peter Geschiere traces how discourses of autochthony are mobilized to limit citizenship and particularly voting rights by majority populations asserting native ties. He cautions readers about the 'perils of belonging'.

Indigeneity and the production of knowledge

Indigeneity is not only a matter of identity and politics but of the formation of knowledge and ways of knowing. In western settler states and Latin America, discourses of indigeneity refer to the ways in which Indigenous identity has survived and may flourish: it is a discourse of being and becoming. Indigenous peoples insist on the fact that they are 'still here' and will continue to live in and protect the lands and waters of their ancestors in new as well as old ways. Discourses of indigeneity connote distinctive relationships to time, for instance in terms of what Anishinaabe scholar, Gerald Vizenor, calls 'survivance'; and even theories of history, for instance in the form of genealogy. The concept of *whakapapa*, in Māori (and related terms in other Polynesian languages), which refers to the layering of generational connections and descent, is a centrally important historical practice.

Indigenous peoples have often been intensively studied over generations by anthropologists and social science scholars. Mobilizing Indigenous rights in relation to knowledge and its production in the last three decades or so, Indigenous scholars have made influential arguments and offered new insights into how research on Indigenous issues, and with Indigenous subjects, should be conducted, who such research does or should benefit, and what are the best methods for carrying out such research.

Perhaps most well-known is the Māori education scholar Linda Tuhiwai Smith's book *Decolonizing Methodologies* which was first published in 1999 and revised in 2012. This book is aimed in particular at Indigenous scholars working within Indigenous communities, of which they may also be a member, and proposes Indigenous-centric approaches to knowledge production and dissemination. A broader 'decolonial' turn, inspired by Latin-American theorists such as Walter Mignolo, Aníbal Quijano, and María Lugones, seeks to 'de-link' the experience and effects of colonialism from modernity, aiming to return Indigenous knowledges to Indigenous peoples.

These arguments carry influence beyond the growing field of transnational Indigenous Studies. Many universities now require researchers to develop particular protocols for conducting research with Indigenous peoples. External funding agencies in Canada, Australia, and New Zealand all offer separate funding streams for Indigenous scholars and/or Indigenous knowledge projects. In other spheres, too, the modern discourse of indigeneity has influenced a reassessment of indigenous knowledge. Most notably, recent reports by the Intergovernmental Panel on Climate Change propose that Indigenous or 'traditional ecological' knowledge is key to solving the climate crisis and must be protected.

The modern discourse of indigeneity, forged in activist networks and the crucible of legal and political rights claims in settler states, is now widely invoked in international organizations as well as in local struggles. The global reach of the discourse is remarkable given that it is usually associated with minority peoples who lack effective access to political power and capital. This fact underscores the effectiveness of Indigenous struggles at multiple scales. The contested politics of indigeneity demonstrate how deeply felt as well as indeterminate such a discourse is, one that is both 'other' to the west and also profoundly entangled within western thought. Rather than being somehow outside of history, a discourse of indigeneity prompts profound questions about relationships between past, present, and future, for a world in crisis.

Further readings

Geschiere, Peter. 2009. *The Perils of Belonging: Autochthony, Citizenship, and Exclusion in Africa and Europe*. Chicago: University of Chicago Press.

Simpson, Audra and Smith, Andrea. 2014. *Theorizing Native Studies*. Durham: Duke University Press.

Tuhiwai Smith, Linda. 1999. *Decolonizing Methodologies: Research and Indigenous Peoples*. London: Zed Books.

34

INFRASTRUCTURE

Socio-legal aspects of a key word of our time

Mariana Valverde

To understand not just what a word means but what practical work it does, it is useful to consider what terms were replaced by it. In the case of 'infrastructure', much of what is covered by this ubiquitous word used to be called 'public works'.

'Public works' have existed in the West since at least the Roman Empire. But it was in the mid-19th century that Western governments began to take increasing responsibility for planning and delivering structures and services that served the public at large, mainly in the imperial metropole but occasionally in 'the colonies' as well. Grand public squares and wide avenues were built in many cities; and networks of roads and bridges (which in many cases had emerged in a haphazard, fragmented manner, often from a group of private investors who would then collect tolls) were either built from scratch or improved.

Railways, invented in the 1820s and mushrooming from the 1840s onward, were initially a private endeavour: but governments often facilitated their expansion by expropriating private owners and literally laying the groundwork. In the early and mid-19th century, a great deal of public money was invested in canals, though railways and then highways made most canals superfluous in short order. Perhaps most importantly, governments (often local rather than national) began to take it upon themselves to provide drinking water and sewage networks for cities. Paris was a pioneering city in this regard.

In the United States, private initiative predominated until the New Deal of the 1930s. President Franklin Roosevelt's radical plan to end the Depression involved massive hiring to build massive projects, such as giant hydroelectric dams. The scale of New Deal public works was unique – until China began its extensive infrastructure program from the 1990s onward.

In general, 'public works' is a phrase that emphasizes not only government funding of facilities aiming to serve the public but also government decision-making. In the public works era, private enterprise flourished: outside of communist regimes, governments contracted private business owners (from architects to road-paving firms) to do most of the actual work. But the decision-making remained in government hands.

By contrast, the original context in which the word 'infrastructure' was coined foreshadows today's 'partnerships'. A term popularized in French engineering works in the 19th century, 'infrastructure' referred to the preliminary land preparation work that had to be carried out before corporations would then construct railways. Private owners and operators of early railways only moved into action once government had done the initial legal and physical work.

Why did this division of labour happen? One reason is that private businesses have never had the power to expropriate other private owners, only governments have. In addition, businesses, then as now, wanted assurances of government support for expensive works that, especially in the early days of railways, were risky: nobody knew in advance if people would abandon stagecoaches en masse in favour of trains. Thus, at its very inception the word 'infrastructure' was not so much an engineering term but a conglomerate of legal and material and social relations. The very word suggested that the 'natural' method of building transportation and utilities is to have governments prepare the groundwork, literally and legally, for private business to then flourish.

Today, this same theory is at work in the 'urban revitalization' projects one sees all over the world. The revitalization agencies and 'urban development corporations' that have given us London's Canary Wharf, New York's Hudson Yards, and touristy riverside complexes everywhere are operating on the same theory as the French 19th century railway magnates. That theory is: give corporations subsidies, paved streets, reliable and drinkable tap water, abundant energy, and maybe a break on taxes, and they will then build projects. Such urban development projects often look pretty; but the jobs they create, once construction is finished, are usually low-paid insecure service jobs (in hospitality and retail). The high-end jobs remain in downtown office towers or else in the suburban 'campuses' favoured by industry.

Today's cookie-cutter urban revitalization projects thus owe a great deal to the assumptions about the roles of government vs private enterprise that were visible in the original French meaning of 'infrastructure'.

If we ask why 'investing in infrastructure' is such a popular refrain today, even amongst conservative, deficit-slashing politicians, we can start to see the different kinds of work – political, cultural, economic, and legal – effected by the current infrastructure mania.

The proponents of 'public works' were fairly unapologetic about the need to levy taxes so governments could provide facilities and services for the general public. However, in the wake of the neo-liberal/neo-conservative revolution in government that began in the 1970s, both levying taxes and spending public money for public purposes came to be regarded with suspicion. Populist moves to freeze or even cut local taxes, without regard for infrastructure needs, were replicated at higher scales. The new spectre haunting both global North and global South countries and international organizations from the 1980s until the 2008 financial crisis was the spectre of national debt, which in turn severely limited spending on infrastructure. After the world financial crisis, however, 'investing' (that is spending public money) on infrastructure became a favoured means of stimulating the economy. Even conservative politicians talk the talk.

To invest in infrastructure is to spend money on (mainly public, though often fee- or toll-collecting) facilities, goods and services. This of course creates debt. But ways can be found to hide or disguise government debt. One popular method of offloading the debt involves creating autonomous 'urban development corporations' and agencies that borrow the money and thus acquire their own debt, which in formal accounting terms is not the government's debt. The debt of the Port Authority of New York, just to give an example, has never been considered as the debt of the two states that created and still govern the powerful agency. The fact that the governments that created it reduced the agency's 'demand risk' to zero by giving it, legally, the monopoly on all cross-Hudson tunnels and bridges, and allowing the agency to keep all the revenue, is not well known. The toll revenue was the most secure future revenue stream ever, and this revenue was then used to continue expanding the 'Empire on the Hudson' to include the major New York bus terminal and countless other facilities and buildings. As is typical for such quasi-government entities, the Port Authority is run by an appointed board, not by public officials. And its books are far less transparent than those of actual cities.

Those features (lack of transparency, lack of accountability to citizens) are replicated in the thousands of newer 'urban development corporations' and agencies that have mushroomed all over the world. But, especially when pressured by populist anti-debt or anti-tax agitation, governments find it very useful to proliferate obscure arms-length agencies.

In addition, ways can be found to shift public debt to the next generation. One popular tactic is seeking long-term private financing, whereby only the interest will be repaid yearly, with the principal being due decades later – when the politicians who signed off on 'the deal' will be, if not dead, safely out of office. All major infrastructure projects in the populous Canadian province of Ontario, for example, are organized, legally, as privately financed projects with a 40-year horizon for repaying the principal. The resulting roads and train lines will all be publicly owned (unlike in many other countries); but the mechanisms used to plan and finance the works greatly empower the private players in the game.

These players are not the road-paving contractors of earlier decades. While in the age of public works contractors were usually local, in the age of infrastructure projects are generally scaled large (financially as well as physically). The large size is seen as necessary to attract the handful of global firms that reap profits from the infrastructure boom and that only deal in hundreds of millions. These include global investment funds, such as the Australian Macquarie bank, giant engineering firms (such as the giant Spanish firms specializing in tolled highways and high-speed trains), global law firms headquartered in New York or London, and, last but not least, the global accounting/consulting firms that charge large sums for certifying the 'deals' (KPMG, McKinsey, Deloitte).

Recent scholarship on infrastructure has pointed out that countries in the global South have been found wanting when measured against a Eurocentric 'infrastructure ideal'. This ideal assumes that the goal is a very large, coherent, and integrated network of efficient transportation routes, with these networks supplemented by a perfectly functioning system of coordinated public utilities. Needless to say this 'ideal' is not realistic even in global North countries. Inhabitants of Amsterdam or Berlin may enjoy something approaching the ideal; but in all countries both public transport and such necessaries as fast internet access are very unevenly spread across the territory.

Scholarship from and about global South cities suggests that the modern 'infrastructure ideal' undervalues or ignores less systematic and less formal solutions to problems. Research in Africa, especially, has shown that local residents, sometimes facilitated by NGO's and 'squatter' community leaders, often have better ideas than the global consultants who go from city to city with pre-prepared templates. For example, in a particular village, one could ask whether all houses need to have indoor toilets connected to water-based sewers – or whether a sufficient number of water-free public toilets that are clean and accessible provide a more environmentally sound and affordable solution. Similarly, the millions of people who have built their own houses on the outskirts of Latin-American megacities and have jerry-rigged connections to nearby electricity poles could be seen as innovators, rather than as 'illegal settlers'. Where electricity has been privatized, legal connections may just be unaffordable. Locally devised solutions in the global South may not replicate the ways in which European cities developed their services and utilities, often with large sums of public money, in the age of 'public works'.

A majority of the world's peoples have severe needs that could be called 'infrastructure' needs: they lack such basics as cheap and reliable drinking water; they are forced into overcrowded unsafe private buses, or they have no access to the internet. However, in the process of planning to meet these basic needs, it may be wise to question both the euro-centric 'infrastructure ideal' (which may be unrealistic in the rural areas of global North countries as well), and the very notion that massive 'infrastructure' projects are the best solution.

If community empowerment were to become central in 'infrastructure' planning, local activists would have to pay close attention to the governance processes (the ones behind grandiose public works, as well as those behind public-private infrastructure projects). The decision-making processes of both public works and public-private infrastructure projects are designed to be invisible to the public; but information can be obtained through freedom of information requests, investigative journalism, and whistle-blowers both within the state and in the private sector.

What politicians and infrastructure builders call 'transparency' often means telling us how much the bridge or train will cost, who will be the contractor, and how many construction workers will be employed. But ordinary citizens are hardly in a position to know whether $200 million is the 'right' price for a major work, or how many construction workers are actually needed for a hydroelectric dam. So, instead of being content with the sort of meaningless 'facts' that visitors to the Hoover Dam or the Eiffel Tower are given by tour guides, the questions that citizens would do well to ask are:

> Who exactly is/was involved in deciding what to build, and where? How are community needs assessed (if they even are), and by what entities? Who decides what will be built and what will remain on the drawing board? Do the infrastructure projects announced by politicians really meet the most urgent needs? Even if the need is clear, is there a way to meet the need, perhaps cheaper and more environmentally friendly, that would never be proposed by the global financing and engineering and consulting firms usually asked to do such work? Who benefits, financially, from the way in which funds are borrowed for these works? Is there an intergenerational injustice in how the debt is structured? Were carbon footprints and other environmental factors seriously considered when deciding whether to 'go big' instead of promoting modest community projects? Who is going to own the facility or service? Will some people lose their houses or their land in the process? How will those losses be costed and remedied?

These are the kind of questions citizens ought to be asking every time they are presented with photos of smiling politicians announcing 'an investment in infrastructure'.

Further readings

Benjamin, Walter. 2010. *The Arcades Project*. Work written 1920s–1930s, never published in the lifetime of the author. Cambridge, MA: Harvard University Press.

Blum, Andrew. 2012. *Tubes: Voyage to the Center of the Internet*. New York: Harper Collins.

Graham, Stephen (ed.). 2010. *Disrupted Cities: When Infrastructure Fails*. London: Routledge.

35

ISLAMIC LAW AND THE STATE

Anver Emon

Depending upon where one lives, calling September 11, 2001 an epic moment in world history is either cliché or hubris. Horrifying in its flash-bang scale and scope, 9/11 sits amidst a distressing spectrum of violence, pogroms, and genocides – the 1994 Rwandan genocide of Tutsis (approx. 800,000-1,000,000 dead); the July 1995 Srebrenica massacre (approx. 8000 dead); and the 2002 Gujarat riots (approx. 1044 dead, 223 missing) – to list a few examples. But unlike these other incidents of violence, the 9/11 attacks unleashed on an enormous scale state-coordinated military and national security campaigns against non-state actors, namely some paradigmatic 'Muslim extremists', in the service of the 'War on Terror'.

For students and researchers, we cannot ignore how 9/11 has had the effect (whether deserved or not) of signalling a sea change in what is researched, who did the research, and how that research was deployed. The security-conscious environment that became the norm ushered in a regime of immigration restrictions, especially in the United States, that limited who could study and produce knowledge, where they could go, and which borders they could cross in pursuit of ideas. 9/11 also informed a more robust policing of the U.S. academy, with 'watchdog' organizations like Campus Watch featuring dossiers on faculty who espoused views deemed contrary to national values. The UK went further with its PREVENT program, which required British academic institutions to control for the possibility of radicalization in the classroom, leading to what many considered a chilling effect in the classroom. Furthermore, the imperatives of national security arising from 9/11 helped to inform the questions for academic study, often in the service of the state. Scholars across a range of American departments, centres, institutes, and clinics, now increasingly examine the measures taken to combat the presumed 'Islamic' threat and those who embody it. In law schools, scholars interrogate immigration and national security rules that guard against an 'Islamic' threat. In psychology departments, researchers advance or critique government programs that counter violent extremism (CVE) by using the 'Islamic' extremist as paradigmatic. In these and other instances, scholars engage, critique, support, or otherwise address state practices in the shadow of varying constructions of the 'Islamic'.

The sheer scope of governance projects (both in the 'West' and the so-called 'Muslim world') that rely on certain constructions of the 'Islamic' or the 'Muslim' demands that we interrogate how we come to define and know what is 'Islam' and who is the 'Muslim'. What is represented as 'Islam' is all too often selectively structured to serve a range of governance projects that take shape across time, space, and scalar variation. Rather than nominally construing 'Islam' and 'Muslims' as passive *objects* of knowledge, I suggest recognizing that these terms serve adjectival purposes. Re-written as 'Islamic' and 'Muslim' in an adjectival sense, both are also rhetorical, socially constructed 'terms of art'.

In addition, the study of Islam (and its Muslim adherents) may benefit from examining how certain constructions of the 'Islamic' and the 'Muslim' are strategically deployed in the service of state governance projects in the West and in the Muslim majority world. For example, debates about Sharia law in the U.S., Europe, and Canada are less about Sharia or law and more about contests about core values of the state (e.g. democracy and rule of law) in light of competing assumptions about what Sharia is.

For example, in an early version of its Shari'a ban (subsequently amended extensively), Tennessee Senate Bill 1028 defined Shari'a as follows:

> Sec. 39-13-904
>
> As used in this part, unless the context otherwise requires:
>
> (1) 'Sharia' means the set of rules, precepts, instructions, or edicts which are said to emanate directly or indirectly from the god of Allah or the prophet Mohammed and which include directly or indirectly the encouragement of any person to support the abrogation, destruction, or violation of the United States or Tennessee Constitutions, or the destruction of the national existence of the United States or the sovereignty of this state, and which includes among other methods to achieve these ends, the likely use of imminent violence. Any rule, precept, instruction, or edict arising directly from the extant rulings of any of the authoritative schools of Islamic jurisprudence of Hanafi, Maliki, Shafi'i, Hanbali, Ja'afariya, or Salafi, as those terms are used by sharia adherents, is prima facie sharia without any further evidentiary showing.

In this passage, we certainly find law in Islam, but it is in the service of a purpose that poses an existential crisis for states in the U.S. who seek to remain free and sovereign. On this reading, it would seem Shari'a threatens an entire way of life. This explains why a year earlier in 2010, Oklahoma's constitutional amendment to ban Shari'a was called, without irony, the 'Save Our State Amendment'.

This is no different from how Muslim majority states deploy Sharia in their domestic environments. Muslim majority states participate in international fora and enter multilateral treaties on a range of topics, including human rights. But in the field of human rights, these same states are notorious for issuing reservations limiting their obligations under an international treaty, out of deference to Islamic law and practices that operate domestically. Perhaps the most notorious example are these states' reservations to the Convention on the Elimination of Discrimination Against Women (CEDAW), where limitations on the scope of their commitment to equality preserves the patriarchy embedded in the historical family law rules they legislate into contemporary statutes. These reservations are no less statements of core value than the Tennessee bill above. In both cases, Islamic law is deployed in the service of state claims to sovereignty and freedom from interference.

Understanding the confluences and divergences across regional, state, and local contexts would allow a more robust appreciation of the shared logics that inform how *Islam* and *Muslim* get construed *and* deployed across regions that are rarely addressed together. By examining these contexts together, researchers can gain a deeper understanding of the relationship between the conditions under which knowledge about Islam and Muslims is generated, and the material consequences arising from how that knowledge is put in practice.

As an example, let's consider insurance. Insurance is a contract that involves purchasers making periodic payments to an insurance provider for a policy that will control against possible future financial challenges (e.g. death, environmental disaster, etc.). A student of Islamic economics will immediately identify this contract as problematic: it involves current payments for a payout at an unknown time, and thereby violates a core Islamic prohibition on speculation or *gharar*. That does not change the fact that people may still want insurance, such as life insurance, to protect against future accidents. Saudi Arabia's government institution of clerics, the Permanent Committee (*lajna al-da'ima*) has pronounced commercial insurance a violation of Islamic law. In *fatwa* 3249, the Permanent Committee states:

> Commercial insurance is a commitment to do something that is not legally obligatory (*fi ʿaqd al-ta'min al-tijari ilzam bima la yalzam, sharʿ*). The insurance company is not responsible for the accident; it only makes a contract with the insurer to protect them against accidents, in case they take place, in return for a sum of money which the insurer pays to the insurance company and the latter does nothing for the insurer to deserve this money. Hence, commercial insurance is *haram* [forbidden].

On this reading, one could easily presume that insurance is unavailable in Saudi Arabia. If we were to end our analysis here, we would have a very narrow construction of the 'Islamic' for domestic political consumption. The problem, though, is that Saudi Arabia also exists in an international context in which migration flows, capital flows, oil wealth, and political multilateralism situate it as a significant political player on the world stage. Consequently, while the state may be able to pacify a conservative demographic by reference to the Permanent Committee's *fatwa*, it must also assure its global partners in trade and labour migration of some degree of egalitarianism among those coming to work in Saudi Arabia, assuming they have enough market power to enter the commercial insurance regime. According to Saudi Arabia's Monetary Authority (SAMA), commercial insurance is indeed a vibrant reality in the Kingdom. Principally seen as a risk-moderating device to promote greater investment in the Kingdom, Saudi Arabia's Monetary Authority glowingly describes its insurance sector as 'a cornerstone of the Saudi financial services sector by providing reliable risk transfer mechanisms, promoting long-term savings, and serving as a conduit to channel funds from policyholders to investment opportunities, thus being a key enabler of a healthy Saudi economy'. According to the Law on Supervision of Cooperative Insurance Companies (Law No. 32 of 1424 AH), insurance in the Kingdom is a regulated activity subject to SAMA's oversight. Insurance companies must operate in accordance with earlier legislation on cooperative insurance, as well as abide by 'the principles of Islamic Shariʿa" (*wa bi ma la yataʿarud maʿa ahkam al-shariʿa al-islamiyya*).

Nothing in SAMA's insurance regulations mention Shariʿa, nor do the regulations invoke the Permanent Committee's *fatwa*. Rather, the *fatwa* and SAMA's oversight of the insurance market in Saudi Arabia, through an aggregated dissonance, contribute to the project of the Saudi state in different ways, depending on who the relevant audience is. The respective audiences of the Permanent Council and SAMA are not likely the same. The Saudi state can say to its own people and nearby Muslim majority countries that it remains committed to Islamic law, whether by referring to the *fatwa* or by the opening provisions of Law No. 32, which caution against any contradiction of Islamic principles. Yet it can also re-assure its international trade partners and foreign investors that any risk in their Saudi investments can be mitigated and controlled by insurance. Maintaining an 'Islamic' milieu and a vibrant economy demands the very inconsistency or dissonance observed here; and in the end, the dissonance remains in the service of the state.

A preference for the adjectival forms of 'Islamic' and 'Muslim' emphasizes how both are, to draw upon Stuart Hall's *The Fateful Triangle*, 'sliding signifiers'. As sliding signifiers, these terms demand that we understand them as replete with philosophical assumptions and material implications, and where both cultural and material aspects inform and are informed by each other. For example, the deployment of 'Sharia' will in part depend on whether the governance project concerns U.S. or Saudi Arabian development aid to Afghanistan, the criminalization/legalization of polygamy in Canada or Jordan, or programs in Indonesia and France to combat violent extremism.

Policy makers pursue a War on Terror by treating 'Islam' as a constant, even a caricature, to rationalize, justify, normalize – and thereby depoliticize – coercive policies, programs, and institutions of the state. In such contexts, the state, whether Tennessee banning Sharia or Saudi Arabia adopting a particular view of Sharia for a domestic audience, is the site of variability, particularity, and difference. However, by recasting the 'Islamic' rhetorically – whether referenced by the left or the right, progressive or conservative, liberal or neo-con – we can better appreciate the relation between the production of knowledge on Islam, and the explanation, if not justification, of state projects of security and surveillance, refugee resettlement, militant democracy, secularism, and the return of history. 'Islam' as object of analysis is instrumentalized in and for the state's knowledge, such as when the Syrian refugee or the migrant labourer is construed as 'only Muslim', and as such informs a range of state policies on citizenship, immigration, and refugee law. By viewing terms like 'Islam' and 'Muslim' as unstable nouns, adjectival modifiers, or more generally, sliding signifiers, students and researchers can bring to the fore the intersection between epistemic assumptions and state practices.

Further readings

Aydin, Cemil. 2017. *The Idea of the Muslim World: A Global Intellectual History*. Cambridge, MA: Harvard University Press.

Davidson, Naomi. 2012. *Only Muslim: Embodying Islam in Twentieth-Century France*. Ithaca, NY: Cornell University Press.

Emon, Anver M. 2012. *Religious Pluralism and Islamic Law: Dhimmis and Others in the Empire of Law*. Oxford Islamic Legal Studies Series. Oxford: Oxford University Press.

Emon, Anver M. 2019. 'The 'Islamic' deployed: the study of Islam in four registers'. *Middle East Law and Governance* 11(3): 347–403. https://doi.org/10.1163/18763375-01103001

Mitchell, Timothy. 1999. 'Society, economy, and the state effect.' In Steinmetz, George. Ed. *State/Culture: State-Formation after the Cultural Turn*. Ithaca, NY: Cornell University Press. pp. 76–97.

Norton, Anne. 2013. *On the Muslim Question*. Princeton, NJ: Princeton University Press.

36
JURISDICTION

Shiri Pasternak

Jurisdiction is a legal mechanism for organizing how political power is exercised, spatialized, and contested. Jurisdiction means the power to *speak the law,* bringing it into existence and defining *who* will be governed, as well as *how* and *where.* It is a fundamentally spatial concept that enacts the governance of law. It is a political, historical, and spatial concept essential for understanding the production of power in society, particularly in settler colonial successor states.

While sometimes dismissed as a simple function of sovereignty – the thing that determines which authority is triggered to act on specific issues – jurisdiction is better understood as the apparatus through which sovereignty is rendered meaningful. In Western political contexts, the relationship between people, place, and the law has changed considerably over time. The meaning of jurisdiction, in particular, has been deeply shaped by imperialism and in relation to territorial expressions of sovereignty. Whereas jurisdiction was understood in Europe for centuries as the legal means to claim authority over people in local places or over those engaged in specific activities, in the late 19th century through settler colonialism jurisdiction became fused with state sovereignty claims to authority over national space. But these territories were not unoccupied, empty space. Therefore, the meaning of jurisdiction has been shaped in equal measure by counter-assertions of jurisdiction grounded in Indigenous legal orders across the globe.

Everywhere we look, the world is governed and shaped by negotiations around jurisdiction; its ubiquity makes it easy to miss. A national border appears to be the hard line where one state's jurisdiction cannot be exercised in another's territory. And yet, there are many ways for state law to reach over into another place: in Canada, U.S. Custom and Border Services is operational, for example, in Toronto's Pearson International Airport. Guantánamo Bay is part of the United States, but based in the country of Cuba, and exists in a realm of multiple, competing, and colluding jurisdictional claims regarding the application of domestic and international law. Jurisdiction is spatial, but it is not necessarily territorial in the way we imagine. And there is often incredible complexity in determining the paramountcy of one set of laws over another.

Jurisdiction is the legal mechanism also used to manage divisions of internal power within states. Where an area of jurisdiction is grey, it leaves open opportunities for governments to claim or evade legal obligations. For example, when a status 'Indian' (a state-designation for First Nations people in Canada) who falls under section 91(24) of the federal head of power for 'Indians and the lands reserved for Indians' requires homecare on an Indian reserve, who should provide the necessities of life to the family? Healthcare is technically a provincial responsibility in Canada under the constitutional divisions of power. A young Cree boy, Jordan River Anderson, perished in Canada waiting for the governments to decide. Lives are often held in the balance of jurisdictional disputes between multiple levels of government over the provision of funding for social services. Jurisdictional powers may be defined by respective authorities, but the world does not easily fall into stark legal categories, creating a wide terrain for contestation.

Finally, jurisdiction determines how individuals, groups, and nations are constituted as subjects. Global migration highlights the way people can be rendered 'illegal' when fleeing states and seeking refuge in countries where they do not hold citizenship. Detentions centres and camps that house refugees can linger when states refuse to absorb asylum seekers for years, even decades, into the citizenry of a nation. People are marked as inside/outside the jurisdiction of host states based on their countries of origin, religion, the colour of their skin, their economic status, and the conditions of arrival in new lands.

Historically, sovereignty has been defined by its claims to an absolute form of political authority and has dominated modern society as the central organizing principle of political order in modernity. But authority is not pre-given to sovereignty. Sovereignty requires legitimacy and conviction of its authority; forming national law is one way in which this legitimacy is sought. And though the common law, or the civil law, comes to take the shape of the state, the fit is never total or complete. The common law has no mystical or transcendental authority that connects it to territory.

In settler colonies, the state's claims to jurisdiction over Indigenous lands assume the authority to inaugurate law where law already exists and presume the new forms that law will take. These presumptions preclude posing pertinent questions about *which* laws should apply on these lands. In *Two Families*, Nehiyaw scholar Harold Johnson explains to non-Indigenous people the authority by which settlers were offered a place in Indigenous territory: 'When your ancestors came to this territory, Kiciwamanak [cousin], our law applied. When your ancestors asked to share this territory, it was in accordance with our law that my ancestors entered into an agreement with them. It was by the law of the Creator that they had the authority to enter treaty'. From this perspective, the condition for shared jurisdiction is the treaty relationship between Indigenous peoples and newcomers that governs the use and settlement of territory. Around the world, contested terrains of authority and plurality define the jurisdictional struggles between competing political orders.

In the settler colonies, the common law's universalist principles of equality were and have been intentionally articulated against the local and particular formations of Indigenous legalities. Citing case law from America and Australia, Lisa Ford traces the transition from a settler legality that claimed jurisdiction over Indigenous bodies only in the case of personal violence toward non-Indigenous people to the period when territorial jurisdiction became a necessary exercise of sovereignty at the turn of the 19th century. Until this later period, an uneasy legal pluralism had existed between overlapping Indigenous and settler social orders. Ford's research shows that the emergence of territorial state sovereignty was introduced in colonial courts through a generalization of the common law as the singular national law.

This fusion of sovereignty, jurisdiction, and territory is a global pattern inhering in the Anglophone settler polities. The assertion of sovereignty over Indigenous lands in the British settler colonies was disengaged for centuries from colonial officials' capacity to *exercise authority* over Indigenous peoples in their territories. Long after the ceremonies of possession, the granting of Royal Charters, and a bewildering parade of imperial performances of power that bore only the slightest relevance to their supposed subjects, Indigenous peoples' social and political orders remained intact, even as they adapted to the influx of European and other settlers on their lands. The ongoing exercise of Indigenous jurisdiction over land, resources, and bodies on their homelands today reveals the continuity of this suspended space between settler assertions of sovereignty and the vitality of Indigenous territorial jurisdiction.

From an Indigenous perspective, jurisdiction is the authority to speak their own laws, articulated as responsibilities rather than simply delegated state rights. Powerful Indigenous movements today clash with settler law on the frontlines of every social, political, and economic issue they face, but most visibly around resource development, where they have the power to grind the economy to a halt. For example, all along the Trans Mountain pipeline that runs from the tar sands in Alberta to the Salish Sea at the Burnaby terminal near Vancouver, British Columbia, First Nations have asserted their inherent laws – those of the Dene, Cree, Secwepemc, Nlaka'pamux, Tsleil Waututh, Squamish, to name a few – to contest the state's authorization of massive, invasive infrastructure on their lands. In Northern Ontario, the Kitchenuhmaykoosib Inninuwug released a Watershed Declaration that nationalized all resources on their homelands protecting the watershed from the impacts of development. In Manitoba, the Sagkeeng Lawmakers Assembly rejected settlement monies from the public hydro facility, asserting their jurisdiction in the face of provincial authority.

Jurisdiction is both a spatial and a legal concept: to render jurisdiction visible, we must place it in the context of geographical studies, otherwise we risk understanding both law and space as non-political categories. A critical legal, geographic perspective insures an interdisciplinary approach to jurisdiction, allowing us to see beyond the standard representation of jurisdiction as a tiered structure that ranks power from the highest to the lowest authority, or in which certain issues are under the exclusive domain of particular authorities. More often, when we really look at the spatial context of jurisdiction, we see a dense patchwork of institutional bodies crowding every place, often governing largely in isolation from each other despite overlaps and contradictions in mandate, authority, and geographic oversight. For example, in the territory of the Algonquins of Barriere Lake, although the Grand Lac Victoria reserve was created to protect wildlife and Indigenous livelihood on their territory, it is superimposed by provincial park legislation that allows for relatively uninhibited resource exploitation, despite the park's creation under the legislative auspices of the Wildlife Conservation Act.

To engage in the question of what it means to decolonize law, we must ask by what authority a law has the authority to be invoked and to govern. Jurisdiction derives its power to allocate authority from many sources. A reconciliation of relations between Indigenous and settler societies requires the radical deconstruction of the authority by which states invoke sovereignty and a re-examination of the jurisdictional orders that underpin Indigenous forms of entitlement to their lands. The source of jurisdiction within Indigenous legal orders is always rooted in place and in the orders of care that renew this legal responsibility for place from one generation to the next. Colonialism was legal in European law, and its principle of discovery remains imprinted on the legal systems of settler colonies today. The contestation of this doctrine and, the questions surrounding the state's authority to liberate itself from earlier law, can be called into question by struggles in the register of jurisdiction.

By disentangling jurisdiction from sovereignty, I do not mean to argue that sovereignty is an illusion: it is the dominant political-territorial ideal of the nation-state, which has had an incalculable effect on Indigenous legal orders. What has been done in its name to protect containerized borders and exclusive claims to authority highlights the violent history of white settlement on these lands. Sovereignty is not simply an assumption made by settler nation-states like Canada that can be dismissed by virtue of competing discourses and assertions of power. But the *legitimacy and legality* of sovereignty can be called into question in the register of jurisdiction.

Further readings

Ariss, Rachel and Cutfeet, John. 2017. 'Kitchenuhmaykoosib Inninuwug First Nation: Mining, Consultation, Reconciliation and Law'. *Indigenous Law Journal* 10 20–35.

Sue M. Hill. 2017. *The Clay We are Made of: Haudenosaunee Land Tenure on the Grand River*. Winnipeg: University of Manitoba Press.

Shaunnagh, Dorsett and McVeigh, Shaun. 2012. *Jurisdiction*. Abingdon: Routledge- Cavendish.

Valverde, Mariana. 2009. 'Jurisdiction and Scale: Legal "Technicalities" as Resources for Theory'. *Social & Legal Studies* 18 (2): 139–157. https://doi.org/10.1177/0964663909103622

37

LABOUR AND EMPLOYMENT

Diamond Ashiagbor

Labour: the standard story

What does it mean to say that someone is 'working', or in 'employment', or to refer to regulation of the 'labour market'?

Many languages contain a distinction between 'labour' and 'work' although, as in English, the distinction is typically not a lexical one but arises from the context; the two words can be used interchangeably. To these terms, in modern times, we must also add the concept of 'employment', with its connotations of a more formal relationship governing the economic activity of work. What is crucial is the *social* and then *legal* meaning ascribed to these terms. According to Hannah Arendt, the central distinction is between the activity of 'labour', which is natural, inherent to 'the human condition', and relating to the private household; and that of 'work' – artificial or socially constructed, and relating to the public realm or the marketplace. In contrast, in the Marxist tradition, that distinction is reversed, with the word 'labour' being associated with a historical category: economic activity which is exploited, expropriated, or alienated; as opposed to 'work', an anthropological category, a fundamentally natural element of human society.

Legal conceptualizations of labour and employment tend to mirror this dichotomy, with legal interventions focused on regulating forms of work which are understood as belonging to the public sphere (as opposed to the household) and which can be commodified as part of the 'market' for labour.

Further, 'labour' and 'employment' have come to be defined primarily by reference to the dominant form in which they have appeared in the industrialized countries of the global 'North' for much of the 20th and 21st centuries. Rooted in the process of industrialization and the historical transition from agricultural subsistence to dependence on wage labour, this employment relationship, in particular, the 'standard' employment relationship which took the legal form of a contract of employment, is usually understood as comprising paid work in the market economy – i.e. not in 'traditional' economic sectors such as subsistence agriculture, casual labour, small-scale trade, or household work. It is also understood as being work which is full-time, stable, with an implicit or psychological contract (i.e. often unspoken expectations about reciprocal obligations, that employees' loyal service will be rewarded with security), for an indefinite duration, and usually done on the employer's premises.

In many industrialized democracies, this employment relationship also became embedded in and dependent upon other areas of social and economic regulation which emerged in the post-war era: the modern 'welfare' or 'social' state, which made provision for mutualization of risk between capital and labour, such as in the form of social security; a role for trade unions (labour unions) in reinforcing such social citizenship for industrial workers, through collective bargaining (i.e. the joint regulation, with the employer, of the terms and conditions of employment such as pay and hours); corporate law and domestic taxation systems which served to reinforce the institution of the vertically integrated firm; and a gender contract premised on a male breadwinner and economically dependent female housewife/unpaid caregiver.

Labour law as a site for struggle?

Relatedly, these dominant perceptions of what labour is or looks like – closely allied to a particular economic history of regulation of primarily Fordist productive relations and the protective capacities of industrialized states – have served to shape prevailing understandings of the purpose, form, and content of labour law or the regulatory frameworks that govern work. Dominant narratives within labour law scholarship thus reflect and give legitimacy to post-war regulatory mechanisms and institutions of labour relations, and also shape which types of employment relationships are deemed suitable for regulation. Thus, traditional regulatory frameworks for governing work relations have taken as their starting point and as their main (sometimes, only) subject of regulation, the post-war ideal type of the 'standard' employee within the 'standard employment relationship', buttressed by institutions of social citizenship.

However, the vast majority of workers globally are located outside this paradigm, and the number of excluded workers is growing. The reality of working life is in fact the simultaneous persistence of non-standard, or 'informal', work in the global South – i.e. work which is beyond the official reach of law, or not covered in practice – and the growth of non-standard and informal work in the global North. If one starts from the conviction of law as a social phenomenon, defining socio-legal scholarship (or a 'law and society' approach) as that which examines law in its social context, then it is essential for any descriptive or normative account of labour and employment to reflect this reality; to ask what instruments or technical means are *actually* governing this economic activity, even if these regulatory techniques do not conform to the paradigm of formal labour law regulating formal work.

As a discipline, however, labour law is sufficiently expansive to reach beyond the confines of a narrow model of 'employment' and of formal regulation which evolved at a particular historical juncture. That is because, unlike other legal subjects such as contract law or tort law – which are considered to be doctrinal, labour, or employment law – is inherently a *contextual* subject, which does not necessarily aspire to *conceptual* unity. The subject derives its coherence, if it has any, from its exclusive focus on a particular context, namely, the economic activity of work or labour. The discipline is often a location for attempts to resolve questions of economic efficiency, redistribution, fairness, equality, justice, development, human rights, and social change. It is imbued with a comparative and an international awareness, since it is through the regulation of work that we try to make sense of a form of economic activity, the performance of labour, typically in exchange for wages, which is experienced globally. Hence, labour law has the capacity to be an agent for social change.

The reality of informality

One of the most significant challenges facing labour and its regulation in market economies is the persistence of informal work and the growth of informalization. With regard to labour markets of developing countries of the global South, informal employment has long been the pre-dominant form, and predictions that work would become formalized as these economies 'modernized' have proven incorrect. Contrary to these predictions, economic development in developing countries has not generated enough 'modern' jobs to absorb surplus labour from the traditional economy. In fact, the informal economy has expanded in developing and industrialized countries, and with it low-skilled, poorly paid, intermittent, and insecure work.

As the International Labour Organization (ILO) acknowledges, there is no universally accurate or accepted definition of informal work or the informal economy, encompassing as it does considerable diversity in terms of workers, enterprises and entrepreneurs. However, as a placeholder definition, informal work can be understood to encompass a broad spectrum of those who labour while remaining outside the protection and regulation of the state – where the law is not applied or not enforced, or compliance is inappropriate, burdensome, or imposes excessive costs, or who have inadequate access to employment protection or social security rights.

Informal work exists in the guise of informal *self-employment*, such as own account workers in informal enterprises (that is, unregistered and/or small unincorporated private economic units), employers in informal enterprises, contributing family workers (in informal and formal enterprises); or as informal *wage employment*, including employees of informal enterprises, casual or day labourers, unregistered, or undeclared workers.

It is important to bear in mind that not all workers who are outside the standard employment relationship are 'informal' in the sense of falling outside the formal reach of the law. For instance, some high-skilled workers from privileged social locations working in the knowledge economy may be excluded from employment law but be covered by commercial law. Nonetheless, for the majority of informal workers, there is a higher likelihood of poverty, lower productivity, and lower income. More than 60 per cent of the world's employed population (2 billion people) earn their livelihoods in the informal economy, with the preponderance of those in informal work located in emerging and developing countries; for instance, the vast majority of employment in Africa, 85.8 per cent, is informal.

The informal economy, according to the ILO, is marked by acute decent work deficits: individuals and enterprises may be trapped in a spiral of low productivity and poverty; informal workers are not recognized, registered, regulated, or protected under labour and social protection legislation, and are not therefore able to enjoy, exercise, or defend their fundamental rights; there are varying degrees of dependency and vulnerability, in particular experienced by women, young persons, migrants, and older workers, including child labourers and bonded labourers.

Noteworthy here is the starkly gendered and racialized experience of informality, in particular when coupled with migrant work. In about half of the countries surveyed by the ILO, the share of informal workers in total employment is higher for women than for men, although the ILO observes that the gender bias in the informal economy is probably *underestimated*. In particular, women are more likely than men to be in those informal activities that are undercounted, such as production for own consumption, paid domestic activities in private households, and home work. They are also more likely than men to be in small-scale economic units where their economic contributions are invisible and therefore not counted. A final, highly significant, element in the persisting patterns of female disadvantage in labour markets and the gendered nature of informal work relates to social reproduction, namely, the importance of unpaid work in making possible much of the paid work that takes place in the labour market, and the fact that most of this unpaid work is done by women. Women are more likely to have responsibility for social reproduction, for instance, remaining the primary providers of child-care, whilst also under increasing pressure to contribute to household income, and thus resorting to non-standard or informal work and 'entering the workforce at the very margins'.

Why is this the case? With regard to developing countries, it is argued that no country has successfully industrialized via export promotion without drawing upon a pool of low-wage female workers. Similarly, most industrialized countries rely on low-wage women workers, a disproportionate number of whom are migrants or racial and ethnic minorities, for increasing amounts of service and production work. Precarious or vulnerable workers are poorly paid and employed in atypical and unstable jobs; women, people from ethnic minorities and racialized groups, and workers with precarious immigration statuses, whether temporary or undocumented, are overrepresented in precarious work arrangements that fall outside the norm of the standard employment relationship.

Even in the global North, the model of 'standard' or formal employment around which much labour regulation is premised was relevant primarily for an elite group of mostly male workers during the period of industrial mass production of the 20th century and has, since the economic crises of the 1970s, been unravelling. Deindustrialization and the rise of the 'service economy', decline in trade union power, and employer preferences for flexibility have meant a shift towards work which departs sharply from the standard. Such a shift towards more flexible work contracts has in large part been facilitated by state deregulation of labour markets. Employers and holders of capital are increasingly organizing production through distancing strategies such as subcontracting, franchising, concessions, and outsourcing, where the 'employer' contracts out work rather than using its own workforce so that work is done not by traditional full time, permanent employees, but by atypical or non-standard workers. This shift is manifested across all industrial sectors, but particularly prevalent in the service sector: retail workers, catering and hospitality workers, home care workers, delivery drivers, museum attendants, security officers – most are working non-standard arrangements, some of which are precarious, many exploitative, the majority without full access to employment rights. Increasingly, labour in industrialized economies is indirectly employed, intermittent, temporary, fixed-term, part-time, 'zero hours', or 'on demand' – for instance, where the employer makes no commitment to provide a set number of hours, and where the work may be governed by a digital platform which mediates between the worker and the end user or customer.

What should be the response of labour law?

What should be the response of labour law to such radical departures from the standard story told about the nature of employment? For developing economies, converting 'informal' into 'formal' arrangements in labour, land, markets, and finance is seen by international financial institutions as the route to economic development, and also the answer to the decent work deficit arising from the persistence of informal work. According to this approach, developing countries should adopt formal legal institutions – in particular, the rule of law and protection of private property – so as to ensure the predictable and effective enforcement of 'background' rules necessary for capitalist economic growth. However, whilst the discourse of formalization has gained increasing purchase, within the ILO as well as the international financial institutions, a parallel discourse of flexibilization which advocates labour market *de*regulation retains its dominance. Orthodox economic thinking, in particular from the IMF and OECD, has long held that the systems of labour market regulation which exist(ed) in many European Union states were the major cause of their high unemployment in the 1970s and 1980s, and that the EU would therefore do best to adopt the more flexible wage systems, decentralized collective bargaining, and labour laws of the United States. Removing such 'labour market rigidities' would arguably foster economic growth by maximizing employer flexibility and freedom to manage. Thus, the standard reform advice issued to developing states, and sometimes codified within lending conditionality of the international economic and financial institutions, advocates formalization of work arrangements, recognizing the importance of *private* law for the operation of private markets. However, these same prescriptions are sceptical about *public or regulatory* law, or state intervention to protect labour rights, with labour market institutions (such as minimum wages or union involvement in wage-setting) perceived to be 'rigid' and thus economically harmful. Thus, *labour market formalization* is urged on developing economies whilst they adapt their institutions to a global template based on a minimalist state, liberalization of trade, privatization of public enterprises and, crucially, *labour market flexibility*: the priority is employer flexibility or 'freedom' from employment protection laws, redistributive social justice, or measures to support countervailing workers' power.

This powerful mantra of labour market flexibility is applied to industrialized and developing countries alike, ensuring that the downward pressure on workers' bargaining power will continue, shifting the risk from employers to workers. For industrialized economies, it may be sufficient to look to a reconceptualization of the *subject* of labour law, by redefining the scope of employment protection measures, to ensure non-standard workers are (re)integrated into historic norms of social citizenship. However, the persistence of informal work in developing economies require us to think more imaginatively about the role of informal *norms* governing informal *work*. The reality of economic activity in developing countries belies a simplistic transfer of models from the global North to the South. Thus, gains for workers may be achieved without necessarily having to resort to formal institutions such as the contract of employment, but instead by reconceptualizing what policy levers or legal forms can be understood as *actually* regulating labour. For instance, policy experiments have succeeded in reaching 'non-standard' workers – own-account workers, economic actors in micro-enterprises, unpaid workers, and domestic workers – by envisaging more inclusive systems of taxation, of social insurance, and other more radical measures which seek to decouple social citizenship and protection from formal work, for instance, through cash transfers and basic income guarantees.

Further readings

Arendt, Hannah. 1958. *The human condition*. Chicago. University of Chicago Press

Ashiagbor, Diamond (ed). 2019. *Re-imagining labour law for development: Informal work in the global North and South*. Oxford, UK; Chicago, Illinois: Hart Publishing.

Fudge, Judy; McCrystal, Shae; and Sankaran, Kamala (eds). 2012. *Challenging the legal boundaries of work regulation*. Oxford: Hart Publishing.

International Labour Organization (2018) *Women and men in the informal economy: A statistical picture*, Third edition, International Labour Office. Geneva, ILO.

38
LEGAL CONSCIOUSNESS

Lynette J. Chua and David M. Engel

'Legal consciousness' has emerged as one of the most dynamic fields of law and society research. It became a central focus for law and society scholars in the 1980s and subsequently underwent periods of elaboration, self-examination, and expansion. Although North American and European researchers published most of the earlier studies, the topic of legal consciousness has now been taken up by researchers working in Asia, the Middle East, Latin America, and Africa.

What do we mean by the term 'legal consciousness'? Law and society scholars have proposed a variety of definitions, but they add up to much the same thing: legal consciousness refers to the ways in which people experience, understand, and act in relation to law. Legal consciousness researchers study not just cognition but also behaviour – the ideologies and the practices of people who are involved with situations in which law could play a role. They explore the absence as much as the presence of law in people's understanding of the social world and their place in it. Therefore, legal consciousness researchers do not merely measure the frequency of invoking law, or knowledge – or ignorance – of legal rules and rights, but also examine situations in which people perceive the law to be irrelevant or do not think about or engage with law at all.

Three schools of legal consciousness research

As legal consciousness research grew in the mid-1980s and proliferated in the late 20th century, it became apparent that researchers were approaching their work in divergent ways, leading to the emergence of three schools, which we call Identity, Hegemony, and Mobilization. The approaches of these three schools are sometimes contradictory but more often are complementary and usually overlap to some extent with one another:

> The *Identity* school views legal consciousness as an ongoing process of constructing the self in relation to law and legal rights. Typically relying on biographical or autobiographical narratives, researchers examine how the relevance or irrelevance of law to a person's experience connects to the process by which that person's identity—or sense of self—takes shape, making legal norms and institutions appear naturally suited in some instances and inappropriate in others. Identities are fluid and people's relationship to law continually shifts and changes. Identities determine how and when law becomes

> active; but law also can shape identities. Moreover, legal claiming can have a paradoxical impact on identities. As marginalized individuals assert legal rights based on an identity protected by the law to win acceptance and inclusion, they may learn that their identity is now perceived as rights-dependent or even stigmatized and oppositional.

The *Hegemony* school treats law as a pervasive and powerful tool of state control that can shape the categories, values, and assumptions of individuals even when it is not applied directly and instrumentally. Researchers using this framework aim to reveal the workings and the often invisible effects of law in the thoughts and actions of ordinary people. They trace the dominance of law and legal institutions in everyday life, and they also examine whether and how individuals resist law's power. According to the Hegemony school, however, even when individuals try to resist the law, they cannot overcome its inescapable reach. Their acts of resistance inevitably operate within law's logic and understandings, rather than outside its framework, and reinforce or leave intact law's superordinate power.

The *Mobilization* school investigates how legal consciousness promotes the role of law and the efficacy of rights in transforming social conditions, particularly to achieve justice or protect disadvantaged populations. Mobilization researchers focus on the relationship between processes of social change and the experiences, perceptions, and actions of individuals who choose to use or avoid the law. Some Mobilization scholars study people's legal consciousness to gauge the extent of social change that has occurred, whereas others study it to explain how and why those people have turned to the law to transform their circumstances. In either case, studying legal consciousness helps Mobilization scholars to view social change through the lens of human agency, thereby augmenting the more typical research on social change that tends to adopt an aggregate and distanced perspective.

At the turn of the 21st century, some scholars suggested that legal consciousness had exhausted its usefulness, while others disagreed and defended its continued use. Since that time, however, legal consciousness research has flourished, and scholars have applied it in increasingly diverse research sites. The critique at the centre of the debate was grounded primarily in the Hegemony school but did not persuade researchers associated with the Identity or Mobilization schools – or even those Hegemony scholars who focused more on the possibilities of resistance than on the irresistible power of law in everyday life. Thus, legal consciousness research did not atrophy but continued to expand, and today scholars from all three schools rely on the theories and methods of legal consciousness research to explore a broad range of social scientific questions, particularly in relation to understudied, marginalized, or oppressed populations.

Future directions and relational legal consciousness

One of the most promising directions of legal consciousness research is what we call relational legal consciousness. Most researchers recognize that consciousness itself is a relational phenomenon. Human beings are social creatures, whose thoughts and actions tend to reflect interactions and experiences involving others. To study legal consciousness, researchers have always had to take into account its relational aspects and, in this sense, the recent upsurge of relational perspectives has been a matter of degree rather than kind.

Given the inescapability of relationality in human consciousness, it may be surprising that individualistic approaches would have any influence at all on this field of research. Yet, concepts of the self as essentially atomistic and autonomous are deeply imbedded in both law and social theory with roots extending back to Enlightenment theory of the 18th century. The image of the autonomous individual as the prototypical legal subject still dominates much of the law and society literature. Nonetheless, the relational perspective on human consciousness also has deep roots, reaching back at least as far as 19th century European and American thought, if not beyond – as in the work of Karl Marx, Emile Durkheim, William James, and others. In the late 20th century, the relational perspective attracted renewed attention among Western scholars across the social sciences and humanities. Relational theory nowadays is enjoying considerable popularity in many academic disciplines and is growing in influence.

The relational approach to legal consciousness has been especially evident in law and society research conducted in Asia, Africa, and other settings outside Europe and North America. There, Enlightenment philosophy had less influence, and highly individualistic understandings of the self were less widespread. Conducting their fieldwork in those settings, legal consciousness researchers discovered that the consciousness of each person is organically connected to those of family members, fellow villagers, or other intimate associates. Thus, legal consciousness research outside Europe and North America has tended to extend or abandon entirely the individualistic framework that characterized earlier research by Euro-American socio-legal scholars working in the field.

We have suggested elsewhere that contemporary legal consciousness research can be imagined on a continuum ranging from studies that view the self as essentially autonomous and independent to those that treat legal consciousness as highly relational or as the co-constitutive creation of two or more minds.

Research located near the individualistic end of the continuum adopts one-by-one or person-by-person methodologies. To the extent that such studies take into account the influence of other persons or social factors, they regard them as independent variables whose significance resides in their impact on the dependent variable of individual thought and action. Many studies from the Mobilization school are located in this region of the continuum, since their primary goal is to explain how and why each research subject decides to invoke or avoid the law. Even when Mobilization school researchers consider the influence of interpersonal relationships, they tend to view such relationships as external to – yet impinging on – the individual's legal consciousness. By contrast, for researchers associated with the Hegemony school, relational concerns are largely implicit in their work. Nonetheless, many Hegemony scholars can be placed in this individualistic region of the continuum, insofar as their studies assume that law's controls and constraints as well as its disempowering ideologies are experienced and reproduced person by person through relationships with external others – government officials and institutions, police, lawyers, judges, and bureaucrats, as well as people and groups to whom individuals may have ties.

In the middle region of the continuum, legal consciousness studies retain the individual as the appropriate object of study but treat other individuals as co-creators of consciousness rather than mere external variables. They place greater emphasis on the porosity of boundaries between individuals. Identity studies are often situated in the middle of the continuum, because they portray individuals' legal consciousness and their relationships as intertwined in mutually constitutive processes. Individuals' legal consciousness shifts and changes as they engage in different social relationships and, in so doing, they manifest different identities. Moreover, Identity studies in this region of the continuum demonstrate that individual legal consciousness is inextricable from the meaning-making activities of a group. Rather than taking the group as a given, they show that individuals can actually create community with their legal consciousness even as the community creates their legal consciousness and sense of self. People participating in such communities produce a legal consciousness that sustains the group and bestows membership on those identified as insiders, who – among other things – share assumptions about the appropriate use and non-use of law.

At the relational end of the continuum, legal consciousness becomes a fully collaborative phenomenon. The life experiences of individuals are associated with collective forms of legal consciousness that transcend the thoughts and actions of any single person. Here, legal consciousness is not formulated person-by-person, but is constituted by their very relationships and exists among and between individual minds rather than within any single person. This co-constitutive model of legal consciousness can take on as many different forms as there are relationships – among couples, families, groups, associations, communities, and so on.

Although legal consciousness scholars have not fully theorized a purely co-constitutive model, some recent law and society writings venture in this direction. They suggest that individuals' legal consciousness is influenced not only by their own experiences, but also by their understanding of others' experiences with the law, implying that legal consciousness can indeed exist at a collective level. It should be acknowledged, however, that the co-constitutive region of the relational legal consciousness continuum is still relatively undeveloped. Consequently, it is not yet clear whether all types of legal consciousness can be regarded as relational or just some types. The co-constitutive process itself has been described in figurative language, but it has not been charted empirically. And where should we assume that consciousness resides if not in the individual mind? Is there a modern version of Durkheim's 'conscience collective' or a kind of computer cloud storage shared by multiple users?

In the most extreme form of a co-constitutive model, individual subjectivity might fade completely into relationships, and researchers might abandon the individual entirely as the relevant unit of analysis. If legal consciousness research were to develop along these lines, researchers could treat dynamic relations involving two or more people as the proper object of inquiry and might view legal consciousness only within such relationships. A significant modification in research methodology would be required. The person-by-person research methods used by most legal consciousness scholars from the early 1980s to the present might have to be abandoned. Instead, researchers might need to devise new approaches that examine how different forms of legal consciousness arise from particular kinds of relationships and social interactions. What those methodologies might look like, and how they might be operationalized, remains to be seen.

Conclusion

Since its emergence in the 1980s, legal consciousness research has provided law and society scholars with new insights into the mutually constitutive relationships among legal practices, the thoughts, actions and decisions of individual actors and groups, and the broader social and cultural contexts in which they live. On the whole, legal consciousness researchers have resisted a singular, monolithic approach. Despite a period of self-questioning at the turn of the 21st century, the field has flourished and has remained one of the most active areas of law and society research. A new generation of scholars is exploring exciting directions in legal consciousness research among marginalized communities and in non-American settings, contributing to insights across the three schools of Identity, Hegemony, and Mobilization.

The growth of legal consciousness research has been accompanied by increased attention to its relational and co-constitutive aspects, partly as a result of influence from relational theory across the academic disciplines and partly as a product of law and society research in contexts where individualistic conceptions of the self are less salient. Thus, legal consciousness research can be reimagined along an individualistic-relational continuum. For now, the more purely relational – or co-constitutive – conceptions of legal consciousness are, the most speculative and least developed. However, as relational legal consciousness receives increased attention, it has the potential to reshape all three schools, leading to formulations of new questions, research designs, and theories about when and how law becomes active.

Further readings

Chua, Lynette J., and Engel, David M. 2019. 'Legal Consciousness Reconsidered'. *Annual Review of Law & Social Science* 15: 335–353. https://doi.org/10.1146/annurev-lawsocsci-101518-042717

Hull, Kathleen E. 2016. 'Legal Consciousness in Marginalized Groups: The Case of LGBT People,' *Law & Social Inquiry* 41(3): 551–572. https://doi.org/10.1111/lsi.12190

Marshall, Anna-Maria, and Barclay, Scott. 2003. 'In Their Own Words: How Ordinary People Construct the Legal World'. *Law & Social Inquiry* 28(3): 617–628. https://doi.org/10.1111/j.1747-4469.2003.tb00209.x

Silbey, Susan S. (2005). 'After Legal Consciousness'. *Annual Review of Law and Social Science* 1:323–368. https://doi.org/10.1146/annurev.lawsocsci.1.041604.115938

39

MIGRATION

Brenda S.A. Yeoh

Migration as a multifaceted economic, social, cultural, and political phenomenon has long been part of human history. Notable migrations in recent history in terms of volume and geographical scope have involved the transcontinental movements of people from the Old World to the New. Millions moved from Britain across the Atlantic to the United States, as well as to settler colonies of the British Empire, including Canada, Australia, and New Zealand. Between the 16th and 19th centuries, the transatlantic slave trade transported enslaved African people to the Americas, again in large numbers. The European expansion of control over trade and territory in non-European parts in the 19th and early 20th centuries generated a large demand for labour met mainly through the migration of non-Europeans. During this period, more than 50 million Chinese migrants and 30 million Indian migrants left their homelands, many to seek work opportunities in European-controlled territories as racialized indentured labour.

Large-scale migrations during the age of empire were disrupted by the world wars of the 20th century, and, with the establishment of nation-states in the post-war era, were further constricted by growing levels of economic and political nationalism. In Asia, for example, border control and institutional barriers were heightened as new nation-states focused on strengthening national identity and consciousness. In this context, immigrants were often regarded as transgressors of the nation's solidarity – even homogeneity – while emigrants who left their 'birth' nations were often seen to have betrayed their homelands.

The closing decades of the 20th century and the dawn of the 21st ushered in an increasingly globalized age of accelerated mobilities of people, ideas, and commodities as well as a heightened sense of interconnectivity among places. The number of international migrants reached an estimated 272 million in 2019, of which 53 per cent are male and 48 per cent female, and almost two-thirds are labour migrants. Migration is increasingly taking on a transnational character where movement is no longer based on permanent uprooting and settlement, but instead is often multidirectional, multinational, complex, provisional, and built on a multiplicity of interconnections and networks sustained between 'home' and 'host' countries. In this context, the original notion of emigration as a form of permanent departure from a home-nation (usually paired with the notion of immigration as a form of full incorporation and permanent settlement) has been partially displaced by transnationalism frameworks that accord more flexibility and interconnectivity to people movements. The rest of this entry focuses on selected emerging trends and key issues that underpin contemporary migrations.

The rise of temporary labour migration

As contemporary life and work cultures become increasingly flexibilized under global capitalism, transience and transitoriness are becoming fundamental characteristics of labour migration flows in many parts of the world. With the polarization of labour markets, the erosion of job security, the rise of global commodity chains and the offshoring of work, subcontracting and the casualization of labour have become key drivers of temporary labour migration. Under these conditions, migrant labour is often stratified by race, gender, and skill categorization. Occupying one end of the spectrum is the hypermobile global elite comprising mainly highly skilled professional migrants and capital-rich entrepreneurs; at the other are low-waged labour migrants, who often face racial discrimination and a gauntlet of restrictions surrounding their access to mobility, social benefits, housing, and job security.

While enforced transience as a governing principle of labour migration regimes rose and fell alongside the 'guest worker' schemes in Europe, labour-importing nation-states in Asia and the Middle East have in recent decades figured temporary labour migration in their social and economic policies. Migrant contract workers make up the majority of the population in Gulf countries such as United Arab Emirates, Qatar and Kuwait. Temporary labour migration regimes gained popularity following the oil price hike in 1973, where oil-rich but labour-short states in the Gulf began to import Asian contract labour to build large-scale infrastructural projects. Intra-Asia labour migration – facilitated by migration brokerage – has also expanded rapidly on the basis of growing demand for low-waged contract labour to take on unwanted, dangerous, and socially devalued jobs, such as construction, domestic, and care work shunned by the citizen workforce in the more developed economies in Asia. These migrant workers on time-bound contracts are frequently excluded or only partially incorporated into labour laws, with minimal access to the full range of rights available to citizens. Rendered easily deportable, their continued socio-legal status in countries of destination depends on the decisions of employers and the vagaries of state calculation. While Europe and 'traditional' countries of immigration such as Australia and Canada selectively provide pathways towards permanent residency, family reunification and eventual incorporation into society as ethnic minorities, low-waged migrants in Asia move primarily as units of labour without family, often treated as 'permanently temporary' and subject to the logic of demand and supply. Even as temporary labour migration helps to relieve labour shortages in destination countries and serves as a safety valve for origin countries grappling with high unemployment rates, many labour migrants continue to face precarious conditions at work, and prolonged separation from left-behind family members.

Migration and development

Over the past two decades, alongside rising uncertainties linked to an unprecedented pace of change in the social, economic, political, environmental, and technological spheres, there is renewed optimism about the relationship between migration and development. This is predicated on the view that the management of migration is key, as seen in the attention given to developing a global migration governance framework through the endorsement of the 2018 United Nations Global Compact for Safe, Orderly, and Regular Migration. Well-managed migration is thought to support development by empowering individuals, bridging cultures, creating wealth, and balancing inequalities while poorly managed migration can undermine development by making migrants vulnerable, and destabilizing origin and destination communities.

The development potential of migration is also underpinned by the growth of international remittance transfers, from an estimated US$126 billion in 2000 to US$554 billion in 2019. Considered a more reliable source than foreign direct investments, it has been argued that international remittances generally assist in diversifying household income risk by helping families in sending communities hedge against the destabilizing effects of local market instability and the lack of social security, whilst working to improve rural livelihoods through upward socio-economic mobility. In this context, major sending states – best typified by the Philippines as a 'brokerage state' in the terms of labour deployment – have increasingly promoted transnational labour migration as an effective development strategy. Notwithstanding the huge volume of remittances flowing back to source countries of the developing world, the verdict with regard to the development impact of remittances is still an open one. While remittances invested in land purchase, business and technology, or in improving health, education, and infrastructure in source countries may have significant multiplier effects, those used to finance consumption purchases may not generate longer-term benefits and instead increase consumerism (e.g. of imported goods) and engender a culture of dependency. For families and individuals, the economic benefits from remittances have to be weighed against debts that may have accumulated in order to finance migration (such as recruitment and training fees charged by migration intermediaries) in the first place. Clearly, migration cannot be offered as an overarching panacea to development without broader structural reform to address local development constraints, such as the lack of credit infrastructure and macroeconomic instability in origin countries.

In weighing the developmental effects of migration, the social implications for sending communities where a 'culture of migration' has taken hold also need consideration. In the case of rural Southeast Asia where feminized migration is becoming common in response to the gender-segmented global demand for domestic and care workers, research has pointed to social consequences linked to the prolonged separation of family members, care deficits in childcare and eldercare, and changing social practices that challenge traditionally scripted roles for men and women. Migration as a household livelihood strategy also trains the spotlight on the 'household' as a site of cooperation and conflict over the management of resources, thereby raising important questions concerning the way gender and inter-generational relations in the family/household mediate the use and distribution of remittances. Research on cultural expectations in relation to remittance behaviour suggests that these familial obligations may press more heavily on women migrants than their male counterparts, and may be associated with conflicting emotions of guilt and gratitude within the moral framework of the family.

Another issue central to understanding the developmental consequences of migration for sending communities concerns the complex question as to whether remittance gains might sufficiently spur development to offset human capital loss. Among the 'emigration threats' feared, 'brain drain' or 'human capital flight' and its impact on developing countries has been an issue of widespread concern given its scale. The depletion of skilled, educated, and talented individuals may adversely affect the development of local knowledge and science, retard economic growth, erode tax revenues, and cripple public institutions and service delivery. In small developing countries with high rates of emigration such as the poorer Caribbean island-states, the sheer volume of emigration often means that any possible positive effects may be outweighed by negative impacts on economic dynamism, the delivery of key public services, and the depletion of the political classes. The hemorrhagic loss of health professionals from African countries has also been seen as a contributing factor to incapacitating health sector reforms and perpetuating a vicious cycle of ill health and poverty.

These negative views have been challenged by those who argue that the emigration of talent is not so much a 'drain' but part of 'brain circulation', and that positive effects for the source countries can include remittance flows, mobilization of foreign capital, exposure to new ideas and technologies, and the creation of diaspora networking that facilitates the exploitation of unique trade and business opportunities. Such an approach to emigration effects depends on the presence of transnational linkages among emigrants and their homelands, including the possibility of return. For example, the massive out-migration from Latin America and the Caribbean to the United States, Canada, and Europe has been accompanied by the development of transnational ties that immigrants maintain across borders between their home and host communities. Similarly, the growth of India's Bangalore and Hyderabad software development and outsourcing centres is tied to business networks facilitated by Indian engineers working in Silicon Valley in the United States.

Migrant-led diversity and integration

As globalization escalates the volume, velocity, and diversity of migration flows, destination countries and cities are experiencing greater degrees of social, cultural, ethnic, and religious diversity in their population composition. Migrant-led diversification has raised challenging questions for national identity and the nation-state framework for resource allocation, particularly where the speed and scale of immigration and the variety of immigrants' backgrounds pose adjustment problems relating to the perception and acceptance of immigrants, and to changing social orders, cultures and identities.

Historical frameworks for accommodating diversity play a continuing role in inclusionary/exclusionary politics. While North American and Australasian settler colonies founded on migration espouse models of multiculturalism that privilege the White/Caucasian subject as the core of the nation, European models pre-suppose homogeneous and insular states. Among post-colonial pluralistic societies such as Indonesia, Malaysia, and Singapore, new waves of migration may either provoke historically encountered conflicts and tensions between ethnic groups, or are more easily absorbed into the fabric of older communities on the basis of accumulated accommodation and integration experiences. In supposedly homogeneous societies such as Korea and Japan, contemporary migration and diversity symbolize a historically significant rupturing of long-held self-representations of ethnically homogeneous nations and communities, posing new challenges to policy-makers, the public and migrants themselves.

As migration takes on increasingly political overtones in the 21st century, becoming progressively weaponized in some cases, the question of how to live together in increasingly diverse communities has become more and more urgent. Assimilationist models (e.g. white Australia policy) requiring one-way adaptation on the part of migrants to fit with the receiving society's national identity and values are becoming untenable in the face of increasing migrant diversity. By contrast, multiculturalism (best illustrated by Canada's policies) allows migrants to retain their cultural identities while expecting a high degree of accommodation by the receiving society. As a policy measure, multiculturalism has been criticized for its ineffectiveness as a counterweight to migrants' exclusion and has also been perceived as a threat to national identity and values in some quarters. The recent shift is towards promoting integration, conceived as a two-way process of mutual adaptation between migrants and receiving societies that is thought to occupy the middle ground between assimilation and multiculturalism.

Combating discrimination, racism, and xenophobia against migrants, however, needs to take into account not just state policies at work but also the politics of space in everyday encounters, particularly in situations marked by a high degree of socio-economic inequality and the continuing salience of racial/ethnic and social hierarchies. Migrant populations are often subject to a politics of invisibilization where they are relegated not only to the socio-political but physical periphery. Urban redevelopment and placemaking projects often aim to reduce the visibility and presence of migrant 'hotspots' in the city. The construction of labour camps and worker dormitories on the outskirts of the city also reflects similar exclusionary policy practice imprinted onto urban space. At the same time, despite – or because of – the lack of integration into the host community, migrants come together in informal but resilient ways in interstitial spaces, thereby cultivating a sense of being and belonging at the local scale, while bypassing the national community to which they lack access. While more inclusive policies in important arenas such as language, education, employment, and residency are crucial to migrant integration, attention to empowering migrants to take charge of their inclusion needs and aspirations are equally vital.

Further readings

de Haas, Hein; Castles, Stephen; and Miller, Mark J. 2020. *The Age of Migration: International Populations in the Modern World* (6th edition). London: Red Globe Press.

International Organization for Migration (IOM). 2019. *World Migration Report 2020.* http://www.iom.int/wmr

Lai, Ah Eng; Collins, Francis Leo; and Yeoh, Brenda S.A. eds. 2013. *Migration and Diversity in Asian Contexts.* Singapore: Institute of South East Asian Studies Press.

Liu-Farrer, Gracia and Yeoh, Brenda S.A. eds. 2018. *Routledge Handbook of Asian Migrations.* Abingdon: Routledge.

40

OWNERSHIP

Persons, property, and community[1]

Margaret Davies

We use the terms 'person' and 'property' in our everyday lives quite happily, without needing to give them a technical meaning. On the whole, we seem to know what the words mean and how to use them.

Lawyers also use these terms routinely and in the quite ordinary course of events. Like many legal concepts, however, they suffer from much ambiguity: both the objects to which they refer and their underlying concepts are uncertain, and are often the subject of political controversy. Indeed, in both cases, it is doubtful whether an 'underlying legal concept' even exists, such is the difficulty of defining them clearly.

The problem is amplified because the everyday usages of the terms 'persons' and 'property' are sometimes at odds with legal definitions – although in this these two words are hardly unique. Having said that, law is embedded in social practice and the distinction between the legal and the non-legal is a convenient fiction, maintained by law. In consequence, there is often both cross-fertilisation of 'legal' and 'everyday' meanings, as well as a certain productive tension between them.

In this chapter, I first introduce some of the difficulties with the concepts of persons and property, what they refer to, and how they are used. Secondly, I explain what I see as the relationship between these two ideas – the fact that in conventional legal thinking, they are supposed to be diametrically opposed, mutually exclusive, but in fact they are inextricably linked. The first two parts draw on existing literatures, but in the third part, I go in a new direction: I try to capture new ways of thinking about property which loosen the property-person nexus without breaking it altogether. These new approaches introduce values associated with community, the environment, and our material futures into the analysis of persons and property.

Persons

What is a person? The everyday use of the term simply points to biological human beings, what lawyers call 'natural persons'. In everyday speech, we have little trouble using the word 'person' even if we are aware that the individual human being is only constituted as a social person through a complex web of norms, values, and practices.

[1]This chapter is an abridged version of 'Persons, Property, and Community' (2012) 2(2) feminists@law.

Lawyers have more trouble with the notion of person. It is usually seen as an abstract term or container which may be filled with quite variable content. In ancient Greek theatre, 'persona' was the mask worn by an actor to denote the character they assumed in the play. In law, each legal person, human or not, is as artificial as the next. The person is simply any entity that bears a right or duty of any sort. The corporation – entirely a construction of law – is a legal person, and so are certain office bearers, such as Ministers of the Crown, who may have dual or even multiple legal status as artificial legal persons and as human beings.

In regard to the so-called natural persons, human beings precede and exceed their legal status because, unlike corporations, they are not entirely defined by law. Importantly, many human beings have limited status as legal persons on account of such factors as young age, mental capacity, citizenship status, criminal record. There is a long history of differentiating legal personhood through categories of gender, heritage, and race.

For this reason, law has been able to find many ways of discriminating against women, sexual and religious minorities, foreigners, and the unpropertied. **Legal** personality is simply shaped to whatever form is desired. Legal rights vary from one person to the next. Furthermore, there is no technical legal difficulty in attributing some legal personality to animals, trees, and ecosystems, as long as they have some guardian to act on their behalf, as many humans do. Thus, there is no essential person underlying the legal concept: the term has been described by legal scholars as 'an empty slot'.

Property

When we think about property the situation is no clearer. In fact, it is probably even more complex. Once again, property is a term in everyday use which appears relatively straightforward. Once again, lawyers often use the term in quite a different way – not to mean things that are owned but rather relationships between persons which distribute rights regarding things. And once again, there can be a startling lack of legal clarity, both about the things that can be said to be subject to property rights claims and about what property means in essence.

Like the person, property can be regarded as an 'empty slot' into which all sorts of rights can be fitted. The law simply identifies and shapes property – always having some intangible or physical thing in mind, but varying the property or the specific property rights according to each case. For example, say that I own three things: a book, a quantity of prescription medication, and a rare object of national significance. I can sell the book itself, but if the author has died within the last 70 years, I cannot reproduce it for further publication. I can destroy the drugs, but I cannot legally sell them. And I can sell and export the heritage item but only with a license. My rights and duties in relation to each object are quite different, even though I have property in all three objects. This is often referred to as the 'bundle of rights' view of property: property is a bundle of different rights for different relationships to objects. Another example is that renting a house gives a person the right to possess or occupy it, while buying the house involves the acquisition of more extensive rights. If there is a current lease over the house, buying it does not give the right of actual possession (until the lease runs out). Importantly, though, property does normally involve a right to exclude others from exercising the same rights as the property-holder. The American pioneer of 'legal realism' Morris Cohen wrote in the 1920s that property in this sense is essentially power over other people, and so a kind of 'private sovereignty'.

Property law manages the boundary between things that can be appropriated by individuals and things that must remain for public use. Some physical things that are traditionally protected from private exploitation include beaches, public parks, outer space, oceans beyond the territorial limit, and Antarctica. But there are also less tangible resources which are also seen as needing protection for the public good – such as built heritage, the atmosphere, and parts of the intellectual domain (as seen in the fact that patents and copyrights have expiry dates).

Law creates and maintains a distribution of property between the community and the individual, but these interests remain in perpetual tension and even conflict. There has been, for instance, much debate over shopping precincts owned by corporations, and the extent to which these businesses should be able to control people's access to such spaces. Although they are private property, such areas may be community spaces, containing libraries, banks, and other necessary services. Questions have arisen that courts have struggled with: should the essentially public function of the space mean that it is treated differently from individual business premises and homes? Is it right that a person's freedom of movement can be quite severely curtailed by private organisations?

Persons and property

As I have explained, there are some similarities in the way that persons and property work in everyday language and in law. These similarities have to do with the fact that there is no central concept of either persons or property; both have long been regarded as legal fictions and highly artificial; and we see highly political debates about their limits and application. Property and the person are effects, not causes, of a multitude of legal relations. At the same time, property and persons struggle to remain in the fictional world of law – they are in a dynamic relationship with the material world of human beings and objects. Persons and property are constantly being invested with new everyday meanings which depart quite radically from law. They are perhaps typical of the many legal concepts that can rarely be separated from what might broadly be called 'the real'.

Bringing property and the person together conceptually does, however, create a difference from other legal concepts because here we see the basis of Western liberalism and capitalism. The connection between persons and property was summarised very succinctly by cultural theorist John Frow when he said that 'the person is at once the opposite of the commodity form *and its condition of existence*'.

The first half of Frow's statement, that the person is the opposite of the commodity form, is connected with modern rejection of slavery. Persons are subjects, and property or commodities are objects. Treating a person as an object would infringe the well-known ethical norm set out by the late 18th century philosopher, Immanuel Kant, that persons should be treated with dignity, as ends in themselves, not as means toward someone else's aims. Slavery also clashes with John Locke's earlier dictum that 'everyone has property in his own person' which 'nobody has any right to but himself'.

The prohibition on owning human beings is an area where property law has trodden only very reluctantly and inconclusively, as seen in recent cases about who owns/controls corpses and body parts, and legal disputes about the status of human DNA.

The refusal of law to countenance ownership of unchanged human tissue can lead to perverse outcomes, especially where the person from whom tissue is removed is not aware that it has been retained or commercialised. This was the situation for the hapless Americans, Henrietta Lacks and John Moore. Lacks' cells were removed from a tumour before she died in 1951, and were the first to be reproduced in an 'immortal' cell line which still exists and is used in the present. John Moore, on his part, had cells taken in the course of treatment at a University of California hospital that were turned into a patented cell line, used in research. He sued, arguing that his property had been in effect stolen from him, but the court rejected his claim.

The irony in both cases is that the human cells in question were not property for the individual, but once the cells were removed and manipulated, they became valuable property for others.

These examples (plus others, such as global human trafficking) illustrate that the supposed dualism separating persons from property is extensively corrupted in practice. And it's not just human bodies being turned into property for others; objects too are often personified and given a value far above that of the least powerful humans. We frequently and quite legitimately invest objects with some personal, cultural, national, or spiritual significance. The object is then more than a thing – it is a thing of special, possibly priceless, value, and may even be invested with its own subjectivity.

Part of the complexity here is that the terminology of property is both compelling and dynamic – it is easy to imagine ownership of all types of things, and the shift from having power over something to owning it is relatively slight. Discourse around property is highly adaptable. In a South Australian case, a man was charged with abducting his own children, and in his defence, among other things, he argued that:

> For a start the children are copyrighted, which means that they are dramatic work produced by using skill … the mother then entrusted me further than that, having me put my surname on them as a trademark, which gives me the ultimate say what's going on with them, until they turned 18 …

The judge declared this claim 'plainly absurd', but it is interesting that the language and the legal tools of property were flexible enough for it to be made.

We now take up the second half of John Frow's statement cited above, namely that the person is the condition for the existence of the commodity form or property in general. Persons are the pre-condition of property because property implies a subject who holds the legal rights and responsibilities of ownership. A deep connection between personhood and property is also found in philosophy, most famously John Locke's claim cited above about every man having property in his own person. An underlying implication of this famous claim is that everything must be either property or persons, but that we adult, free humans are both subject and object to ourselves. Persons are thus fundamentally depicted as owners, and our skills and knowledge are constructed as our individual property, in a highly individualistic and competitive framework, as the political theorist C.B. McPherson influentially argued in his work on 'possessive individualism'.

But Locke's views on self-ownership were not only the basis for the liberal individualist notion of personhood; they were also the basis for justifying 'us' owning other things. Locke says that if we take something – especially land – and change it, we acquire a right to it. In the present day, this view underpins copyright and other forms of intellectual property. But in the 17th century, when Locke was writing, the key context was colonial expansion. Indigenous people's relationship to their land was seen as non-productive, non-creative, and thus not leading to property claims; this justified white settlers seizing Indigenous land simply by fencing it and cultivating it.

Law moving on

I indicated at the outset that property has long been understood by legal theorists as based on social relationships between people. However, because of the strong philosophical ties to individualism, a pervasive view of property has emphasised the priority of individual rights over broader community interests.

For some time, however, there has been what might optimistically be called the beginnings of a paradigm shift in the meanings and the extent of property. Looking at the excesses of capitalism might put this in question; perhaps what I am calling a paradigm shift is simply an 11th hour reaction to these excesses.

Nonetheless, I think it is possible to discern in law, in scholarship, and in many forms of activism some pressures and changes in the idea of the person, the idea of property, and the ways in which they relate. Increasingly, we see the values of community, place, environment, and the common future reflected in discourses and indeed in the law.

To begin with the person, the liberal ideology of the bounded and autonomous self has been challenged by more relational ideas of the person. In legal scholarship, one finds this in feminist work criticising notions of the person based on the model of the rational self-determining individual, developed by authors such as Jennifer Nedelsky and Ngaire Naffine.

When we look at property, we also see a heightened concern for the relationships in which property is situated, for example, in debates about 'the commons'. Promotion of commonly owned resources (belonging to communities, not necessarily to the state) has been hampered by the economic orthodoxy embodied in the 'tragedy of the commons' argument, which argues that common resources will necessarily degrade over time because individuals will have no incentive to manage those resources with care. However, in 2009, Elinor Ostrom was awarded the Nobel prize in Economics for her work on the commons. Ostrom's work challenged the 'tragedy of the commons' narrative and provided accounts showing that well-managed common resources, such as fisheries, forests, and pastures can be more efficiently managed than private resources if there are strong co-operative bonds among users and solid management principles.

Numerous property law scholars have also turned their attention to urban, rural, and global commons, as well as the intellectual property public domain and of course, the environment. In this literature, individual rights and private interests are only part of a more complex picture that includes the interests of a multitude of communities, including non-human life. To give just one example, we can see commitment to community-oriented property in the open-access licences that are available for the distribution of intellectual property.

This re-valuing of community is not confined to activism, or to the Internet. Law itself has arguably been moving in this direction for some decades, as environmental, heritage, and planning regulations have been strengthened. Such regulations often appear to individual property owners as a subtraction from their rights. But it is arguably more accurate to see such regulation as evidence of an alternative conception of property embedded in law.

As part of this rebalancing of rights and responsibilities, property in land is sometimes now imagined through the language of stewardship or custodianship. Stewardship implies that an owner owes a duty to current and future users of a resource, and in a sense, also a duty to the resource itself. Stewardship is not well integrated into property law, but it is making its way into law mainly through environmental law.

Concluding thoughts

Recent scholarship has seen the development of a more complex notion of the person, as well as a greater emphasis upon community and environmental imperatives, in the definition of property rights.

We are moving away from a social and legal imaginary based on boundaries, self-containment, and individual control, to a consciousness which is relational, contextual, and deeply social. The strong nexus between persons and property must now be seen as mediated by values associated with the commons, the public domain, and the numerous communities within which we find ourselves. It is simply no longer possible to remove shared interests from questions of identity and ownership.

Thinking about property and persons relationally means that the image of the quintessential owner shifts from autonomous powerful individuals to connected individuals and groups, including marginalised people. Recent jurisprudence on native or Indigenous title in Australia and Canada recognises that in the case of Indigenous land, it is the group that is the relevant interest bearer, not a collection of individuals.

We can also see a different approach to distributional justice. In relation to something I own, we can ask, what is the distribution of rights between me and the community?

> But it is not possible in this or any other context to speak of a single undifferentiated community. Communities are multiple, contingent, dynamic, and overlapping. Therefore, no general principles about property will hold for all situations.

Finally, persons and property are artificial legal constructs whose meaning cannot be governed by law alone. They circulate through many layers of social relationships and networks. They constitute and are produced by material and symbolic economies. Uncovering a few threads in these relationships begs a multitude of other questions.

Further readings

Bhandar, Brenna. 2016. 'Status as Property: Identity, Land, and the Dispossession of First Nations Women in Canada'. *Darkmatter Journal* 14: 1–20.

Fennell, Lee Anne, 2011. 'Ostrom's Law: Property Rights in the Commons'. *International Journal of the Commons* 5(1): 9–27. http://doi.org/10.18352/ijc.252

Underkuffler, Laura. 1990. 'On Property: An Essay'. *The Yale Law Journal* 100: 127–148.

41

OWNERSHIP OF INTANGIBLES

Intellectual property and the contested commons

S. Ali Malik and Rosemary J. Coombe

How do you typically start your day? While 'wellness experts' advise that we meditate, exercise, and eat a healthy breakfast, many of us begin our day by reaching for our phones, seamlessly entering our technologically mediated digital cultural worlds. Spotify notifies you that your favourite DJ has just released new tracks and mixes. You receive an email from HBO commanding that you 'cease and desist' downloading the series finale of *Game of Thrones*. Putting that off, you scroll through your Instagram feed and find stories about the controversial practice of donning Native American-style headdresses at music festivals and accusations of design plagiarism against fast-fashion retailer Zara.

To finish your presentation for class, you copy and paste some photographs from Google Images to illustrate it and save your file to a flash drive. The morning is moving on, so you get dressed and bolt through the door into the world that just minutes ago you were curating with the swipe of your finger. Already issues about the ownership of intangibles have shaped your day.

Disputes and concerns about property in intangibles (e.g. data, software, cultural products, and plant and human genetic resources) say a lot, both about how we live within digitally meditated worlds and how we engage with the cultures of others. Humans everywhere express their collective and individual identities through cultural forms; they hold strong opinions about what kinds of things should be private properties and which should remain social-held goods. Individuals and corporations create, use, and distribute various forms of intangible property, such as the goods protected by copyright, trademarks, patents, publicity rights, design rights, and database protections. The legal systems that govern the use of intangibles or non-physical forms of property, known as intellectual property (IP) laws, structure and regulate the way we entertain ourselves, obtain our news, choose our fashion, and get access to education, medicine, and art.

We begin this chapter by showing how 'information capitalism' is structured by an IP system that creates exclusive property rights that enable owners to exclusively control and profit from the circulation of goods we call intangibles. We then illustrate how IP facilitates the 'commodification of culture' between groups and societies, resulting in accusations of piracy and cultural appropriation. Similar struggles surround the potential for biotechnology to transform forms of life into products for sale and create new restrictions on access to drugs, medical technologies, therapies, and agricultural goods. Finally, we consider how controversies about the ownership of intangibles in the global South take form in a human rights context that emphasizes Indigenous and biocultural rights.

The political economy of information capitalism

The economic drive to expand and integrate world markets through the management of transnational flows of information enabled by the rapid speed and power of information and communication technologies are characteristics of what we call information capitalism. Product bar codes, for example, allow flows of product information to travel across countries in seconds. You can purchase goods in foreign countries using your personal credit card without leaving home. Digital platforms facilitate live exchanges of branded news, entertainment, and expertise. Proprietary databases of information about consumer preferences and industrial capacity as well as patented 'just in time production' methods enable corporations to easily move manufacturing plants and quickly fill stores and online orders around the world. We refer to this as economic globalization.

As economic globalization has intensified since the late 1980s, industrial production activities have increasingly moved to the global South where labour costs are low, while the economies of the global North have shifted emphasis to services, branding, and information-based goods. By designating what informational resources are recognized as property and the extent of owners' rights, IP restricts the ways citizens can use informational goods. Consequentially, IP can limit consumer creativity and constrict freedom of speech. Informational resources that would otherwise be public goods and free to use, become expensive. The power of corporations who trade in such goods has dramatically increased since they pressured the World Trade Organization to mandate that all countries in the world enact legislation to protect IP and enforce the rights of foreign IP holders.

IP rights include copyrights, trademarks, patents, integrated circuit topographies, and rights of publicity, but also geographical indications, plant varieties, designs, and databases. Such legal protections turn musical compositions, artistic expressions, software, films, logos, means of manufacture, fertilizers, pharmaceuticals, medical treatments, computer chips, celebrity images, food names and even seeds into commodities with market value. But where does the information that goes into such creative work come from and how do we gain access to it? And who benefits from it? How would anything new and creative come into being unless there were cultural resources available for us to transform? Liberal democratic societies rationalize IP as a social bargain in which the creator or innovator is rewarded with a limited term monopoly over the creative or innovative 'work' protected by IP in exchange for eventually donating that work to society when the monopoly expires.

Informational resources lacking IP protection reside in the 'public domain' and are thus available for everyone's use. These include cultural goods in which IP rights have expired and intangibles that have yet to be crafted into anything legally recognized as a human/corporate creative work. From traditional stories, literary and musical genres, myth, folklore, digital code, open source software, wild plant germplasm, chemical elements, generic tools, methods and conventions of manufacture, the creation of IP, and thus the future of information capital requires ongoing access to what we might call 'raw' cultural goods. These building blocks for cultural and industrial transformations are everywhere. Consequently, we tend to see them as part of a cultural 'commons', naturally governed by the social values of accessibility, openness, sharing, cooperation, and reciprocity. When terms of IP protection get longer, as they have recently done, and fewer resources thus become socially available to us when copyrights and patents expire, activists complain that our cultural commons are being unduly contained. How long, after all, should Disney be able to own Mickey Mouse?

Within a single society, divisions between proprietary and non-proprietary goods tend to be relatively unproblematic, or at least widely understood, although they may be contested. For example, in the early era of electronic development, we witnessed a widespread movement to use

Creative Commons licenses (a form of contract) to create open source software that could not be captured by corporate proprietary interests. Whole new industries can be stunted if too many early innovations are immediately protected as information goods, thereby limiting the resources available for the work of others. Even in contemporary industrial societies, we see some efforts to maintain public domain resources for future creative development.

If we look historically at the conditions of early modern development, however, a more complicated picture comes into focus. Globally, the concept of the 'common heritage of mankind' assumed a singular public domain held collectively by humankind. Empires, fur traders, explorers, travelling botanists, and eventually countries, corporations, and international institutions advanced the creation of global knowledge banks in which all traditional knowledge and arts (TK), as well as plant and genetic resources from communities around the world were collected. TK refers to the folklore, innovations, medicinal and agricultural practices, and creative expressions of Indigenous and local communities, many of which predate the creation of modern nation-states; these are also resources on which communities depend for social reproduction. Collectively possessed TK evolves through its transmission from generation to generation.

Much of what we consume today, both materially and culturally, derives from the application of TK. Historically, TK provided the 'raw material' necessary for facilitating scientific research and stimulating modern invention. Archives of cultural and material forms of knowledge, practices, and genetic resources provided the ingredients from which modern technology, medicine, and various arts were forged. Folk art, stories, oral poetry, tones, and musical scales were collected by Western researchers in collections that continue to provide a common pool of resources for creative sampling.

'New' music fuses previously created forms with the creative efforts of current artists. American jazz and rock and roll derived from African and African-American traditions; Drake's music draws heavily upon Dancehall and other Caribbean sound styles. Music emerges through circulation, remixing, collaboration, and reciprocal patterns of influence.

The introduction of IP rights such as copyright puts limits on such flows of creativity and influence. Copyright privileges certain creators as authors of musical works who have absolute rights over their reproduction, performance, and circulation which limits their use by others for several decades. Today, very little of our own societies' musical creativity becomes part of the common pool of cultural resources on which creators can draw without expensive royalty fees. This pool further shrinks as copyright becomes globally enacted and enforced. Corporations obtain additional rights as owners of trademarks – arrangements of symbols, characters, designs, or sounds that function as brands which distinguish one company's goods from others in the marketplace. Such trademarks do more than simply represent the company's reputation. Over time, they embed themselves in our everyday social lives; the Nike 'swoosh' or the Jordan 'jump man' have become part of popular culture. Yet, their corporate owners can limit our capacity to use them for our own purposes and/or appropriate such expressive activities for their own IP-protected advertising campaigns.

When copyright and trademarks rights are widely infringed, however, corporations tend to incite moral panics over the growing prevalence of 'piracy.' In the last 20 years, new technologies have enabled consumers to engage in novel ways of embedding IP-protected cultural products into their daily practices. Illegally downloaded music, films and television shows, counterfeit trademarks and fashion imitations have provoked diverse crack downs on people often accused of terrorist activities. In campaigns to root out illicit copies, 'fakes' and 'knock-offs', industrial actors insist upon concepts of ownership and originality that are at odds with social realities of cultural production and diffusion. In short, as we have seen historically, so-called acts of 'piracy' may be integral to cultural (re)production.

We can now see the crux of debates over cultural appropriation; if the basis from which contemporary art is created, refashioned, repurposed, or improved upon is a common cultural pool, much of which has been contributed by peoples and groups outside of modern industrial societies, then who can legitimately claim ownership of such goods? How do we balance the rights of creators to freely express themselves and to capitalize upon their creations with the rights of the peoples who make legitimate claims that their own knowledges, practices, and cultural contributions were unacknowledged, and often simply seized in compiling the cultural pool upon which today's creators draw? How do we create the conditions for knowledge, information, and art to be 'free' for new creative use, while still maintaining incentives for new forms of innovation, acknowledging industrial investments in producing, and publicizing intangibles in markets while respecting the cultural rights of communities who are protective of their TK and cultural heritage?

Resisting information capital

In an era of rapid advances in biotechnology and computational processing, all natural and genetic resources hold the potential for becoming 'information' in new technological developments. Patents are exclusive IP rights granted for an invention, which is a product or a process that entails a new way of doing something. To be awarded a patent right, the inventor must prove the innovations' novelty and its usefulness and show that it is a non-obvious step in the development of a practice. For example, your DNA is not patentable because it is a naturally occurring substance. But what happens if a segment of your DNA is taken in a medical setting and then altered by a pharmaceutical company to provide you with a personalized medicine? What if, instead, the resulting patented medicine is useful to others who can afford to pay huge sums for it but is not made available to you or your family at a price you can afford? The medicine might not exist without you, but your genetic information is not seen as your work or innovation, but simply part of a common pool of natural resources from which new creations can be made.

Controversies about the propriety of patenting life forms abound, not only in medicine but also in agriculture. Modern scientific plant breeding draws upon the knowledge and skills of farmers from around the world, accumulated as 'common heritage' resources. New biotechnologies enable companies to quickly edit the genetic 'information' held in seeds. Just as families are upset when their cell lines become the basis for patented treatments which boost corporate pharmaceutical profits, farming communities in the global South may be outraged when the plant biodiversity resources they have cultivated and contributed to modern agriculture are subjected to genetic profiling for the corporate production of patent-protected engineered seeds. The conversion of plant genetic resources into information capital seems especially unjust when the new technologies are unaffordable to source communities or their purchase of Monsanto seeds renders them dependent upon the same corporations' costly patented fertilizers and pesticides (contributing, in a notorious instance, to a debt-suicide crisis amongst Indian farmers). The introduction of genetically modified plants into the areas from which traditional crops came may also endanger the original plant varieties as Indigenous peoples in Mexico realized when genetically modified corn made its way into their family gardens. Traditional healers have also contributed a great deal of plant knowledge to modern medicine, often alerting pharmaceutical industries to the active compounds in plants and their potential usages for human treatment. Should traditional cultivators and medicinal practitioners have rights to recognition and compensation when corporations appropriate their knowledge and skills in producing new seeds, plant varieties and medicines?

A variety of social movements emerged in the late 20th and early 21st century in resistance to the power of information capital. Such movements insisted upon the importance of maintaining diverse protected commons in cultural and natural resources, from human and plant genetic materials to digital software and creative cultural works. Coalitions of farmers, artists, environmentalists, feminists, software developers, food and health activists, and Indigenous Peoples created new alliances around these issues. Many of them oppose the expansion of proprietary biotechnologies and the mass commodification of human life. Advocates look to defend place-based (or 'local') modes of cultural production and distinctive place-based products against the onslaught of mass-manufactured goods. Indigenous Peoples' movements assert their stewardship of plant and animal biodiversity, while rural farmers, aligned under 'slow food' banners, identify themselves as protectors of traditional artisanal techniques. Both groups promise to cultivate resilient and sustainable landscapes and livelihoods.

To mark goods as coming from protected commons, ironically, such groups have increasingly turned to forms of IP similar to trademarks, such as appellations of source like Champagne, denominations of origin like Kalamata olives and collective and certification marks, such as Fair Trade and Small Producers symbols, which identify goods as coming from sustainable conditions and community traditions of practice rooted in local ecologies. Historically, these were common in Europe, such that producers of a sparkling wine could only call it 'Champagne' if its production occurred in the Champagne region in France using specific, carefully regulated production methods. Other examples include Gruyere cheese, Basmati rice and Kashmiri Pashmina. Consumers are often willing to pay a premium to ensure that they are purchasing authentically produced, local goods rather than mass-manufactured generic products coming from unknown global sources.

Increasingly, however, consumers also want to know that such goods are ethically and equitably produced. In short, we have seen an emerging concern with respect for human and environmental rights in production practices. All too often, goods that are marked as coming from specific and authentic conditions of origin using geographical indications do not guarantee either an environmentally sustainable conditions or respect for human equality in conditions of production.

The protection of Tequila as a geographical indication, for example, was celebrated for extending such IP protection to the global South, but the expansion of production the geographical indication enabled drastically reduced biodiversity in the region and excluded the holders of TK in the cultivation of mezcal. Authentic South African Rooibos tea, famed as Indigenous to the Western Cape region in South Africa, is grown in a region populated only by white farmers with large landholdings and landless 'coloured' farm laborers, descendants of Khoisan Indigenous peoples whose lands were seized by Boer settlers. They earn pitiful wages for doing seasonal work on these same lands today. Khoisan/coloured people gain no benefits from the appropriation of their TK of this crop by their white overlords and the country's dominant black population is most excluded from its production. In India, Darjeeling tea was the first registered geographical indication. It is produced in a region largely populated by people whose ancestors were shipped in by British plantation owners as indentured labour from 'tribal' regions and neighbouring countries. For centuries, they were denied citizenship or labour rights and subject to beatings or imprisonment if they attempted to escape these conditions of virtual bondage. Even today, most of these laborers are women, members of ethnic minorities who have few rights and are still held responsible for repaying the debts of their ancestors who were forced to compensate employers for the costs of their transportation and medical care. How might we ensure that such workers see better working conditions and realize their fundamental human rights? How might they have their interests represented in plantation management as a consequence of the additional revenues generated by the 'Darjeeling' mark?

Indigenous Peoples have unique interests in contemporary IP struggles. Their cultural heritage was deliberately subject to destruction and dispossession in the making of modern states and a global economy. Their territories were occupied, their cultural heritage looted and placed in imperial museums, their ancestral music recorded and stored in institutions to which their descendants have no access, their religious rituals were criminalized and ritual regalia seized; the plant genetic resources they cultivated were appropriated for use in global food and medical systems. Rights-based claims to cultural heritage are central to the collective struggles of many Indigenous people to assert their continuing existence as a people and to articulate new forms of engagement with (and resistance to) global markets.

Insisting upon the inter-connectedness of the biological and cultural realms of their livelihoods, Indigenous communities make territorial claims as stewards of plant, animal, marine, and soil biodiversity as well as TK. The concept of biocultural diversity recognizes the inter-relationship of nature, culture, and knowledge; community biocultural rights are now acknowledged in environmental law. The International Institute for Environment and Development (IIED), an environmental NGO working with the Peruvian organization, Association Andes, has promoted the development of a 'Biocultural Heritage Indication' labelling scheme for Indigenous communities to use as collective trademarks that would function like Fair Trade or Slow Food designations. They endorse their extension in other mountainous farming communities to mark goods and services produced in accordance with Indigenous principles of environmental stewardship, gender equality and customary law. This final example suggests new frontiers for the ownership of intangibles, in new networks of commitment to agricultural, environmental, and rights-based social change.

Further readings

Coombe, Rosemary J. 2020. All authored and co-authored publications are located at https://yorku.academia.edu/RosemaryCoombe.

Hettinger, Edwin C. 1989. 'Justifying intellectual property'. *Philosophy and Public Affairs* 18: 31–52.

Kundani, Arun. 1999. Where do you want to go today? The rise of information capital. *Race & Class* 40 (2/3): 49–71. https://doi.org/10.1177/030639689904000205

Perry, Bronwyn. 2000. 'The fate of the collections: Social justice and the annexation of plant genetic resources'. Zerner, Charles, ed., *People, Plants and Justice: The Politics if Nature Conservation*. New York: Columbia University Press. pp. 374–400.

42

FROM REPRODUCTIVE RIGHTS TO REPRODUCTIVE JUSTICE

Rachel Rebouché

Introduction

Since the mid-1990s, 'reproductive rights' have been the focus of United Nations documents, a priority for agencies like the World Health Organization, and the subject of judgments issued by national and international courts. The advocacy by women's rights activists helped place reproductive rights, including abortion rights, on the agendas of human rights bodies and organizations. And, presently, gender equality and women's empowerment are central to international conversations about reproduction, fertility, and population.

The proliferation of reproductive rights at the international level also benefited from high-profile national cases decided in the 1970s and 1980s, which interpreted constitutional rights as requiring state permission for individuals to use contraceptives or to have an abortion. Decisions in the global North, like *Roe v. Wade* in the United States or *Morgentaler v. R.* in Canada, interpreted rights to privacy or equality or autonomy as protecting reproductive choices. More specifically, these iconic cases spurred social movements to tether reproductive rights to the pursuit of women's human rights.

Of course, the implementation of reproductive rights depends on health systems and healthcare resources; rights without services may not be worth very much. So, although reproductive rights announce important, universal values, improving the reproductive health of individuals and populations requires the technical and practical work of offering accessible and affordable healthcare. But healthcare resources, at the local and global levels, are not evenly distributed. Those committed to the advancement of reproductive rights have recognized that people routinely have their reproductive health needs, over the course of their lives, under-funded and under-served. Scholarship and advocacy campaigns have therefore adopted a 'reproductive justice' framework to address systemic and structural inequalities. This essay focuses on the foundational commitments of the reproductive justice movement, and how those commitments differ from previous thinking in the field. To make these distinctions clear, this essay will rely on examples of how scholars and advocates apply reproductive justice principles in the United States.

Chief among the distinctions between 'reproductive rights' and 'reproductive justice' is a primary concern with the experiences of marginalized groups. Reproductive justice is a movement founded by people of colour who recognized the relationship between racial oppression and state coercion of reproductive choices. Moreover, reproductive justice looks beyond campaigns for constitutional rights to abortion to reach a range of issues that affect people over the course of their lives. Reproductive justice advocates situate themselves as a broader force for social justice that responds to people's needs through grassroots organizing and community-based interventions. The tenets of reproductive justice therefore share a concern for people's lived experience of law.

In the U.S. context, the move from reproductive rights to reproductive justice reflects significant shifts in thinking about how law and reproduction intersect. However, like any social justice movement, reproductive justice is evolving and faces the inevitable difficulties of realizing its ambitious goals. This essay concludes with reflections on one such challenge, namely, how rhetoric, legal reform, and activism can inspire a globally responsive vision for reproductive justice.

Movement priorities

Reproductive justice, as distinguished from reproductive rights, has gained significant momentum in the United States. In increasing numbers, non-profit groups have dropped 'pro-choice' from their materials and incorporated 'reproductive justice' into their organizations' names. In fact, reproductive justice writings express a concern that the phrase 'reproductive justice' has been co-opted by groups that do not share its commitments. This section describes four such commitments – the recognition of race and intersecting identities as fundamental to reproductive justice, the inclusion of priorities outside of abortion, a retreat from litigation as the predominant source for change, and attention to community or local engagement.

Reproductive justice founders describe meeting at the 1994 International Conference on Population and Development (ICPD), an event at which the statement '[r]eproductive rights are human rights' gained international prominence. Advocates who attended the ICPD were inspired by the human rights movement generally and the ICPD's inclusion of poverty, gender equality, and the empowerment of women specifically. They noticed, however, that the concerns of their communities were not represented in the statements of government delegates. Non-profit groups advocating for people of colour, organized by the Sister Song Women of Colour Collective, strategized to advance an agenda that would harness the spirit of inter-sectionality of the ICPD on behalf of people of colour in the United States. The Sister Song Collective included groups representing Native American, African-American, Latina, and Asian-American communities.

These activists identified the ways in which mainstream reproductive rights organizations had not concentrated on the racial and income disparities that perpetuate inequality and impede the delivery of reproductive healthcare. Indeed, the pro-choice movement's focus on preserving constitutional rights to abortion ignored problems of arguably more salience to women of colour. Reproductive justice emphasizes the connection between present discrimination and the legacies of slavery which forcibly controlled the fertility of women of colour and low-income women. Dorothy Roberts, for example, describes the past and present population control measures imposed on women of colour and the racism that historically underpinned laws governing reproduction. Yet matters such as the involuntary sterilization of women of colour and the underfunding of reproductive healthcare were not focal points for the mainstream reproductive rights movement.

Thus, at the core of reproductive justice is a focus on racial injustice as well as on how overlapping forms of oppression shape one's wellbeing and experiences. Of particular note is the activism of Loretta Ross, who worked for the first reproductive justice organization for women of colour (the National Black Women's Health Project, founded in 1983) and directed a program focused on women of colour at the National Organization for Women (NOW). Ross's co-founded the Sister Song Collective to address the repeal of the federal ban on abortion funding, for example, which had not been a priority of NOW and other mainstream organizations.

Consistent with the theme of inclusion, reproductive justice supports an expansive agenda for reproductive health. As noted, abortion rights have been the focus of the U.S. reproductive rights or pro-choice movement. Reproductive justice begins with the premise that concentrating on abortion law represents only a narrow slice of the issues important for people's reproductive choices and health. Reproductive justice initiatives address reproduction across a life span, including, but not limited to, pre-natal and post-partum healthcare, the availability of sexual education in schools, subsidized or free contraceptives, and access to safe assisted reproductive technologies, and affordable childcare. Zakiya Luna and Kristin Luker, in distinguishing reproductive rights from reproductive justice, define 'reproductive justice [as] equally about the right to not have children, the right to have children, the right to parent with dignity, and the means to achieve these rights'.

Many of the interests and rights described in the preceding paragraph cannot be fully protected or won in courts. Indeed, courts may not have the tools to implement remedies for systemic social or structural problems, and they do not make political and policy decisions that affect directly the delivery of health services. As Lindsay Wiley writes, 'Access to healthcare – not merely as a matter of the "right to choose" contraception or abortion, but as a matter of the general affordability, availability, and cultural appropriateness of a wide range of health services for women and families – is a priority issue for the [reproductive justice] movement'. In this vein, the reproductive justice movement is sceptical of what court-driven reform, which historically has been the bedrock of U.S. abortion rights, can accomplish. The reproductive justice movement emphasizes that negative rights, in which an individual has a right to be free from government interference, support people living with certain privileges, who can depend on governments to protect their rights. Said another way, a constitutional right to privacy is only valuable to those who can afford to exercise privacy and autonomy under a state's laws. That freedom may not be available to those vulnerable under law, such as people subject to racial discrimination, receiving state assistance, or new to a country. Accordingly, reproductive justice advocates consider the multiple avenues by which people meet their reproductive health needs, emphasizing tailored law reform and localized interventions.

In sum, the agenda for reproductive justice seeks to meet an array of interests for diverse communities of pregnant people. One way that reproductive justice writings articulate these commitments is through evoking the power of international human rights. Reproductive justice materials note that human rights law addresses a range of reproductive health matters, and human rights advocacy urges states to adopt positive measures to meet women's economic and social rights as well as their civil and political entitlements. Jael Silliman, Marlene Gerber Fried, Loretta Ross, and Elena Gutierrez – authors of an early manifesto for reproductive justice – wrote, 'Linking civil, political, economic, sexual, and social rights' can 'bridge [] the gap between having legal rights and lacking the economic resources to access those rights'. In the same passage, they chastise 'the mainstream movement' for being 'unfamiliar with the Universal Declaration of Human Rights (UDHR) and international treaties that protect women's reproductive rights'. Although the turn to international and comparative sources of legitimacy is important, an account of how and why human rights strengthen reproductive justice claims could help bridge U.S.-based activism with the work of advocates across other countries.

Movement trajectories

Co-founder of the Sister Song Collective, Loretta Ross, noted above, wrote that reproductive justice demanded the 'protection of women's human rights to achieve the physical, mental, spiritual, political, economic, and social well-being of women and girls' and a cause 'embedded in a human rights and social justice framework'. In the United States, advocates and lawyers have looked to human rights law to fill gaps in U.S. constitutional and statutory protections. A human rights approach has been particularly important to reproductive rights groups that have confronted the shortcomings of constitutional protections in U.S. courts and the animosity toward abortion in U.S. legislatures. Human rights advocacy has been one way to protest U.S. exceptionalism, or the U.S. resistance to signing international treaties and declarations such as the Convention on the Elimination of All Forms of Discrimination Against Women.

Reproductive justice advocates might ask, however, if human rights arguments have the capacity to redistribute income and societal resources. Human rights advocates know all too well the challenges of implementing socio-economic rights, especially rights to healthcare and to effective healthcare systems. Human rights law may not be the best tool for addressing the thornier problems in the delivery of reproductive health services, particularly at the local level. And human rights strategies often cannot provide answers to questions about infrastructure or resources – how communities and countries decide to allocate social goods and social assistance. Resource distribution is an intensely place-specific issue. Human rights can sometimes elide differences among places and populations, which might obscure the experiences of various populations that reproductive justice seeks to bring to the fore.

Reproductive justice's reliance on community and grassroots mobilization engages these practical and political considerations; namely, given the movement's focus on anti-subordination, advocates focus on why some populations and individuals lead healthier lives than others. But for reproductive justice to be truly global in its reach, it faces the challenge of translatability. The U.S. conception of reproductive justice will not and should not be the same for other jurisdictions. Reproductive justice offers a template for strengthening solidarities between causes for social justice and alliances among disparate jurisdictions. At times, human rights can help galvanize decision-makers and communities to seek common change. In other instances, different tactics and frames will be necessary to achieve distributive goals.

Further readings

Luna, Zakiya and Luker, Kristin. 2013. 'Reproductive Justice'. *Annual Review of Law and Social Science* 9: 327–352 https://doi.org/10.1146/annurev-lawsocsci-102612-134037

Price, Kimala. 2010, 'What is Reproductive Justice?: How Women of Color Activists Are Redefining the Pro-Choice Paradigm'. *Meridians* 10(2): 42–65 https://doi.org/10.2979/meridians.2010.10.2.42

Rebouche, Rachel. 2017. 'Reproducing Rights: The Intersection of Human Rights and Reproductive Justice'. *UC Irvine L. Rev.* 7: 579–610.

Silliman, Jael; Gerber Fried, Marlene; Ross, Loretta; and Gutierrez, Elena R. 2004. *Undivided Rights: Women of Colour Organize for Reproductive Justice*. Cambridge, M.A.: South End Press.

43

SETTLER COLONIALISM

Sarah Hunt

Over the past several decades, scholars in a range of fields have documented socio-legal processes inherent to processes of colonization. A central component of this scholarship is clarifying the enduring nature of settler colonialism, as distinct from other colonial and post-colonial formations. Settler colonial societies, as opposed to societies that are run overtly as colonies, are comprised of ongoing structures and systems that have become separate from an external governing structure, such as an imperial ruler. Settler colonial societies build their power on the ground in the places they seek to claim, through establishing legal processes by which settler colonists can actively acquire and control land and resources, as well as assert the lawful right to make these claims as settler subjects. This body of cross-disciplinary scholarship illuminates the assertion that settler colonialism (as it exists in Canada, Australia, the United States, and elsewhere) is not marked by a single historic moment of colonial conquest, but rather is structurally embedded. Where colonialism is focused on economic accumulation by people acting on behalf of a distant imperial power, and, where post-colonialism is understood to mark conditions where these imperial powers have since vacated, settler colonists do not leave but remain on the lands they have claimed as their own. Further, in claiming lands previously occupied and governed by Indigenous peoples – and often still occupied by Indigenous peoples – settler colonialism is marked by the reconfiguration of these lands through socio-legal imaginaries which dispossess the original occupants via ongoing assaults on their worldviews, bodies, and ways of being.

As such, any systems of law which existed prior to the establishment of settler colonial societies are rendered illegitimate or no longer authoritative, as Indigenous people become instead subjects of newly formed settler colonial legal regimes. In settler colonial countries such as Canada, the United States, Australia, and Aotearoa/New Zealand, it has been argued that historic processes of dispossessing Indigenous peoples of their lands and subjugating them to settler colonial law are not merely situated in the past, but rather continue to structure everyday life in those countries.

While settler colonialism is understood by many to centre on land – specifically on territorial claims – socio-legal analysis reveals that settler colonial power seeks to rework not only relationships to land but all areas of social life. In particular, settler colonialism establishes authority, and naturalizes that authority, by reconfiguring Indigenous life across a range of sites and scales. Within these demonstrations of authority, law is often experienced by Indigenous peoples as a source and expression of violence, as a key tool of settler colonial states and state actors. A cross-scalar analysis emphasizes not only the structural nature of settler colonial power, but its visceral, embodied, and material nature, as evidenced in frequently violent reconfigurations.

Key to the establishment of settler colonial sovereignty, and the dispossession of Indigenous peoples which is necessary to establish this claim, is the reconfiguration of land into distinct settler spaces and Indigenous spaces. Thus, the creation of settler colonial nations first entails the imaginative work of categorizing Indigenous people as racialized subjects incapable of claiming legal authority or tenure, such as by codifying classifications of 'Indians' or 'Aborigines'. These racial classifications work alongside the mapping out and dividing up of land for Indigenous inhabitants ('Indians') and settler colonists. In the formation of settler colonial states, new geo-political spaces – those of demarcated reserves, towns, provinces, and spaces of industrial development, resource extraction and production – were superimposed onto the existing land-based relationships and geo-political formations of Indigenous peoples, their boundaries closely monitored and policed. Settler colonial power tended to operate in favour of settlers and the expansion of capitalist ideologies of settlement, and law was used to construct, enforce, and normalize power relations literally on the ground through separating racialized spaces from settler spaces. Thus, as materializations of the force and power of law, arbitrary social boundaries became legal realities.

The establishment of reserves or reservations was used in the creation of North America (with similar spatial strategies developed in other settler colonial contexts) in order to confine Indigenous peoples and their territorial claims. While in some settler colonial contexts, settler governments negotiated land use in treaties with Indigenous nations (such as the *Treaty of Waitangi* in Aotearoa/New Zealand), these treaties were often not formed with a clear set of meanings shared by all parties, and/or the terms of the treaties were never honoured. Regardless of whether or not treaties were signed, the process of clearing the lands of Indigenous peoples and confining them onto lands reserved for their use often entailed the deployment of legal force, including outright violence. Through these processes, Indigenous people became trespassers in their own homelands and, in many contexts, remain constrained by these racialized spatial formations to this day.

A central claim made by scholars of settler colonialism is the assertion that settler colonialism operates via an eliminative drive which seeks to kill or erase Indigenous peoples, their claims to land and ways of being in order to assert and naturalize state assertions of sovereignty. However, scholars of Indigenous studies have nuanced this claim by revealing the ways in which settler colonialism depends on, and assumes the presence of, Indigenous peoples and their relationships to land. In other words, although Indigenous people are rendered racialized subjects of settler colonial systems of power, settler colonial subjectivity is always formed in relationship to the Indigenous racial 'other' and, thus, in some ways, depends on Indigenous presence. Moreover, the establishment of settler colonial spaces is always formed in relation to its outside – not only the social realities external to its own national borders but also those 'outsider' spaces within its borders – spaces characterized by the logics of the frontier.

The differentiation of lawful space from unlawful space was central to the frontier landscape out of which settler societies emerged, a process which is also evident at the scale of Indigenous bodies and homes. It has been argued that the frontier remains the most recurrent common sense myth about Indigenous people in Canada and in the US. The key features of the frontier include imagining North America and other settler colonial lands as empty, unoccupied wilderness with rich resources freely available for the taking, or the rightful wealth owed to those representing the interest of civilization and progress. Settler colonial frontiers are characterized by encounters between opposites, situated across boundaries of difference: civilization and savagery, man and nature, good and evil, and, of course, white and 'Indians'. Conflict is seen as inherent to the frontier, as nature, evil and the racialized 'other' must be overcome, subdued, destroyed, contained, or assimilated (often through friendship or the benevolent 'care' of settlers, not only violence) in order to realize the values of self-reliance, democracy, freedom, and 'will of God' upon which settler colonial society will be built. In this way, it is evident that the 'Indian' and 'Indian space' are necessary components of the frontier.

Indigenous scholars have argued that, within settler colonial frontier imaginaries, Indigenous bodies are constructed as abnormal or criminal through highly gendered colonial logics, such that they come to occupy degenerate spaces at the edge of the nation. If Indigenous people step outside of the marginal roles assigned to them, they become criminalized or targeted for disappearance, as evidenced in the high numbers of Indigenous women who have been killed or disappeared across Canada and the United States. They are not simply to be destroyed once and for all, but they are situated at the edges of the newly forming society and in opposition and relation to the identities of settlers.

One frontier figure is that of the 'traditional' stoic Indian who lives off the land and who finds contemporary life confusing, confounding, and beyond their grasp; unable to adapt; unable to progress; and unable to return to the romanticized past. Thus, the story goes, settler subjects are required to help usher these sad Indians into a new, real industrial economy. Rather than being limited to a fixed set of figures or places, the flexibility of the frontier imaginary is key to its hegemonic power, via its ability to transform into a particular historical worldview that is reinforced and reproduced in everyday dimensions of life. The frontier myth structures the limit of settler colonial social, legal, and historical imaginaries.

Heidi Kiiwetinepinesiik Stark powerfully illustrates this in her 2016 article 'Criminal Empire: The making of the savage in a lawless land', in which she examines the ways in which Indigenous resistance in the 19th century was treated as criminal in order to avert attention away from the illegality of the US and Canada. Within frontier zones in which struggles to obtain control over Indigenous peoples and their lands was in flux, Stark demonstrates that the brute force of guns and violence typified at the frontier eventually shifted via the assertion of criminal jurisdiction over Indigenous people. No longer was brute force needed – legal procedures and legal technicians could assert control over Indigenous peoples' lands, mobility, and virtually all aspects of life, and do so in the name of the law. As Stark discusses, case law pertaining to Indigenous people in the 19th century primarily focused on land title and criminal law. Simultaneously, state law failed to protect Indigenous people from settler violence nor punish settlers for murder or other atrocities against Indigenous people. Law itself thus became a direct threat to Indigenous nations, suppressing culturally distinct legal orders while failing to provide protections afforded to citizens of newly forming nations.

Further, with the assertion of settler colonial legal regimes, Indigenous resistance to colonial dispossession and violence could be categorized as criminal activity such that Indigenous lands could be seized lawfully and leaders could be killed under the rule of law. Criminality was produced through settler colonial ideologies of race and gender. For example, in some stereotypical representations, Indigenous men were portrayed as part of the 'wilderness' that had to be 'tamed' through law and order, while women were viewed as either virginal territory in need of conquest or inherently cunning and deceptive, requiring a civilizing force. The introduction of measures of control were portrayed as necessary and for the good of residents of these supposedly uncivilized, lawless spaces. The bodies of Indigenous men and women at the frontier thus necessitated the creation of newly forming spaces of law and order, and newly forming roles for settlers to maintain the security of those spaces, such as Indian Agents surveilling Indigenous homes and police surveilling public streets of newly built towns and other settler spaces. The imagined boundary between spaces of civility/lawfulness and wilderness/unlawfulness continues to typify the frontier.

Within settler colonialism, frontier rationales necessitated the creation of reserves, reservations, or other spaces set aside for Indigenous people. Stereotypes of criminality were (and are) commonly used to justify the need for separate spaces for Indigenous people and settlers, as well as to justify separate legislation that applies only to Indigenous people within settler states. In the Canadian context, the *Indian Act* of 1876 included provisions which criminalized Indigenous spiritual, cultural, and legal practices, such as the sun dance or potlatch, and, over the years, included a range of other provisions specific to Indigenous people, such as criminalizing leaving reserves without permission, mandating all Indigenous children attend church-run residential schools, forbidding the drinking of alcohol, and preventing Indigenous people from hiring a lawyer to make a claim against the government. Patriarchal ideologies of settler colonial states were inherent to these legislative measures, with provisions that disadvantaged women (i.e. only allowing men to hold elected leadership positions). Similarly, the Australian *Aboriginal Protection Act* of 1869 gave the colonial state sweeping powers over Indigenous peoples lives, including their places of residence, marriage, and dictating who was and was not 'Aboriginal'. While many of these race-based legislative provisions have changed over time, settler states continue to use legal tools to manage Indigenous family life, mobility, and land use, as well as determine who is considered Indigenous and how Indigenous communities are governed.

In settler colonial nations across the world, legal measures introduced to ban the cultural, spiritual, and political practices of Indigenous peoples and enforced through criminalization and incarceration have been deemed genocidal in intent. Despite global recognition of the violence settler colonial states have imposed and continue to impose on Indigenous people, there are few legal mechanisms through which to hold settler colonial states to account. It has been argued by such Indigenous theorists as Glen Coulthard and Audra Simpson that Indigenous nations have turned increasingly to settler states for recognition of the validity of their rights. With little progress being made within state mechanisms for recognition of Indigenous rights, land title, or self-governance, however, Coulthard, Simpson, and others call for customary Indigenous laws, systems of governance, and everyday practices to be asserted on their own terms regardless of the way settler states continue to impede or criminalize these activities. Significantly, settler colonial states such as Canada, Australia, New Zealand, and the United States initially voted against the 2007 *United Nations Declaration on the Rights of Indigenous Peoples* which was intended to provide international rights provisions for Indigenous peoples globally. This demonstrates that settler states continue to avoid responsibility for the true nature of structural violence and denial of Indigenous sovereignty inherent in their formation.

Further readings

Blomley, Nicholas. 2003. 'Law, property, and the geography of violence: the frontier, the survey, and the grid'. *Annals of the Association of American Geographers* 93(1), 121–141. https://doi.org/10.1111/1467-8306.93109

Coulthard, Glen Sean. 2014. *Red Skin, White Masks: Rejecting the Colonial Politics of Recognition*. Minneapolis, MN: University of Minnesota Press.

Harris, Cole. 2004. 'How did colonialism dispossess? Comments from an edge of empire'. *Annals of the Association of American Geographers* 94, 165–182. https://doi.org/10.1111/j.1467-8306.2004.09401009.x

Morten-Robinson, Aileen. 2007. *Sovereign Subjects: Indigenous Sovereignty Matters*. Crows Nest, Australia: Allen & Unwin.

Simpson, Audre. 2014. *Mohawk Interruptus: Political Life across the Borders of Settler States*. London: Duke University Press.

Stark, Heidi. 2016. 'Criminal empire: The making of the savage in a lawless land'. *Theory and Event* 19(4).

44
SEXUALITY

Brenda Cossman

Sexuality is often associated with a person's sexual orientation – heterosexual, gay, bisexual. But, sexuality includes more that sexual identity; it describes the range of human sexual desires, sexual practices, sexual pleasures, and sexual identities. A person's sexuality may be related to sex or gender identity, but sexuality is a broader concept referencing the full range of sexual feelings, behaviours, preferences, and activities.

Emergence of sexuality studies

Sexuality first became the focus of serious scientific study in the West in the late 19th and early 20th century. In 1886, Richard von Krafft-Ebing published *Psychopathia Sexualis,* often considered the birth of sexology as a scientific discipline. In England, Havelock Ellis undertook similarly ground-breaking scientific analyses of issues such as masturbation and homosexuality in his *Sexual Inversion* (1897). Alfred Kinsey, and then Masters and Johnson, would further develop the scientific study of human sexuality in post-war America.

Critical scholarship, located within the humanities, emerged in the 1970s. Michel Foucault, in his ground-breaking *The History of Sexuality, Volume One*, was first published in 1974 and translated into English in 1976. Foucault argued that the emergence of sexuality as a distinct sphere of life and identity was a modern phenomenon, with its origins in the 17th and 18th century. He argued that sexuality should not be understood as natural or biological, but rather as a discursive object. In contrast to the widely held belief in the 'repression hypothesis' that is, the idea that sexuality in the West was repressed from the 17th century onward, he argued that this period actually saw a '... veritable discursive explosion' of discourses and discussions about sexuality. Indeed, he argued that the scientific studies of the sexologists were amongst these, seeking to unearth a 'truth' about sex and sexuality.

Foucault initiated a new era of both the historical and critical study of sexuality. Critical sexuality studies blossomed, across a range of disciplines and theoretical frameworks. In the 1970s and 1980s, feminists began to articulate theories of sexuality, tied to questions of the oppression of women. Catharine MacKinnon developed her theory of sexuality as a key site of women's oppression. In her highly influential articles, she argued that women are constituted as subordinate sexual objects in and through sexuality. By extension, she critiqued various sexual practices, from pornography to rape as the mechanisms by which sexuality worked. MacKinnon's work would cast a long shadow not only over feminist understandings of sexuality, but also over law.

But, other feminists and critical theorists during this time would develop very different approaches to sexuality. The period witnessed a deep clash of perspectives on the question of women and sexuality known as the 'sex wars'. On one side, radical feminists pursued an anti-pornography politics, that directly linked the sexuality of pornography to pervasive violence against the women. On the other side were a group of more sex-positive feminists who were worried about the sexual prudism of the anti-pornography critique, and its failure to offer a more positive vision of sexuality as a source of pleasure in women's lives. During this time, Gayle Rubin would write a highly influential article entitled 'Thinking Sex' which introduced the idea of a 'charmed circle' of sexuality. Inside the circle is normatively acceptable sexuality – marital, monogamous, heterosexual, private, reproductive. Outside the circle was the unacceptable or 'bad' sex – non-marital, in groups or alone, in public, homosexual. Rubin argued this hierarchy of sexual behaviours and acts is reinforced by a series of dominant ideologies, like sexual negativity, wherein sex is seen as a dangerous and destructive force.

There began to be a proliferation of gay and lesbian sexuality studies during this period. Following on the emergence of the gay and lesbian liberational movement, often marked as beginning with the Stonewall riots in New York City in 1969, the gay and lesbian sexuality activists began to tell previously un-told historical, political, and socio-logical narratives on homosexuality. By the 1990s, a break emerged between these gay and lesbian studies of sexuality and what would emerge as queer theory. Writers like Judith Butler and Eve Sedgewick argued for an approach to sexuality that was neither tied to gender nor gay and lesbian identities. Instead, they advocated for a non-essentialist approach to sexuality that broke down the binaries between homosexuality and heterosexuality, and explored the multiple ways in which sexualities were constituted.

Sexuality in law

While sometimes considered to be a deeply private dimension of intimate life, Western law has long been deeply implicated in the regulation of sexuality. Gayle Rubin's work gestured towards the many laws that have regulated sexuality, promoting and reinforcing dominant norms of sexuality – heterosexual, marital, monogamous. The history of the legal regulation of sex in the West was one that channelled sex into marriage, by effectively prohibiting any and all sex outside of marriage. Despite the liberalization of Western laws with the move to separate law and morality in the latter half of the 20th century, law continues to regulate sexuality in significant ways. Legal regulation of sexuality explores the broad range of laws that impact on issues of sex, sexual behaviours, and sexual identities. While criminal law has featured prominently, socio-legal studies of sexuality also explore the way other types of law have also impacted the legal regulation of sexuality.

One issue often associated with law and sexuality are the laws regulating gay men and lesbians such as the criminalization of homosexuality, the rise of gay and lesbian rights, and same-sex marriage. The laws impacting sexual orientation or sexual preference have a long history. British criminal law long criminalized sodomy, the similar provisions were adopted in the criminal laws of the British colonies. The first Canadian Criminal Code of 1872 criminalized sodomy. In the late 19th century, the offence of gross indecency was added to capture a broader range of sexual activities, since it was often difficult to establish whether the act of sodomy had actually taken place. These laws stayed in force until the 1960s, with the rise of both early gay rights organizations and a liberalizing shift in attitudes towards the criminal law. Homosexuality was partially decriminalized in England in 1967 and in Canada in 1969, although it remained criminalized in many parts of the world. In the United States, sodomy laws were found unconstitutional in 2003. Following decriminalization, gay and lesbian rights advocates moved towards securing human rights protections preventing discrimination on the basis of sexual orientation. This was a shift away from the focus on criminal law to a consideration of how civil law, particularly, human rights and constitutional law, might be used to promote gay and lesbian rights. In the jurisdictions where these rights were secured, advocates turned their attention to relationship recognition in same-sex marriage. Marriage was long defined in exclusively heterosexual terms.

While LGBT issues are an important subset within the legal regulation of sexuality, they do not capture the full range of laws impacting sexuality. Law regulates sexual practices in many ways. Modern criminal law has prohibited many sexual practices and activities. Criminal law has sought to regulate and prohibit non-consensual sex, with rape and sexual assault laws. Rape laws have typically been directed at prohibiting non-consensual sexual intercourse, while sexual assault laws are defined more generally as the intentional sexual touching of another person, without that person's consent. While non-consensual sex is criminally prohibited in Canada and other countries, there is much variability in the definitions of consent, and in the application of the law. Rape and sexual assault trials often revolve around the question of consent, and many feminist critics have argued that rape myths have been brought to bear on this law. According to these rape myths, women with sexual histories are either thought to be more willing to consent and/or subsequently lie about the absence of consent. The criminal law of rape and sexual assault remains a highly contested area of law, all the more so in light of the #MeToo movement.

But modern criminal law has regulated many sexual practices and activities beyond those that are non-consensual. For example, criminal law regulates and prohibits elements of sex work and prostitution. Some jurisdictions have explicitly prohibited the sale of sex for money, while others have prohibited many of the activities surrounding prostitution, such as communication for the purposes of prostitution and/or running an establishment in which prostitution occurs. While some jurisdictions have decriminalized prostitution, more recently, there has also been a turn to the so-called Nordic model, which criminalizes the purchase but not the sale of sex. According to its advocates, it is intended to reduce the demand for sexual services, but not punish the sex workers themselves. There has also been a significant international campaign against sex trafficking. Although the language of trafficking suggests non-consensual sex work, the campaign tends to conflate trafficking and sex work in general. The criminal regulation of sex work and prostitution, including the Nordic model, is highly controversial, and rejected by most sex worker rights advocates.

Modern criminal law has also criminalized the expression of sex and sexuality through obscenity and indecency laws. Criminal laws in Canada, the UK, and other countries have sought to prohibit obscene materials, although the standards for defining obscenity have been vague. Initially the law was used to prohibit any sexually suggestive or explicit materials on the basis of their ability to corrupt morals. While the definition of obscenity continued to be vague, these laws are no longer enforced as broadly, as standards of acceptable sexual representations have liberalized over time. In contrast, indecency laws were directed to live acts or performances, from theatrical performances to sex in public spaces. While they too were initially directed to any sexually suggestive or explicit sexual performance or acts, these laws have also been liberalized over time. In Canada, for example, the courts now require proof that the alleged indecency causes harm. But, the law is still used to prohibit sex in public spaces. In the aftermath of the HIV/AIDS epidemic, Canada and other countries moved to criminally regulate HIV transmission. Either through laws directly targeting the intentional or reckless transmission of HIV, or existing sexual assault laws, the criminal law has been brought in to punish individual's for non-disclosure of their HIV status. It too is a highly contentious area of criminal regulation, with studies suggested that these laws increase the stigma of HIV status, without any demonstrable benefits to public health.

Together, the web of criminal laws continues to draw lines between sex that is legitimate and sex that is not, in ways reminiscent of Rubin's charmed circle. But, sexuality is not only regulated through the criminal law. Marriage and divorce law have played a role in channelling sex into marriage and promoting marital monogamy. Marriage was traditional reserved for monogamous, heterosexual couples, and only sex within marriage was legal. Adultery laws helped to police the monogamous nature of marriage. Long after adultery decriminalized, modern family law in Canada and other countries has long allowed individuals to obtain a divorce on the basis of their spouse's adultery. In more recent years, family law has moved towards a recognition of more diverse family forms. Some have moved to a greater recognition of family forms outside of marriage. Others have recognized same-sex couples. Yet, monogamy continues to be a privileged family form, and non-monogamous relations are not recognized. Marriage continues to be defined as the union of two persons 'to the exclusion of all others'. In many countries, couples – same and opposite sex alike – can have sexual relationships outside of marriage. But, they cannot have more than one conjugal, marital relationship. Polygamous and polyamorous relationships are not recognized, and in some cases, criminalized.

The recognition of adult relationships outside of marriage cuts across a broad range of laws, where relationships are recognized by the state for the purposes of imposing benefits and obligations – from taxation law to immigration law. In order to be legal recognized, a couple must be marriage-like, and legal definitions relying on concepts of conjugality still tend to privilege a particular kind of relationships, namely, those that involve a sanctioned sexual relationship. While a few jurisdictions include non-conjugal adult relationships, most continue to focus on the existence of a sexual relationship (alongside other indicia such economic interdependence and emotional intimacy) to determine whether the couple will be legally recognized.

Further readings

Foucault, Michel. 1980. *The History of Sexuality, Volume 1: An Introduction*. New York: Vintage Books.

MacKinnon, Catharine. 1983. 'Feminism, Marxism, Method, and the State: Toward Feminist Jurisprudence'. *Signs* 8 (4): 635–658. https://doi.org/10.1086/494000

MacKinnon, Catharine. 1982. 'Feminism, Marxism, Method, and the State: An Agenda for Theory'. *Signs* 7(3): 515–544. https://doi.org/10.1086/493898

Rubin, Gayle. 1992. 'Thinking Sex; Thinking sex: Notes for a radical theory of the politics of sexuality'. Vance, Carole S. Ed. *Pleasure and Danger: Exploring Female Sexuality*. London: Pandora. pp. 267–293.

45
SOVEREIGNTY

Shaun McVeigh

Like many terms that are used in political, legal, social, and cultural thought, sovereignty – a quality of being sovereign and an attribute of rulership – is more often viewed today as a cluster of contested ideas and disputes than as a single concept. The academic disciplines that study sovereignty, like political and legal theory and social and legal studies, do so with differing concerns and methods.

As a contemporary concern, the understanding of sovereignty is understood here as form of authority and a practice of deliberation, rulership, and government. For many living in territorial states, sovereignty is an idea and ideal that encapsulates what is most important about such forms of association and government. It has been a central concern of the European and then International project of state-building and maintenance that started in the 16th century ('states system'). The assertion of sovereignty has been addressed through the political and languages of foundation and action: revolution, liberation, co-existence, and self-determination. Others have contested the form and claims of sovereignty altogether.

1.

The language of sovereignty first came to prominence in 14th century Europe. In Medieval political and legal thought, one central concern was to ensure a proper relation between Divine authority and the worldly power of the Prince. Within this hierarchy, the authority of the sovereign, their duties, and the practice of rulership were judged as a part of a divinely ordered world. In the 15th century, sovereignty became associated with the power of a supreme ruler or Prince over a territorially defined city or nation.

In the complex entwining of spiritual and temporal authority of the Catholic church and Holy Roman Empire, sovereignty was not the only language of law and government. Theologians and jurists also considered many other matters of authority, power, and dominion (property). For example, drawing on the understanding of the Catholic church, jurists devoted considerable effort to theorising the corporate form of the sovereign as divided between a continuing legal form of the sovereign – for example, the Crown – and the mortal transient body of a living sovereign. They also developed sophisticated accounts of jurisdiction as the authority to declare the law and to judge and act in the name of the law. Sovereignty is, in many respects, enmeshed with the workings of jurisdiction.

Since the 17th century, sovereignty has become associated with the formation of territorial states. In this period, often dated from the ending of the 'Wars of Religion' and the political treaties of Westphalia in 1648, the central concern became the establishment of a political order free from higher religious authority. As a concern of government, sovereignty became associated with the exercise of supreme political authority over a population within a geographical territory. It also became associated with the creation and maintenance of forms of secular and social relations. Since the end of the 19th century, the sovereign territorial state has become the dominant international form of political organisation.

The elements of modern state-based territorial sovereignty can be summarised briefly. First, sovereign authority was viewed as a unified source of authority free from superior external authority. Second, that sovereignty and government were public and political. It was concerned with the arrangement of the conditions of life lived in association and conducted through public institutions such as the executive, the legislature, and the judiciary (as opposed, say, to private corporations). Third, as an expression of the public authority of the state, sovereignty was exercised by and conducted through public office rather than familial or personal authority. Finally, sovereignty expressed a public relation between ruler and ruled, the sovereign, and citizen or subject.

The formulation of sovereignty in terms of public law and political power may seem too law-focused. However, it sheds light on disputes about the authority of the sovereign territorial state. In one tradition, drawn from Jean Bodin (1530–1596) and Thomas Hobbes (1588–1679), sovereignty has been understood as the exercise of a supreme authority or representative office subject to no other. For Hobbes, this Supreme Authority was established by covenant between the isolated humans of pre-political society who (isolated humans who agreed...) agreed to be governed through the transfer of their rights. Sovereign Will in this account was expressed as the command of the ruler. Freedom or liberty was understood as what lies beyond and is not subject to sovereign command. In other versions, sovereignty was expressed in terms of the will of the people, the collective. For example, for Jean-Jacques Rousseau (1712–1788), it is the constituent power of the people as a unit, the General Will, that is sovereign.

In both Hobbesian and Rousseauian versions of the social contract, sovereignty emerged from the foundation pact, or social contract, binding people together. It was expressed or constituted in the legal form of right – law in general, 'droit' in French. As a political language, popular sovereignty helped shape the great constitutional revolutions of rights undertaken since the 18th century in, for example, France. Such accounts of sovereignty were given institutional form in many 19th and 20th century state constitutions.

Along with concern for the internal authority, relations between sovereign territorial states have also been addressed by foregrounding sovereignty. From the 17th century relations between states, largely established through war, trade, and diplomacy, were formalised as a body of international law in the 19th and 20th centuries. It established the sovereign territorial state as the central actor in an international domain. Since the establishment of the United Nations in 1948, the consolidation of the international system of states and the development of the supra-national governance of world peace, population and trade, has deepened such projects by establishing the states system against older forms of (European) empire.

The grand claim made for sovereignty, then, is that its formulation has helped address the conceptual foundations of modern political thought, the jurisdictional form of its institutions, and the scope of territorial government.

2.

In the first section, sovereignty was considered as giving shape to forms of political and legal authority. This section turns to the ways in which sovereignty has been understood in relation to government. For political and legal theorists, sovereignty has both constituent and constituted elements that are in uneasy relation. For many, the importance of political thought lies with the maintenance of deliberative authority and rulership. However, set the concerns of sovereignty and government too far apart and sovereignty loses its effective relation to action and conduct. Set the relation too closely and sovereignty cannot represent, legitimate, or shape government. Socio-legal scholarship has helped make visible the ways in which sovereignty has been understood from the standpoint of the government of populations and the conduct of social life.

From the perspective of government and administration, sovereignty is most readily considered in terms of relations of power and knowledge. Where sovereigns have ruled territorial states through the exercise of a monopoly of coercive power, sovereignty has often been considered a matter of will and command. However, government in complex societies and economies, is also typically conducted through techniques of police, administration, welfare (social security), and forms of discipline that shape institutional and individual conduct and conscience. In such circumstances, sovereignty is expressed in plural legal forms and governmental practices. Law and Society scholarship has also emphasised the diverse institutional arrangement in which government conducted including both state and non-state entities such as corporations, hospitals, and schools. Similarly, the economic, social, statistical, and ascriptive knowledge through which government is practiced – all now processed and interpreted through big data – has shaped the different forms of life of citizens. As the objects, scale, and mode and manner of government varies, so do the expressions of sovereignty.

Two brief examples can illustrate some of the ways in which sovereignty and government interact. In many respects, national security, that is, securing the existence of the state and nation, particularly its territorial integrity, is the pre-eminent task of a sovereign. However, the mixing of national security with the concerns of regulation, order, and social well-being is common given the shared knowledge and means of government. The concerns of sovereignty and security are now raised in the face of a climate emergency, the contested exploitation of natural resources, food scarcity, pandemics, and mass movements of people across borders. Responses to these concerns mix direct sovereign action and national survival with the effective government of the overall health and welfare of the population.

In a slightly different register, the concerns of government can also shape those of sovereignty. The many ways in which the concerns of international and transnational governance have shaped the understanding of sovereignty as relational rather than simply self-constituting (more a matter of 'interdependence' than 'independence'). Examples can be found in supra-national organisations such as the European Union that stress multi-layered co-ordination of government rather than shared or federated sovereignty as a mode of cooperation between nations.

Likewise, the emergence of human rights has had significant effect on the ways in which sovereignty as the representation of the constituent power of the people has come to be viewed. The state signatories of the United Nations *Universal Declaration of Human Rights* (1948) might be taken as accepting mutual obligation in their recognition of a moral measure of government within sovereign territorial states. However, it is also the case that the acknowledgement of human rights has become a part of the constituent founding of the People.

3.

Accounts of sovereignty and the status of the sovereign territorial state have never been without challenge. Here, three contemporary challenges are addressed relating to: the political and legal understanding of sovereignty, the capacity to maintain forms of political sovereignty, the forms of association and government and rival sovereign existences.

The contests of the importance of sovereignty are as varied as the claims for sovereignty. For example, it has been contested that supreme authority is not necessary for shared deliberation; that the modern state is better understood in terms of social administration; and, that insurance and risk, rather than authority, contract, and subordination are a better model for understanding the political bond. Since the 19th century, for example, it has been argued that the state should be understood as an instrument of planning and social co-ordination rather than popular will. In such an account, the authority of the state rests on forms of public service. Contesting sovereignty altogether, philosophers and activists in the Anarchist traditions have imagined political and legal orders without sovereigns. By contrast, Legal Pluralists have emphasised the ways in which there are multiple sites of authority, and even sovereignty, through which relations of law and cooperation are created both within and without the nation.

A second contest of sovereignty relates to the capacity of the sovereign territorial state to exercise authority and govern. In part, this can be understood in terms of the contests between states. However, transformations in computer-based technologies and knowledge and the power of private corporations has meant that it is increasingly difficult for sovereign territorial states to control the movement of capital, labour, and goods. The sovereign territorial state has become a small part of many different networks of transmission and exchange. Whether such transformations are understood as the waning of state sovereignty or the intensification of the elite control depends on how closely sovereignty is tied to popular state sovereignty.

A further formulation of sovereignty can be seen in Indigenous political and legal thought. Many First Peoples establish their authority through a relation to, and custodial knowledge of, the land which gives them their law. Sovereignty, as responsibility for the conduct of self and others, is understood in relation to distinct cosmologies, non-human and human relations, as well as rights and responsibilities. In these accounts, the language of sovereignty bears a complex and critical relation to sovereignty as understood in the International states system. One part of this concerns the contest of recognition of First Nations with the International states system both as a part of struggles for decolonisation and self-determination as well as the assertion of rival forms of International ordering. The more complex engagements of sovereignty elements relate to the understanding of the responsibility for the conduct of lawful relations and the material practice of government, since these concerns are not circumscribed by state or civil institutions. Sovereignty here might be understood as a practice and expression of responsibility and relationship more readily than superiority or will.

As a focus for the exercise of authority and rulership, sovereignty remains a vital point of engagement. Whether the International states system remains central or adequate to those concerns remains contestable.

Further readings

Black, Christine. 2011. *The Land is the Source of the Law*. New York: Routledge.

Brown, Wendy. 2010. *Walled States, Waning Sovereignty*. New York: Zone Books.

Foucault, Michel. 2009. *Security, Territory, Population: Lectures at the Collège de France, 1977-1978*. New York: Picador.

Loughlin, Martin. 2010. *The Foundations of Public Law*. Oxford: Oxford University Press.

Sassen, Saskia. 2006. *Territory Authority Rights: From Medieval to Global Assemblages*. Princeton: Princeton University Press.

46
SPACE AND BELONGING

Sarah Keenan

For many activists and socio-legal scholars, a key challenge is how to plan and think outside of liberalism and its focus on the rights and freedoms of individual subjects, and their formal equality before the law. One way of not only revealing liberalism's failures, but also building a constructive counter-analysis of law and society is to shift the focus away from the legal subject and on to the broader spaces in which that subject exists. Doing so reveals how political issues usually considered as unrelated (for example, private property law and immigration law) are both practically and conceptually connected.

Law focuses on the actions of subjects – the obligations they have to each other, the harm they cause each other, their relationship to the state, their rights of access and movement – while paying relatively little attention to the physical and social spaces in which those actions occur. By paying little attention to the spatial contexts of persons and their acts, law adopts an implicit understanding that space is something pre-existing, inert, and singular in meaning. Legal judgments, executive powers, legislation, and legal commentaries tend to treat space as something to be planned over, built on, cultivated, bought, sold, and/or protected; a blank canvas or platform to be acted upon. This understanding of space also assumed that it is cleanly separate the subjects who occupy and move through it.

Legal geography challenges the implicit legal conception of space as the background to law, and is concerned with revealing the active role of the uneven spaces that law operates through and produces. This approach necessarily adopts an understanding of space as something which is not only socially meaningful, but also active rather than static. Geographer, Doreen Massey, offers a useful conceptualisation of space as 'the simultaneity of stories so far' – the simultaneity of multiple and very different stories of subjects, streets, mountains, communities, and empires; stories which are, importantly, unfinished. These stories are practised, embodied, and relational – to understand space as 'the simultaneity of stories so far' is not to reduce space to narrative, but to capture its multiplicity, its vitality, and its interconnectedness. So, space is not fixed because it is the constantly shifting set of processes that both physically and conceptually constitutes the world, whether that be 'the whole world' or the different worlds that attach to particular individuals, groups, or networks. Space is constantly being (re)produced, in no fixed direction.

Massey's understanding of space challenges not only the idea of space as a static surface upon which law evenly operates, but also the related idea that the legal subject is a whole, discrete individual moving through the world, powered simply by their own agency and essentially disconnected from the space around them. Subjects not only exist in space, but are partly constituted by their relationships with space. For example, subjects can only be owners, citizens, and students in relation to particular spaces.

A particularly significant relation between subjects and spaces is that of belonging, which manifests in many ways, both legally and socially. As legal theorist Davina Cooper has explored, belonging can signify property ownership, membership of a community, a relationship to place, an emotional attachment, and/or a behaviour or identity that 'fits', or is 'at home'. Cooper shows how questions of belonging are relevant not only to ownership of land and resources but also to community membership. Relations of belonging help to structure the identity and status of 'insiders' and 'outsiders', and to thereby determine who can represent the past and present of particular locations. Belonging thus connotes a sense of propriety, of the proper. To belong is to fit smoothly, or without trouble, into either a conceptual category or a material position. It is necessarily a relational term; an object/subject/practice/part that belongs cannot exist in a vacuum, it must belong *to* or *with* something else.

Cooper uses belonging as a way to understand property relations. She argues that property practices involve a number of intersecting dimensions of both ownership and membership, but that belonging is the most important part of property practices. Cooper considers belonging in two ways: firstly, the relationship whereby an object, space, or rights over it belong to a subject ('subject-object'), and secondly, the constitutive relationship of part to whole whereby attributes, qualities, or characteristics belong to a thing or a subject ('part-whole'). Both types of belonging implicate social relations and networks that extend beyond the immediate subject and object of property; property is instead understood as a network of belonging in which the subject is embedded. The first understanding of belonging (subject-object) fits with legal and conventional understandings of property as ownership, which tend to revolve around the subject's right to exclude the world from the subject's object (e.g. the car (object) belongs to the owner (subject), who has the right to exclude the rest of the world from that car).

The second understanding of belonging that Cooper describes (part-whole) is a departure from traditional and legal understandings of property, but resonates strongly with feminist and critical race theories of identity. Critical race theorist, Cheryl Harris, for example, has made the influential argument that whiteness is property. Analyzing centuries of American law since British colonisation, Harris argues that whiteness is an exclusive, protected legal category which confers on those who have it an array of benefits including, historically, protection from being enslaved and colonised. Although slavery and colonialism are no longer facilitated by law (or at least, not in the explicit ways that they were prior to abolition), the argument is that law continues to protect the expectations of the descendants of those who benefited from these processes (i.e. white people). In the US and the UK today, white people descended from those engaged in colonisation and the slave trade are protected from deportation through immigration and nationality laws, and their land and financial wealth are protected by private property laws.

Applying this analysis of property as a network of belonging, whiteness can be understood as property because the property-holder is embedded in certain social relations and networks of belonging. A white person can enjoy the privileges of whiteness because he or she *belongs to* the various social relations and networks that constitute whiteness. As sociologists have shown, those relations and networks are complex and far-reaching. Whiteness, like all identity categories, is socially constructed through historically specific fusions of political, economic, and other forces. And whiteness in turn constructs daily practices and worldviews. That is, whiteness is productive of subjectivities. So, while whiteness can be understood as belonging to the white subject, as Harris argues (whiteness as property in the sense of subject-object belonging), the white subject also belongs to the complex relations and networks that form whiteness (whiteness as property in the sense of part-whole belonging). This analysis suggests that in order to understand the varied social powers of property, both subject-object and part-whole belonging must be considered. Ownership and membership are always intertwined. Having is always linked to being, property to identity. In policy terms, this analysis means that if the normative goal is to challenge the way whiteness (or another identity category) operates as a structure of exploitation and oppression, then it is the relations and networks that form whiteness which must be undermined rather than the narrower project of giving rights to the individual subjects who belong to them (e.g. liberal measures such as anti-discrimination law and hate crime law).

Drawing on critical geography, we can analyse the spatiality of embedded relations of belonging. If we accept, as Massey argues, that space is an active agent of material life rather than its backdrop, then the space in which those relations of belonging are embedded must also be a constitutive part of what makes them into property. In order to constitute property, the set of networked relations that Cooper describes must form a space that holds up relations of belonging: conceptually, socially, and physically supporting them. Relations of belonging are held up when the wider social processes, structures, and networks that constitute space give force to those relations. By this, I mean that they are recognised, accepted, and supported in ways that have a range of effects and consequences. In policy and strategy terms, this means that there must be a multiplicity of tactics and demands, going beyond legal recognition of particular relations of belonging as property. As has been pointed out by many Indigenous scholars such as Glen Coulthard and Irene Watson, recognition fails to acknowledge or address the violence inherent in colonial orders. Instead, recognition brings new relations of belonging into those orders and thereby depoliticises them. Because the concept of 'holding up' is about the multiplicity of processes, structures and networks that constitute space, it is directly concerned with these orders.

In the aftermath of the Brexit and Trump victories, understanding these electoral victories through the framework of space and belonging can provide useful insights. In trying to piece together why people voted for Brexit and Trump, many on the left were quick to blame 'identity politics' for taking attention away from material inequalities and thus leaving behind the 'white working class'. This analysis relies on a false opposition between questions of identity, which is seen as trivial, self-interested, and individualistic, and those of materiality, which is seen as class-based and transformative. If we understand identity not as a fixed end product of seemingly unbreakable social structures, but rather as a relation of belonging between a part and a whole which occurs not in the abstract but rather in relation to particular spaces, then identity is necessarily material. 'Identity', and whiteness in particular, must not be seen as a distraction from issues of material inequality, but rather as an essential aspect of those issues. To belong to national spaces is a material relation, one which involves questions of physical safety from both street harassment and state deportation, and resource allocation in terms of access to a living wage and essential services. Analyses of Trump and Brexit which see these electoral victories as being about two groups of individuals: the 'haves' and the 'have nots', are not useful unless they acknowledge that whiteness is something to be had – a relation of belonging that functions as property – in the U.S. and in Britain. The appeal of Trump and Brexit was not only about the material promises each campaign made to improve imbalance in the distribution of wealth, but also the promises made to reshape national space such that white men in particular would feel a sense of unbridled ownership and membership reminiscent of eras when both nations were more explicitly white supremacist. Both campaigns used language that made it clear who would belong in the national space and who would not.

The Trump and Brexit victories did not occur because white people are 'bad' or uniformly racist in mindset, but because they are embedded in spaces that have been shaped by the white supremacist practices of colonialism and slavery, which are still today the material and social basis of the U.S. and Britain (and Australia and Canada and elsewhere in a world shaped by European colonialism). There has never been a formal redistribution of property by way of reparations from that which was violently taken through these practices – instead it has been held up as the property of those who have inherited or bought it from those who acquired it through slavery and colonialism. Both the Trump and Brexit campaigns were able to succeed because they occurred in spaces that, despite gains in formal equality terms (e.g. anti-discrimination laws) and a level of cultural shift, continue to hold up exclusive relations of belonging between white people (men in particular) and the national space.

Focusing on space and belonging requires us to broaden the focus of analysis away from the individual subject and onto the broader networks of relations in which the subject exists. Rather than understand law as operating on individual subjects with space in the background, as liberal discourse requires, law becomes one of the many active forces which constitute any particular space, and the subject becomes a networked figure, constituted by material and social relations. Social and legal questions become less straightforward, as shifting spaces and relations is more difficult than disciplining individuals, but this complexity matches that of the world, and makes for rich research.

Further readings

Cooper, Davina. 1998. *Governing Out of Order: Space, Law and the Politics of Belonging*. London: Rivers Oram Press.

Keenan, Sarah. 2015. *Subversive Property: Law and the Production of Spaces of Belonging*. London: Routledge.

Massey, Doreen. 2006. *For Space*. London: Sage.

47

SUPPLY CHAINS AND LOGISTICS

Galit A. Sarfaty

Recent tragedies – including the collapse of a garment factory building in Bangladesh killing more than 1,000 people and a string of suicides at an Apple supplier's manufacturing plant in China – have highlighted the risks of outsourcing production to suppliers with poor working conditions. These abuses range from corporate complicity in labour rights violations to indirect support of corrupt or oppressive regimes. As multinational companies are becoming more dependent on global outsourcing, there is a need for regulation to ensure responsible supply chains. Governments and the public are now demanding more information on the origins of a company's products. For instance, are such practices as human trafficking, forced labour, or child labour used in the production of a consumer good? Calls for supply chain transparency are part of a larger movement to hold corporations liable for adverse human rights practices that occur in their overseas operations.

Human rights abuses have occurred within global supply chains, in part, because a significant governance gap exists in the reach of domestic and international law with respect to the regulation of corporate activity abroad. The governance gap is particularly pronounced in conflict-affected areas, where host states lack the political capacity, rule of law, and/or will to enforce human rights norms and provide redress to victims of human rights violations. Host states are concerned primarily with attracting foreign investment, which may mean turning a blind eye to domestic law violations or abstaining from passing human rights regulations that could (in what has been called the "race to the bottom") prompt foreign companies to shift their business elsewhere to a jurisdiction with fewer regulatory burdens. States in so-called 'weak governance zones' are usually plagued by corruption and may be unable or unwilling to prevent or stop human rights violations within their borders (and may sometimes be implicated in those violations themselves). This governance gap leaves companies not legally accountable for human rights violations that occur extraterritorially. Closing this gap is a critical challenge.

The nature of global supply chains shapes the ways in which they operate as relational infrastructures, or networks that facilitate the flow of goods, people, or ideas and allow for their exchange over space. Global supply chains frequently include multiple layers of suppliers that may be difficult to trace and therefore regulate. In addition, power asymmetries between buyers and suppliers can shape the regulatory effects of supply chains. Furthermore, supply chains can be quite fluid and unpredictable even within the same industry. Their dynamic nature not only makes it challenging for a company to maintain an updated list of suppliers and sub-suppliers, but it also complicates regulatory efforts as there may be a temptation for firms to shift their supply chains away from the regulated regions.

Supply chains as relational infrastructures

By analyzing global supply chains through the lens of infrastructures, we can better understand their regulatory effects and political implications. Supply chains connect disparate places, objects, people, and ideas, and enable commodities to be manufactured across 'logistics space' (rather than a singular space). As anthropologist Brian Larkin observes, 'Their peculiar ontology lies in the facts that they are things and also the relation between things'. In this way, supply chains serve as relational infrastructures and operate as systems. Moreover, logistics is more than just a story about distribution; it is an assemblage that recasts geographies of law and violence and transforms practices of citizenship. It is therefore profoundly political despite its claims to pure technicality.

Supply chains include multiple layers of suppliers and feature power asymmetries between buyers and suppliers, which can shape the regulatory effects of supply chains. For example, when a company has a high degree of control over its direct suppliers and the power to switch among suppliers, it can more easily monitor and influence their behaviour. In such instances, lead firms are better able to drive coordination, enforce agreements, transmit environmental and human rights norms, and conduct due diligence along their supply chains. Alternatively, if lead firms face high switching costs with respect to their suppliers, lack the purchasing power to hold suppliers captive, and/or cannot transmit tailored incentives that motivate cooperation, they have less influence and more difficulty exercising due diligence.

As with other types of infrastructures, supply chains exhibit a tension between aspiration and failure. Supply chains promise modernity, progress, and profitability through a rational, depoliticized network of outsourcing production. There is a sense of possibility through enhanced mobility of labour and greater efficiency via subcontracting – thus creating what anthropologist Anna Tsing calls 'supply chain capitalism'. Yet despite hopes of global integration and opportunities for economic and social development, supply chains have an underbelly that is lined with relations of power, inequality, exploitation, and expropriation. Human rights abuses abound, including poor working conditions, child labour, forced labour, and displacement of local communities. Victims come from the most vulnerable groups, such as women workers, migrant workers, children, and residents of rural or poor urban areas. They often lack access to whistle blower protections, complaints mechanisms, or legal recourse.

Central to supply chains and other infrastructures are the role of experts and the use of technologies. Experts apply practices of accounting, management techniques, and financial instruments to maintain webs of relational contracts between suppliers and to mediate exchange among people and objects across vast distances. There is a datafication of supply chains, as auditors rely on data analytics, performance indicators, and other statistical tools to conduct supply chain mapping and risk assessments. While such techniques imbue a technocratic rationality into decision-making and hide behind a guise of neutrality and objectivity, they are in fact embedded in politics. They display governmentality by serving as a technology of power that constitutes populations and makes individuals calculable and therefore governable.

While corporations attempt to exert control over actors within supply chains through technologies, management tools, and other administrative techniques, there is a process of doubling that occurs within supply chains (as with other infrastructures). Doubling is when systems and practices operate in variance with their purported objective. In addition to the tension between aspiration and failure described above, supply chains (as well as the laws that purport to regulate them) also exhibit a tension between control and detachment. Corporations attempt to exert control over certain issues (e.g., prices, marketing, logistics) but relinquish responsibility over other issues (e.g., labour arrangements, environmental/human rights practices). While contractual commitments, labelling practices, and auditing procedures may be standardized, labour practices are anything but. When it comes to compliance with human rights standards, corporations fall back on a 'fiction of contractor independence' despite their control over other features of the supply chain. Below we will see the process of doubling in the regulation of supply chains, as laws that purport to control actors and promote transparency are actually fostering detachment through a chain of outsourcing.

Supply chain legal pluralism

There is a layering and intersection of multiple legal orders – which create a supply chain legal pluralism – that coordinate and compete to promote human rights accountability within supply chains. While voluntary codes of conduct, third-party certification systems, and international soft law (non-binding rules that have normative force) operate in this field, there remains a significant governance gap in their reach with respect to the regulation of corporate activity abroad. International standards and self-regulation have generally failed to shape corporate behaviour because they lack independent monitoring and enforcement mechanisms – thereby 'lacking teeth'. Yet, recent domestic legislation is emerging as an alternative method for regulating the extraterritorial human rights abuses of corporations and for achieving corporate accountability. Governments are now demanding more information on the origins of a company's products.

An ever-growing body of domestic regulations reflects the growing trend towards supply chain transparency. Supply chain-related laws represent a shift in international regulation from the prevailing model of 'transnational new governance', which is dependent on voluntary standards adopted by private actors and international institutions, towards the use of domestic law. This shift is particularly notable with regard to the field of business and human rights, which has traditionally been driven by international soft law that is non-binding. Recently passed and proposed regulations in a variety of jurisdictions require companies to publicly disclose information on their global supply chains, including due diligence measures that they have undertaken to prevent human rights violations by third-party suppliers.

While a growing body of domestic laws address social, environmental, and human rights risks in supply chains, the complexity of supply chains complicates their implementation. The recent regulations require companies to conduct (or at least report on) human rights due diligence within their supply chains. Due diligence typically involves a company identifying actual or potential risks associated with its activities and relationships, and taking steps to mitigate those risks. While firms already conduct this process on various business activities, applying it to supply chains presents unique challenges.

Global supply chains frequently include multiple layers of suppliers, which may be difficult to trace and therefore regulate. Since companies often rely on first-tier suppliers to identify and audit those in the second-tier, who in turn identify and audit the next tier and so on, comprehensive monitoring by the company may not be possible. Usually, companies can locate first-tier suppliers, but those suppliers in the lower tiers are not so visible. Chains can comprise many tiers – when tracing the source of conflict minerals, for example, there are often seven or eight layers in the supply chain between the original artisanal mine and the final packaged good in the consumer sector. With so many layers, it can be difficult to identify all suppliers, let alone conduct due diligence on them.

In response to complex, multi-tiered supply chains, we see complex, multi-tiered regulation. Because of the nature of supply chains, regulation of outsourcing is being outsourced to the regulated parties themselves, who must secure compliance by their suppliers. This model of outsourcing stands in contrast to examples in existing literature on the privatization of public law and the development of 'public-private governance'. Scholars have focused on the outsourcing of regulation to non-governmental organizations, auditors, and other third parties that take on standard-setting, implementation, and enforcement functions traditionally occupied by regulatory agencies. Yet, supply chain regulation exemplifies a distinct model of outsourcing that dilutes the host state's governing authority.

While outsourcing regulation to companies is a means by which states can indirectly regulate firms in other countries, this practice raises accountability concerns when private actors are performing functions that are fundamentally public. Like other forms of public-private governance, there are risks to public values when private actors perform functions conventionally reserved for government agencies. Supply chain laws pose a greater risk to accountability given that the regulated entities themselves are sometimes ill equipped to monitor and enforce compliance by third parties. Companies may then outsource compliance to private parties such as consulting firms or suppliers who audit the tiers of suppliers below them. Third-party auditors frequently conduct human rights due diligence and implement supply chain policies on behalf of companies. They utilize legal technologies such as human rights risk indicators and social auditing to shape the behaviour of large firms and their suppliers. Third-party auditors are thus serving as translators as they interpret and implement supply chain policies on behalf of companies, which are in effect outsourcing their compliance obligations.

A chain of outsourcing has developed, one involving layers of monitoring and enforcement, and often competing systems of incentives. This chain of outsourcing regulation presents a number of implications for companies and the state. It creates a new role for companies as not just regulated entities but regulators themselves. Companies are setting standards for their suppliers and developing procedures for suspending or terminating firms that do not comply with their sourcing policies. While corporate codes of conduct have been on the rise for the past decade, supply chain regulations raise the bar for the level of due diligence that companies require in their codes. Assigning private parties the role of implementing regulation is concerning given the lack of government oversight and transparency to the public.

Scholars of public-private governance have recognized that with increased efficiency and innovation come costs such as lack of transparency, limited oversight, and weak accountability. The accountability costs are particularly prevalent in supply chain regulation, which becomes further removed from the state as it gets outsourced first to companies and then to other suppliers and private consultants, thus creating second-order regulation, third-order regulation, and so on. Supply chain management could therefore be analyzed as an 'ethic of detachment', in the words of anthropologist Jamie Cross. As part of this ethic of detachment, there are 'multiple nodes of authority' so responsibility is redistributed across a network of actors. The limited accountability along supply chains is especially disconcerting given that the purpose of these regulations is to enhance corporate accountability to the public and hold companies responsible for human rights violations in which they are complicit, whether wittingly or unwittingly.

This chain of outsourcing regulation presents a number of further implications for companies and the state. It requires that firms locate and regulate their suppliers, a task that they may currently lack the resources and expertise to perform effectively. Furthermore, when numerous layers of monitoring and enforcement are present, competing systems of incentives may develop. In addition, companies are setting increasingly rigorous standards for their suppliers and developing procedures for suspending or terminating firms that do not comply with their sourcing policies. While corporate codes of conduct have been on the rise for the past decade, supply chain regulations raise the bar for the level of due diligence that companies require in their codes.

Scholars are increasingly studying the accountability concerns that accompany regulatory outsourcing in this context as they address such questions as: Are multinational corporations properly equipped to monitor and enforce compliance by third parties? To what extent are multinational corporations simply 're-outsourcing' to auditing and consulting firms to carry out this task, and what are the full implications of such a practice? Given the complex and nuanced link between global supply chains and human rights, it is important to apply a multidisciplinary approach that borrows insights from a variety of fields and generates knowledge of interest to governments, business leaders, and the broader public.

Further readings

Cowan, Deborah. 2014. *The Deadly Life of Logistics: Mapping Violence in Global Trade.* Minneapolis: University of Minnesota Press.

Larkin, Brian. 2013. 'The Politics and Poetics of Infrastructure'. *Annual Review of Anthropology* 42:327–343. https://doi.org/10.1146/annurev-anthro-092412-155522

LeBaron, Genevieve and Lister, Jane. 2015. 'Benchmarking Global Supply Chain—The Power of the Ethical Audit Regime'. *Review of International Studies* 41:905–924. https://doi.org/10.1017/S0260210515000388

Tsing, Anna. 2009. 'Supply Chains and the Human Condition'. *Rethinking Marxism: A Journal of Economics, Culture & Society* 21:148–176. https://doi.org/10.1080/08935690902743088

48
TERRITORY AND LAW

Nicholas Blomley

A territory is a bounded, contiguous section of the Earth's surface that is claimed or controlled by a particular group, nation, or state in order to restrict or control spatial access for specific periods. It is material, as it entails land or ocean. It is functional, organized, and controlled for particular ends. It is very often symbolic, linking the territory to some form of identity. Territory is clearly geographic, entailing the marking out of bounded spaces. It is also temporal, operating at specific times, for example. Territoriality is the strategic use of territory to achieve certain ends.

Territory differs from many other forms of spatial organization. It is not simply a process of drawing a boundary. The boundary must seek to control others – usually people – for it to be a territory. In this sense, territory entails power. A territory organizes power relations through a set of spatial markers – particularly the boundary. The boundary marks a unit of space, and designates an inside and an outside.

Territories can be found in all societies, and at all scales, although they take different forms. Some territories are permanent (for example, property territories), while others are temporary (for example, the way passengers organize their personal territories on a bus). Some are formal (for example, state territories, which require legal documents such as visas and passports to access), while others are informal (for example, a family picnic at the beach). Some territories aspire to universal significance, affecting everyone (like state territories), while others work only for particular groups (like gang territories in the inner city). Territorial arrangements are present in the global North, and in the South. Some territories, like the US/Mexico border, seek to regulate flows between North and South to the advantage of the former.

Arranging relations between people by the use of territory can appear very straightforward. It is sometimes assumed that laying claim to territory is a natural process, like that of animals. Territorial arrangements, like those claimed by nation-states, or the desire to defend personal property, are thus assumed to be natural. However, such a view is reductionist, limiting a wide array of experiences and relationships associated with territory to a narrow set of functions. It also naturalizes a set of social and historical arrangements which are relatively recent, like nationalism and its ties to territory.

Human territoriality is more complicated and diverse. While it may have functional characteristics, it also entails communication, meaning, identity and – most importantly – the workings of power. Power may take many forms. It may be negative or positive. It may advance or slow the workings of dominant social interests. These different forms of power can be organized territorially. Space can be organized so as to enforce instrumental or 'power-over' forms of power. A prison, for example, is designed to spatially contain designated people. Associational power, or the power produced through the action of people or institutions pooling their resources to secure certain outcomes, can also be territorialized. Indigenous claims to territory are an example of this.

Territory can also serve to render power impersonal, thus obscuring the hierarchical relations it so often produces, because of the tendency to think of space as natural or given. When a police officer tells a homeless person that they have to move on, because they are in the wrong place, they are relying on a territorial form of control that displaces attention from the power of the police officer to the territory itself. Territories and their boundaries appear to be doing the controlling.

As these examples suggest, territory is central to many forms of legal control. Western law, in particular, has long relied on a set of territorial understandings regarding law. Medieval European lawyers interpreted Roman law to argue that legal jurisdiction should apply to a bounded territory. This is different from claims to property. The legal territory, which marks the spatial extent of legal rule, is not a possession, but the object of political rule. Law applies over the things that take place within it. This marked a crucial shift from the idea of the personality of law, which governed members of a defined community, regardless of their location, to the territoriality of law. From this emerged the tight connections between territory and state sovereignty with which we are familiar. Law takes place in relation to the territorial envelope of the state. Concepts such as security become tied to the defence of the territorial state. Different forms of law within the boundary of the territorial state, such as Indigenous law, become seen as a threat to state sovereignty. Because of its roots in state sovereignty, modern law is therefore deeply territorial.

Law, as a form of power, also often works through a variety of territorial strategies. Some of these are through 'everyday' forms of social practice. While not technically law, they nevertheless rely on legal understandings and commands. When people lay claim to street parking spaces they have cleared of snow, becoming very offended if others ignore the sign or chair they have put in the space, they perform such everyday forms of law.

In so doing, they echo formal legal practices. Territorial forms of law can be found in many contexts, serving multiple legal functions. For example, law often tries to use territorial forms of power at attempts at creating forms of desired behaviour. Modern societies often use spatial strategies of partitioning or enclosing spaces to create forms of social discipline and surveillance. Reserves, or residential schools, set aside for Indigenous people and their children by settler societies, often entailed attempts at regulating spatial movement in an attempt at creating 'civilized' forms of culture, economy, and behaviour. Criminal law in Canada and other countries controls the spatial movement of people released on bail for petty offences through various forms of territorial control. In Canada, these 'red zones' attempt to control behaviour by territorially excluding alleged drug users from areas of drug supply. Such forms of territorial control are temporal, operating only until a person on bail appears before the court.

Legal territories can also be used to expel unwanted people from particular places. Homeless people who camp out on city-owned land for extended periods of time can experience the power of trespass, the act of unlawful entry on another's land. Originally denoting any crime less than a felony, trespass has been tied to land, and framed in sharply exclusive and territorial terms, such that merely crossing a territorial line, without any damage, constitutes an offence. It is routinely used by the state not only to enforce its ownership interest in public property, but also in the delegation of private powers to police in many US cities, which allows them to 'trespass' homeless people not just from private property, but from libraries, recreation centres, public transportation, university campuses, hospitals, and social service agencies. Although public property is technically open to all, the combined effect of these forms of territorial control can be that there are very few spaces that homeless people can effectively be present.

Private property is also highly territorialized. The legal rationale here is that the benefits of private property can only be realized if unwanted people are excluded from using or benefitting from a parcel of land. Law provides a series of powers to private owners to defend their land against others. In some extreme cases, owners are permitted to use lethal force against trespassers. More conventionally, owners can call upon the state to defend their land, through agencies such as the courts, land-titling offices, and the police. Such exclusive claims to pieces of land are sometimes treated as natural expressions of human behaviour. However, like territory and sovereignty, the connection is a recent one, at least within Western societies. Feudal property, for example, was essentially a set of relations between people that allowed them conditional rights to access and use land. Even the land of elites was often used by poorer people. Different territorial arrangements, in other words, were at work. It was only in the early modern era that the law began to identify a single owner, who was assumed to have exclusive rights to land. Other changes, especially the development of modern map-making, played a crucial role in re-defining the territorial understanding of land. Even now, despite dominant myths regarding property and land, territorial parcels of private property are often legally defined as open to others. It is a well-established principle of law, for example, that owners who serve the public, through establishments such as malls or restaurants, have an obligation to allow access to their property without inappropriate discrimination.

Legal punishment or sanctions are also often territorial. The institution of the prison is, of course, designed to territorially regulate people, containing offenders within bounded spaces, and limiting access to this space to non-offenders, for designated periods of time. The prison relies on architectural layout and technical means, such as cameras, as well as the temporal and spatial regulation and movement of inmates through programming, and prisoner counts. More intensified forms of territorial control within such spaces – notably the use of solitary confinement – individualize territorial regulation with very punitive effects. The rise of mass incarceration in many legal systems means that an increasing number of people – often poor and racialized minorities – are caught up in these forms of territorial control. Even outside such spaces, many inner city neighbourhoods are so intensely policed, often through forms of territorial control, that they have effectively become extensions of prison.

The significance of territory to forms of social power and dominance expressed in law is clearly evident when we look at the resistance to such forms of social control. These often challenge the territorial dimensions of law. Large scale occupations of state or private land by landless peasants in the global South directly challenge prevailing territorial norms, for example. Indigenous people controlled by the reserve system, or criminalized people in the inner city, often find ways of avoiding the territorial controls law places upon them through subterfuge and quiet forms of resistance. The early modern commoners expelled from the land they regarded as available to them continued to break through the enclosure hedges designed to keep them and their cattle away from land that the lord now claimed as his own.

Resistance can also be expressed through creative alternative uses of territory. Indigenous people defending their traditional lands against resource extraction or pipelines often use blockades to stop traffic or work crews, physically demonstrating their territorial claims to sovereignty. Squatters attempting to occupy empty houses not only defy territorial property norms, but also use them to their advantage, claiming that by fixing up abandoned or under-used housing, they have a better right to exclusive use than the absent or negligent formal owner.

These forms of resistance also often challenge the ideological dimensions of territory, in particular, the manner in which territorializing law may have the effect of depersonalizing a set of power relations. Tenant organizers, for example, challenge the apparently impersonal nature of eviction – a highly territorial form of expulsion – through creative strategies designed to reveal the punitive and violent consequences of eviction on real people, and by naming and shaming the landlords who use eviction as a tool for profit extraction.

Territory, therefore, is clearly central to the work of law. To understand how law works, and how it is contested, it is vital to think carefully about territory, and its multiple functions. Territory is not simply an outcome of law. It is a means or resource through which law is communicated, enforced, legitimized, and challenged.

Further readings

Blomley, Nicholas. 2016. 'The territory of property'. *Progress in Human Geography* 40(5): 593–609. https://doi.org/10.1177/0309132515596380

Ford, Richard Thompson. 1999. 'Law's territory (a history of jurisdiction)'. *Michigan Law Review* 97(4): 843–930. https://doi.org/10.2307/1290376

Foucault, Michel. 1977. *Discipline and Punish: The Birth of the Prison*. London: Allen Lane.

Herbert, Steve and Beckett, Katherine. 2008. '"This is home for us": Questioning banishment from the ground up'. *Social and Cultural Geography* 11(3): 231–245. https://doi.org/10.1080/14649361003637661

Sack, Robert D. 1983. 'Human territoriality: A theory'. *Annals of the Association of American Geographers* 73(1): 55–74 https://doi.org/10.1111/j.1467-8306.1983.tb01396.x

49

THE TRANSNATIONAL LAW OF HUMAN TRAFFICKING

Prabha Kotiswaran

'Trafficking' is a term with a complicated genealogy. For most people, it conjures up the figure of the innocent young woman forcefully moved across borders to work in prostitution. Yet, the term found definition in international law only 20 years ago. Of course, the problem of the movement of human beings across borders for economic exploitation is hardly new. As far back as the 19th century, abolitionists like William Wilberforce fought long and hard against the transatlantic slave trade, stigmatising it by calling it 'traffic' rather than trade. Indeed, at the turn of the 20th century, feminists drew on this abolitionist heritage to mobilise against what they called the 'white slave trade' involving young girls and women being coerced into prostitution, leading to a series of international instruments which finally culminated in the 1949 UN Convention for the Suppression of Traffic in Persons and of the Exploitation of the Prostitution of Others. The word 'traffic' was used to refer commercial movements characterised by illegality and/or immorality (as in drug traffic, people trafficking, and liquor traffic). The rise of the automobile in the 1910s and 1920s led United Nations' bodies to investigate ways of minimising auto-related accidents by imposing norms, but these were initially labelled as norms regulating 'circulation', not traffic, possibly because of the negative connotations that the word 'traffic' had acquired.

The conceptual shift from 'traffic' to 'trafficking' happened in the mid-1990s when anti-trafficking advocates were in search of a term to describe the irregular movement of Beninese and Togolese children to Gabon to work as live-in domestic workers. The term 'trafficking of children' was adopted for Article 3(a) of the Worst Forms of Child Labour Convention in 1999 to imply a type of slavery or practice similar to slavery.

Shortly thereafter, in 2000, the UN Protocol to Prevent, Suppress, and Punish Trafficking in Persons, Especially Women and Children (hereafter the UN Protocol) and the Protocol on Migrant Smuggling were both introduced to supplement the UN Convention against Transnational Organised Crime. Trafficking was defined for the first time in the UN Protocol. Since then, trafficking has been used to describe labour exploitation in scenarios ranging from the production of 'slave chocolate' in Cote D'Ivoire and cotton from Uzbekistan to exploitation in war-time captivity in Nigeria, bonded labour in Pakistan, fishing boats in Thailand, Qatari construction sites with Nepali workers, the brick kiln industry in India, and Unilever's supply chain in Vietnam. Invoking the spectre of transatlantic chattel slavery, modern day neo-abolitionists term these new forms of exploitation as 'modern slavery' and claim that there are 40.3 million 'modern slaves' around the world.

Experts who drafted the UN Protocol never imagined that the term trafficking would one day be used to denote exploitative working conditions. Consider the definition of trafficking under Article 3 of the UN Protocol, which requires three criteria to be satisfied for adults: a mode of action (recruitment, transportation, transfer etc.), the means by which it is obtained (threat or use of force or other forms of coercion etc.), and the purpose for which it is obtained (namely exploitation). These were initially meant to require 'movement', that too, across borders. Indeed, the 2000 Trafficking and Migrant Smuggling Protocols were thought to be a response to the large-scale movement of persons in the wake of globalisation, one which facilitated states' border control agendas, in turn explaining their rapid ratification. To illustrate, the UN Protocol was adopted in 2000, came into force in 2003, and has been accepted by 178 countries to date.

The UN Protocol has, however, been poorly institutionalised. According to the UN Office on Drugs and Crime, the guardian of the Protocol, conviction rates are low and the criminal justice system response is stagnating at a low level. There are several reasons for this, including the irreducibly political concepts of coercion and exploitation that are at the heart of the definition of trafficking and its consequent malleability.

Starting in the 1900s, international law has evolved in relation to each of the forms of exploitation listed in Art. 3, resulting in a division of labour amongst UN agencies. The UN Protocol brought these definitional terms together without resolving issues of conceptual overlap and competition between these agencies. Moreover, there are ideological differences on the scope of the **forms of exploitation** listed in the Art. 3 definition as well as the **appropriate regulatory** response to them.

One can identify three phases in the development of anti-trafficking law and policy spanning the past 19 years. These are:

1. a phase between 2000 and 2009, when sex work was the main focus of anti-trafficking law and policy;
2. a phase between 2009 and 2014, when attention to labour trafficking rendered visible the competing frames of 'modern slavery' and 'forced labour'; and
3. from 2014, when legal interventions were framed explicitly in terms of slavery and forced labour.

The existing international sex work debates greatly influenced the negotiations of the UN Protocol, even as governments and civil society actors, including feminists and sex work advocates, adopted conflicting positions over the normative status of sex work. Feminists calling themselves neo-abolitionists claimed that sex work was always coerced, by definition, while sex worker groups insisted that sex workers retain some modicum of agency when they undertake sex work voluntarily. These differences resulted in compromise terms in Art. 3 such as 'the abuse of a position of vulnerability', a means element that is thought to cover voluntary sex workers as well.

Unsurprisingly, as countries began to articulate their anti-trafficking response, they defaulted to amending anti-sex work national criminal laws. Thus the initial years of the UN Protocol were marked by sex work exceptionalism, by which I mean (a) the characterisation of the sale of sex for money as an egregious violation of human dignity and as an exceptionally harmful activity and (b) the overwhelming association of trafficking with trafficking for sex work and with sex work itself. The Bush Administration in the United States in particular, interpreted trafficking as primarily concerned with forced migration for sex work. A robust sex panic developed. This has been extensively documented by feminists, who have argued that much like the 'white slavery' panic at the turn of the 20th century, anti-trafficking regimes seek to allay fears about globalisation through a yearning for a familiar race and gender order wherein women's migration is discouraged.

By 2007, scholars and activists agreed that the implementation of the UN Protocol was over-inclusive in targeting voluntary as well as coerced sex workers, and that it was under-inclusive in addressing the labour exploitation of millions of workers. In 2005, the ILO argued for an expansive understanding of trafficking to include both sexual and labour exploitation. By 2009, under the Obama administration, the U.S. State Department in its annual reports began to focus on labour trafficking, not only sex trafficking and sex work. Soon, critics of the anti-trafficking framework put forward what they called the 'labour paradigm of trafficking' and trade unions and workers' groups came to form networks with conventional anti-trafficking NGOs.

From around 2012, in international circles, trafficking became increasingly reframed in terms of both slavery and forced labour, without so much emphasis on sex work. In 2014, a significant Protocol and Recommendation revised and updated one of the most highly ratified ILO Conventions, Convention No. 29 Against Forced Labour, in recognition of the fact that 90 percent of the forced labour in the world today is exacted by private sector employers. Meanwhile, Kevin Bales, a sociologist, coined a new term 'modern slavery'. The Walk Free Foundation founded by mining magnate and philanthropist Andrew Forrest was influenced by Bales and released the Global Slavery Index in 2013 and subsequently in 2014, 2016, and 2018, ranking countries in terms of their 'modern slavery' problem.

The term 'modern slavery' has found purchase with governments, with the UK and Australia passing anti-modern slavery laws with provisions on supply chain transparency to ensure corporate responsibility for labour abuses committed in supply chains abroad.

The relationship between trafficking on the one hand, and forced labour and modern slavery on the other, is fuzzy and contested. Different organisations claim the primacy of their preferred terms. Thus, a section of the ILO claims that forced labour is the umbrella term of which trafficking is a sub-sect, while the Walk Free Foundation claims that modern slavery is the umbrella term, subsuming both forced labour and trafficking. For the UNODC, trafficking is the umbrella term. These contestations at the international level have been re-played at the domestic and local levels.

Given the controversies, and the disagreements about the very meaning of words, it is unsurprising that there are many, inconsistent, regulatory responses to various phenomena labelled as 'trafficking'. There are at least four major approaches here: the criminal law approach, the human rights approach, the labour approach, and the development approach. The criminal law approach views trafficking as an exceptional aberration in otherwise normal circuits of commerce and exchange, thus warranting the use of the criminal law. The human rights approach seeks to mitigate the harshness of the criminal law regime by bolstering the human rights of victims of trafficking.

The labour approach rejects the criminal law view of trafficking. It understands coercion and exploitation as spanning a continuum of social scenarios, so that the difference between the exploitation of workers and trafficking is a matter of degree not of kind. Advocates of the labour approach call for addressing structural systems of subordination and exploitation in labour sectors, as well as highlighting the problems caused for workers by certain overzealous prosecutions of 'smugglers' or 'traffickers', whereby workers are often deported or otherwise punished.

Finally, the development approach refocuses attention on the role of the state in effecting redistribution and problematises the hegemonic growth-oriented understanding of development in the UN Sustainable Development Goals and its target 8.7, which is devoted to eradicating trafficking, forced labour, and modern slavery. The criminal justice and human rights approaches both view trafficking through an individualist, liberal frame while the labour and development approaches meanwhile consider the root causes and the structures of patriarchy, capitalism, and racism. The criminal law approach, however, remains the most influential in national legal systems around the world to the detriment of structural interventions.

Further, the number of states that criminalised both sex and labour trafficking in domestic laws increased from 10 percent in 2000 to about 73 percent in 2013, which would suggest a reduced focus on the nature of the work performed and a greater focus on the conditions under which it is performed. However, anti-trafficking law and policy have reinforced cultures of sex work exceptionalism. Numerous countries around the world have adopted the so-called Swedish model, which criminalises customers of sex workers in the hopes that this will help reduce the demands for sex work and thereby minimise and ultimately abolish the sex sector alongside trafficking for sex work.

Yet, other states have stepped up the enforcement of anti-sex work laws against voluntary sex workers on the pretext of countering trafficking. As Anne Gallagher, an international lawyer and champion of the UN Protocol notes, 'Trafficking for sexual exploitation receives the lion's share of criminal justice attention and resources in most, if not all, countries'. Hence almost 20 years since the UN Protocol was negotiated, the myth of the innocent girl duped into prostitution against her will continues to colonise our imagination on trafficking whether at airports, shops, and sports events or through primetime TV shows, Hollywood movies, and the mainstream media.

In conclusion, while some claim that the 'trafficking' campaign has firmly placed exploitation on the international political agenda, anti-trafficking laws continue to be misused against sex workers, migrants, migrant brides, and sexual minorities in many countries. The jury is still out on the value of the transnational anti-trafficking legal and political order in addressing extreme exploitation. Even if these laws were to be implemented properly, they focus on a sliver of exploitative work conditions while legitimising vast swathes of globalised working arrangements that are deeply exploitative. Yet, popular media and culture keep promoting the idea that 'trafficking' is a single phenomenon with identifiable 'bad guys' and 'virtuous victims'. This stereotype continues to justify the use of criminal laws against vulnerable people.

Further readings

Chuang, Janie A. 2014. 'Exploitation Creep and the Unmaking of Human Trafficking Law'. *American Journal of International Law* 108(4): 609–649. https://doi.org/10.5305/amerjintelaw.108.4.0609

Global Alliance Against Traffic in Women. Edited by Mike Dottridge, 2007. *Collateral Damage: The Impact of Anti-Trafficking Measures on Human Rights Around the World.* www.gaatw.org/Collateral%20Damage_Final/singlefile_CollateralDamagefinal.pdf (accessed March 17, 2017).

Kotiswaran, Prabha. 2019. 'Trafficking: A Development Approach'. *Current Legal Problems* 27(1): 375–416 https://doi.org/10.1093/clp/cuz012

Shamir, Hila. 2012. 'A Labor Paradigm for Human Trafficking'. *UCLA Law Review* 60: 76–137.

50

WATER DISPUTES ACROSS BORDERS

Tamar Meshel

Freshwater is an essential, irreplaceable, and finite natural resource. Of the total amount of water on Earth, only 2.5% is freshwater and only around 30% of this water is available for human use. Freshwater also has unique hydrological and geographic qualities. The movement of water through a watercourse is but one phase of what is known as the 'hydrologic cycle' – water evaporates from the sea and land into the atmosphere, falls as rain and snow onto the earth, reappears in watercourses, and then drains back into the sea. One consequence of this hydrological cycle is that the amount of water on earth and in its atmosphere is finite and fixed, although its availability may change. Another consequence is that water is in constant motion between several major reservoirs: the atmosphere, oceans, rivers, lakes, soils, glaciers, snowfields, and groundwater. This implies that water cannot be easily subjected to human control.

Key from a socio-legal point of view is the fact that much of the world's freshwater crosses state boundaries. There are 148 states whose territory falls within international river basins and more than 30 states that are located almost entirely within such basins. For example, the Danube and the Nile river basins are shared by 19 and 11 states, respectively. Countries sharing freshwater resources are therefore inter-connected by the flow of the water. The use of water in one part of a transboundary lake or river may affect its use in another part, making states sharing such resources inter-dependent.

The immense diversity in size, nature, purpose, and location of cross-border freshwater resources should also be noted. These resources range from enormous systems such as the Congo, Amazon, and Ganges, to minor streams. Many are located in arid parts of the world, so that they flow on the surface only intermittently, while others are in water surplus areas. Some have been integrated in domestic uses and productive processes for their entire lengths, while others remain untouched. Some pass through developed and wealthy nations, while others flow through resource-poor and developing states. Some may be used principally for drinking and household purposes, others for irrigation, and still others for industrial or hydroelectric production.

In addition to these unique hydrological and geographical qualities of freshwater, it is increasingly becoming a scarce resource for several inter-related reasons: depletion and degradation, industrialisation, and inefficient use. Depletion and degradation result from the main uses of fresh water: agriculture, industrial use, and domestic use, as well as from global warming. The use of freshwater for industrial purposes has also been increasing dramatically, with harmful environmental consequences and conflicts between uses. National water needs have expanded since the 19th century from an almost exclusive focus on navigation to other water uses such as fisheries, irrigation, and hydropower production. As countries industrialise, more freshwater is needed for energy production. The availability of freshwater is further limited by its inefficient use, the results of which are evident, for instance, in the declining water levels of the Aral Sea, Lake Chad, the Colorado River, and the Rio Grande.

These hydrological and geographical characteristics of freshwater as a shared natural resource, coupled with the growing demands of competing water uses and increasing shortage, contribute to the complexity of cross-border freshwater disputes and may hinder their resolution.

Even in the face of the inter-dependence and inter-connectedness of shared freshwater resources crossing political borders, states sharing such resources often behave in a self-interested manner, for instance by unilaterally diverting rivers, constructing dams, or polluting, which may result in violent conflicts.

Most vulnerable to such conflicts are the Middle East, North Africa, and sub-Saharan Africa regions, which are the driest areas in the world and contain a large number of rivers and other fresh water sources that cross borders. Ongoing disputes in these regions concern the Nile River, the Jordan River, the Euphrates and Tigris Rivers, and Lake Malawi. Asia has also seen its fair share of cross-border freshwater disputes, mostly attributed to the lack of agreements governing the use of dozens of rivers shared by countries such as India, Bangladesh, Nepal, and China, as well as ineffective agreements and mistrust among the post-Soviet countries of Central Asia. There are ongoing water-related conflicts around the Indus River, the Ganges River, the Helmand River, the Mekong River, the Syr Darya basin, and the Aral Sea. In South America, socio-economic issues, water quality and management problems, and competing demands for human consumption and for industrial uses or irrigation have all been frequent causes of cross-border disputes. Previous and ongoing disputes in this region concern the Uruguay River, the Silala River, and the Lauca River. Finally, in Europe and North America, water quality and quantity issues have also given rise to cross-border freshwater disputes such as those around Lake Lanoux, the Danube River, and the Meuse River in Europe, and the Colorado, Rio Grande, Columbia, and Red Rivers in North America.

Cross-border disputes over the use of freshwater resources are particularly common in regions where relations among states sharing water resources are historically conflicted and where such resources are scarce. Such disputes are particularly complex: water issues are indivisible from divisive issues relating to economic, political, and national security interests. Ultimately, cross-border freshwater disputes are rooted in an, at times inevitable, clash between the necessity of sharing water and states' pursuit of sovereign interests, whether political, strategic, economic, or cultural, which may hinder the resolution of these disputes and potentially exacerbate them.

The political aspects of cross-border freshwater management and dispute resolution have become so critical that a distinct discipline for the study of 'hydro-politics' has emerged. This field of research draws from many other disciplines, including geography, development studies, sociology, anthropology, political science, and law.

Geographical characteristics of shared freshwater resources contribute to the complex politics of cross-border freshwater dispute resolution. In the case of transboundary rivers, there are two main geographical configurations, namely the through-border, or successive river, and the border-creator, or contiguous river. The former provides for an asymmetric relationship between upstream and downstream states since one country occupies a geographically superior position. The latter creates a relationship based on a geographically symmetric relationship between two states whose common border is situated on a shared river. This physical geography of a river is important to cross-border freshwater dispute resolution since it influences the use and development of the shared resource, as well as states' respective interests.

From a development studies perspective, disputes between states regarding access to freshwater may result from the different stages of economic, technological, and hydrological development of upstream and downstream states sharing rivers. This is illustrated, for instance, in Iraq and Turkey's respective development of the shared Tigris and Euphrates Rivers. Downstream Iraq began such development in the 1950–1970s, while the ongoing irrigation and hydropower dams developed by upstream Turkey only began in the 1990s. A similar pattern can be seen between Israel and its neighbours with regard to their respective development patterns.

However, while downstream states controlled their hydrological developments earlier, they also became dependent on the continued usage of these rivers, and therefore subject to any water development by their upstream neighbours. This raises concerns regarding the potential downstream impacts of such upstream developments, which may lead to conflicts. Similar concerns may arise concerning potential upstream impact from activities undertaken by a downstream state.

Disputes between states sharing freshwater resources can also be examined from a social science perspective, as these disputes often relate to socio-political conceptions of freshwater. For instance, freshwater has been viewed by some countries as freely available, ignoring the need to manage shared resources efficiently. Freshwater is also frequently seen as a resource that should be provided by the government exclusively for its citizens and not for sharing with neighbours. Another recent development is the recognition of local rivers as legal persons, with all accompanying rights and liabilities. The New Zealand Parliament as well as an Indian High Court, for instance, have noted the spiritual significance of these rivers to the local population.

Finally, law has an important role to play in the resolution of cross-border freshwater disputes. International water law's three core principles – no significant harm, equitable and reasonable utilization, and the duty to cooperate – aim to ensure the effective use, development, and protection of international watercourses. This body of law can provide states with well-defined rights and obligations to help resolve cross-border freshwater disputes. It can also assist in the objective resolution of these disputes by facilitating harmonious interactions between states. Indeed, once a cross-border water dispute has arisen and each state believes that its vital interests are at stake, peaceful resolution can become extremely difficult in the absence of objective and uniform legal principles that limit the states' unilateral claims to water flowing through their territory.

In conclusion: there are hundreds of international river basins and transboundary aquifer systems in the world. Hundreds of freshwater-related disputes have occurred in the last century alone, many concerning such cross-border resources. These freshwater disputes tend to involve complex political, social, legal, geographic, and hydrological issues, and if left unresolved, may deteriorate to violent conflict, as has occurred in the case of the Jordan River dispute between Syria and Israel, the Senegal River between Mauritania and Senegal, the Nile River between Egypt and Sudan, and the Euphrates-Tigris River dispute between Turkey and Syria, just to name a few. The fact that freshwater resources are rapidly depleting while human dependency on them continues to grow suggests that disputes between states over such resources are likely to arise with increasing frequency in the future. It is therefore crucial to understand and account for all aspects of these complex disputes if we are to resolve them effectively and peacefully.

Further readings

Benvenisti, Eyal. 2002. *Sharing Trans-boundary Resources – International Law and Optimal Resource Use*. New York: Cambridge University Press.

Boisson de Chazournes, Laurence. 2013. *Fresh Water in International Law*. Oxford: Oxford University Press.

Dinar, Shlomi. 2008. *International Water Treaties: Negotiation and Cooperation Along Transboundary Rivers*. London; New York: Routledge.

Leb, Christina. 2013. *Cooperation in the Law of Transboundary Water Resources*. Cambridge: Cambridge University Press.

51

WATER JUSTICE AND INDIGENOUS PEOPLES

Pooja Parmar

With increasing scarcity of fresh water, access to potable water has emerged as a major issue in many parts of the world. As communities, water experts, governments and rights activists seek answers, the issue is framed in multiple ways. For some, the issue is about population growth, rapidly changing lifestyle, consumer rights, sustainable water resources development, and regulating the market, while others are concerned about national security, governance practices, the rule of law, and human rights. Some are engaged in questions about property in water (for e.g., who can or should own water?), while for others critical answers lie in determining how humans should *live with* water. With the intensification of the global water crisis, demands for water justice that have emerged around the world raise questions that are rooted in layered histories of oppression and resistance, as well as of social, political, and economic exclusions in particular places. Worldviews and experiences that shape contemporary water politics are diverse, and therefore the struggle for water is inevitably also a struggle over meaning. I truly understood the significance of this insight, and more importantly, the violence in not recognising it, particularly in the case of Indigenous struggles, in a village in Kerala, India that became the site of a powerful movement for water justice.

Plachimada and a few other nearby settlements in Kerala have been home to *Adivasis* for generations. While formally recognised as Scheduled Tribes, like many other Adivasis in the region, these communities identify as Indigenous peoples. In 2002, several members of the local Adivasi community began a sit-in protest outside the gates of a Coca-Cola bottling plant. The plant had started operations on land an Indian subsidiary of The Coca Cola Company had acquired with the goal of using groundwater for the production of its beverages in the area. In the absence of any effective regulation of groundwater, ownership of land gave the company the right to extract any amount of water for its use in accordance with the principles of the Common law, a legacy of the British colonial rule in the subcontinent.

Within a short time, changes in quality and quantity of water became noticeable to the residents. After the Company refused to do anything about the increasing scarcity or the pollution, several members of the Adivasi community decided to begin public protests. The sit-in, sustained primarily by Adivasi women, who sat in a protest hut day and night, lasted over a decade. Between 2002, when the protests began, and 2017, the year Coca Cola declared it would not resume operations in that unit, there were many protest marches and rallies in and around Plachimada. An international water conference was held there, and there were several rounds of litigation at different levels of the Indian court system. Reports were published and the dispute was analysed from a range of social, legal, and scientific perspectives. As the dispute gained international attention, it garnered widespread support from a wide range of committed people and groups. This support helped sustain the momentum and also shaped the government's responses to the protestors. But a closer look at the narratives of this quest for justice reveals something important about the manner in which meanings change as claims arise and travel – as they are presented and re-presented. Two examples illustrate this.

The meanings of water

Water sustains life. It is necessary for life. Access to water is therefore important. A pre-occupation with quantifying per capita 'needs' for water can therefore be justified. Most of us are satisfied if we have access to sufficient quantity of water as long as the water is of the quality that meets our requirements. At a time when we have realised that freshwater is not infinite, many of us also support or advocate for sustainable water use practices. This is one way to understand water, but it is not the only one.

In Plachimada, *Adivasis* believe they share the place *with* water (among other forms of life). Water is therefore not a commodity to be consumed but rather a life form alongside others. An elder in the village spoke about living and dead water as if she was referring to a person. In her view, a place cannot survive without water, and in saying that she, of course, understands the necessity of water for life. But conceiving water as a life form with its own place in the world, this elder also suggests a relationship with water that is not centred on human needs. In suggesting that their water was dying because of the activities of the Company, the elder revealed a violence to water itself. In her world, this violence is a wrong, independent of the injustice any lack of water might visit upon human and non-human life forms in the place. This conception of water inevitably informs the politics of *Adivasis* in the region. It is not, however, reflected in any mainstream political, social, or legal narratives about the dispute. As a result, some of the decisions made as the protests continued were misunderstood and even dismissed as unreasonable by people the *Adivasis* describe as 'outsiders' or 'settlers'.

One such decision was the *Adivasis'* rejection of the Company's offer to bring in water in trucks to meet the needs of the locals soon after the litigation began. What the Company, the courts, and many supporters of the protests failed to appreciate is the meaning or place of water in the life of the community. These are people for whom water is not simply a single thing meant to meet human needs. In conversations with me, they did not speak about water in general, but rather always referred to particular waters: lorry [truck]-water, well-water, canal-water, dam-water, tap-water, Company-water. These distinctions mattered to them, and were always present in their responses to questions about the dispute. An interlocutor who is not familiar with the worldviews that inform these meanings could easily miss these important differences. 'Company water' was unacceptable to the community because it was associated with an entity that had disrupted the various relations that existed in the place, including those with water. Their complaints could thus therefore not be met by trucking in water. It is not an adequate or just response to their claim. They wanted the existing underground water to be respected.

Similarly, the Adivasis' dissatisfaction with the 'lorry water' subsequently provided to residents by the government also requires an appreciation of the distinctions they make between the various waters. The necessity of waiting for the water trucks (sometimes at the cost of losing the day's wages), increased use of plastic pitchers (because they are cheaper than the metal ones and more durable than traditional clay pots) to store water for days, and the differences and disputes that can arise within the community due to the scarcity have all altered their lives. Moreover, for people who have always relied on wells near their homes for water, the uncertainty of the source of lorry-water is a source of anxiety and an unwelcome change. Trucked-in water is water they cannot trust. They'd rather walk longer distances to fetch well-water from other places. The well-water is important because it connects them to the land.

Water and the histories of dispossession

There are always multiple ways of understanding a movement and the politics of and for justice involved. However, what the Adivasi struggle in Plachimada shows quite clearly is that disputes over water are misinterpreted if they are seen as singular events, unconnected with other struggles and events. Adivasis, like Indigenous peoples in many parts of the world, remain on the margins of the social, economic, political, and cultural life in India. Many have been dispossessed and displaced (often more than once) in the name of national and local 'development' and are up against powerful corporations and a powerful state. Despite Constitutional and statutory protections, their traditional economies, cultures, knowledge systems, and ways of life are being destroyed rapidly even as a majority of them struggle to live on their traditional lands.

The stories of early and more recent displacements Adivasis in Plachimada narrate to 'outsiders' clearly speak to the fact that their protests are not only for access to water, but also against further dispossession. The struggle for water in this place is therefore inseparable from their struggle for land and for the recognition of their relationship to the land and everything it sustains. They repeatedly demanded that the Company leave so that they would not have to. The meaning of their struggle for water here therefore cannot be understood without acquiring some knowledge about the layered histories of dispossession and exclusion in colonial and post-colonial India. Some of these histories can be learned from the official records in state archives, but others that are not in any official records emerge from the stories shared by elders and are even reflected in contemporary social relations of power and oppression in the place.

The Adivasis' struggle for water is also a struggle for a right to say 'no' to a company that extracted excessive amounts of water and the state that granted it license to operate in the area. It is a signal to the rest of Indian society that demands that Adivasis continue to bear the burden for the 'progress' of the nation are unacceptable. Their struggle is to keep their own ways of being alive. Whatever the protests stand for to their many supporters, for the Adivasis, the struggle is clearly also for self-determination. Their insistence on not leaving, and on demanding water that is *from* that land relates to who they are and how they understand their lives as an inseparable part of the place and everything else that belongs there, including the land, water, rocks, and trees. Translating their demands into a simple right to enough water, a right to fulfil a basic need, misrepresents their cause. Neither the language of a right to water, nor that of environmental rights can fully represent what the opposition to Coca-Cola's operations meant to those who began the protests and refused to back down for a decade and a half. Unfortunately, many who spoke on behalf of the Adivasi protestors did not speak about land. This is not surprising given we can often only hear what is within the realm of our comprehension. Sometimes the interlocutors chose the narrative of water rights because it is a more powerful language. It is also less threatening than the language of dispossession and reparations for historical wrongs. Some therefore likely chose to not speak about land over concerns for losing an audience that might be complicit in the dispossession and displacement.

Conclusion

The Adivasi struggle that unfolded in Plachimada for over 15 years is unique for its situatedness in a particular place, but it is also connected to Indigenous peoples' struggles for land and self-determination elsewhere. There are numerous such struggles ongoing across the world, and as we face the prospect of running out of something that is essential for life and non-substitutable, more such disputes and calls for justice are inevitable. As with any threats to life, marginalised communities will continue to suffer the most harm with increasing scarcity and pollution of water. Stories like that of the struggle in Plachimada offer critical insights into understanding the problem and useful responses to it. Treating such protests as simply demands for access to potable water, as many of the Adivasis' supporters in Plachimada eventually did, is in itself a violence – one that is rarely recognised or addressed. As both the government and the Company learnt, trucking in water from other places is not the solution. A meaningful resolution of such disputes over water requires us to look more closely and understand claims within a wider context of social realities and layered histories.

Further readings

Baviskar, Amita ed. 2007. *Waterscapes: The Cultural Politics of a Natural Resource*. New Delhi: Permanent Black.

Curran, Deb; Napoleon, Val and Morales, Sarah. 2020. *Nutsa'maat: Cowichan Water Law*. Victoria: ILRU, UVIC.

Parmar, Pooja. 2015. *Indigeneity and Legal Pluralism in India: Claims, Histories, Meanings*. New York: Cambridge University Press.

Sundar, Nandini. 1997. *Subalterns and Sovereigns: An Anthropological History of Bastar, 1854-2006*. 2nd edition. New Delhi: Oxford University Press.

Xaxa, Virginius. 1999. 'Tribes as Indigenous People of India'. *Economic and Political Weekly*. 34 (51): 3589–3595.

52

WHITE SUPREMACY

Jemima Pierre and Aisha M. Beliso-De Jesús

In recent years, the concept of white supremacy has been associated with extreme racist groups and ultranationalists, as well as high profile acts of associated racial terrorism, particularly in Western countries. Some examples are: the massacre of nine African-American worshippers at Mother Emmanuel AME Church in South Carolina (USA), the violent white nationalist march in Charlottesville, Virginia (USA), the mosque shootings in Christchurch, New Zealand that killed 51 people and injured 49, the Hanau, Germany attack that killed nine people and wounded six others, and the shooting deaths of eleven congregants in a synagogue in Pittsburgh, Pennsylvania (USA), among many others. There has also been a renewed rise of right-wing movements, politicians, and governments who espouse and advocate for ethno-nationalist and white supremacist policies. In Europe, we have seen a resurgence of ethno-nationalist parties, neo-Nazi movements, and other radical stances, often under the guise of populism, against non-white, non-Christian local and immigrant groups. In the United States, the 'Christian Identity Movement' – in which members believe in an impending Armageddon and a battle between whites and non-whites – is the dominant religious viewpoint of white supremacist groups. Latin America has also seen the rise, through the undemocratic usurpation of power, of right wing and Christian fundamentalist white supremacist regimes, such as those in Brazil and Bolivia.

The concept of white supremacy also conjures up well-known and specific historical examples. *Jim Crow* segregation laws in the US, and the attendant terrorism of the Black community by Ku Klux Klan; the *apartheid* system in Southern Africa, that allowed for the economic, political, cultural, and social domination of the majority Black population by a racist white minority; the Aryan theories and violence perpetuated by Hitler's Nazi party against Jews and other non-Aryans – these all serve as the standard references of white supremacy. But it was the emergence of Western racial science in the early 19th century, promulgating the idea of white racial superiority and non-white racial inferiority, that provided the ideological and intellectual scaffolding for the enactment and justification for the brutalities of European expansion and military and political domination of the rest of the world. Through Western racial science, 'evolution was made to prove that Negroes and Asiatics were less developed human beings than whites; history was so written as to make all civilization the development of white people … brain weights and intelligence tests were used and distorted to prove the superiority of white folk' (DuBois, 1946:

37). The result was the complete domination of the world by Europe (and its satellites in North America).

White supremacy is, as philosopher Frances Lee Ansley contends, 'A political, economic, and cultural system in which whites overwhelmingly control power and material resources'. This broad global system includes widespread 'conscious and unconscious ideas of white superiority and entitlement', as well 'relations of white dominance and non-white subordination' reenacted 'daily across a broad array of institutions and social settings'.

It is important to understand that the concept of white supremacy extends beyond the recognized understandings of race, racialization, and racism. White supremacy certainly depends on the construction of the idea of race as a hierarchical relationship of power based on presumed biological and cultural difference, as well as related practices of racism, the valorization of whiteness and the denigration of non-whiteness. But we should also recognize that the workings of race depends on racialization processes where racial meanings are often in flux and can be malleable and shifting in various contexts – even as whiteness retains its power position. The articulations of white supremacy, therefore, may shift depending on cultural and political contexts. There remains, nevertheless, a set of specific power dynamics inherent in the construction of race: that is, the hierarchical categorization of 'white' as racially superior. The point is to be explicit in naming whiteness as the clear power position in any analysis of race, racialization, and racism. In naming whiteness and analyzing white supremacy, it is just as important to stress the dominant role of racialized-as-white people in creating the material realities of inequality, racial oppression, hierarchies, and 'white-framed interpretations', but also, the 'white-imposed community norms; scientific and medical categorizations; residential, educational, or occupation segregation; and the racial images and ideologies of the media, popular culture, and science' that assure white privilege and power. Race is always a description of a social, historical, cultural, and political position, and as James Baldwin reminds us, 'whiteness is a metaphor for power'.

To examine white supremacy is thus to point to the presumed *power* and privilege of whiteness, and to analyze how whiteness is structured in and through our institutions, our disciplinary theories and methods, our everyday relations, as well as through global economic and political processes, processes that are always racial. It is to understand how, as Rev. Martin Luther King once reminded us, the doctrine of white supremacy is 'embedded in every textbook and preached in practically every pulpit … [it is] a structural part of the culture'. At the same time, and in keeping with sociologist, W.E.B. DuBois's analysis of the 'complete domination of the world by Europe', ideologies and practices of white supremacy are transnational and global.

The structures of white supremacy were established through the history of European expansion, the colonization of the Americas and dispossession of First Nations, the enslavement of Africans, the governance, classification, and ordering of peoples based on presumed racial difference, as well as a capitalist economic system dependent upon this difference. Thus, the current international power system emerged in the 15th century through violent 'conquest, colonization, and universal legitimacy of European – and racialized white – power'. The impact of this power system continues to have global cultural, socio-political, and economic reverberations. For example, so-called 'Third World' nations are part of a global economy that is dominated by white capital and white international lending institutions, as well as Western intellectual, political, and juridical traditions and practices structured as white – ensuring the position of the non-white populations of the world at the bottom of all socio-economic and political hierarchies. The cultural projects of the white West dominate global experience – examples of which include worldwide practices of chemical skin bleaching, the western development regimes that buttress a 'white saviour industrial complex', education policies in former colonial (neocolonial) spaces, among others.

To account for the long history of European conquest of the world and the establishment of white supremacy as the 'central organizing logic of western modernity', is to also recognize the shared history of colonial invasion, extraction, and elimination that link Africa and Asia, to other areas such as the Middle East and Latin America within global structures of race and power. Thinking through white supremacy within a global framework thus allows for us to connect 'indirect' and 'direct' rule under formal colonialism and the continued impact of racialization processes in the 'post-colonial' contexts of Africa and Asia. For example, even though South African *apartheid*, is considered the unique case of white supremacy on the African continent, apartheid was the norm rather than the exception in colonial Africa, as Mahmood Mamdani pointed out. It existed as a form of institutional segregation marked by racial difference called 'associations' by the French and 'indirect rule' or 'suzerains' by the British and Portuguese. In this way, the white supremacist colonial order – both its settler and non-settler variants – is an important frame of reference for understanding the making of nations and empires, new systems of capital and labour, new concepts of culture and identity, and new constructs of humanity. Notions of 'Western liberal modernity', 'democracy', and 'morality', emerge within the context of violent conquest and justified by an intellectual and political apparatus – what has been called the joint march of liberalism and white supremacy. Disciplines such as anthropology played a key role in the development of racial science, helping to create the racio-biological taxonomy along with the 'whiteness-above-all' ideology that turned 'physical difference into relations of domination', which continues to shape our modern racial worldview.

Black and feminist scholars of colour have long documented and analyzed white supremacy as an important and global system of white racial dominance. But they have cautioned that white supremacy does not work alone; it is the modality through which many social and political relationships are lived. For example, capitalism, patriarchy, and racism works hand in hand with white supremacy, a set of relationships that cultural critic bell hooks refer to as 'white supremacist capitalist patriarchy'. And there can be no symmetry of race and gender subordination when cis-gendered women are differently positioned within white supremacy than non-white (and cis- and trans-) women and men. The same can be said about the relationship of white domination to class, ethnicity, nationality, and sexuality, among other factors.

In this historical moment, as acts of white supremacist violence continue to proliferate, we caution the reduction of white supremacy only to 'white nationalism'. The logic of white supremacy as white nationalism is both flawed and obscuring. It is flawed, in that it assumes that white supremacy was not already built into modern Western-ruled world. It is obscuring because it shifts understandings of white supremacy away from contemporary systems of power, and to white supremacy as only reflected in specific acts of material violence. In order to intervene in the flawed and obscuring logic of white supremacy as 'white nationalism', we argue that scholars and activists should not simply focus on race, racialization processes, or racism alone, but also attend to the specific structural dynamics inherent in the hierarchical construction of race and of 'whiteness' at the top power position. To do so means: 1) establishing as given the long history of European expansion for the political, intellectual, cultural, and ideological sedimentation of presumed white superiority; 2) shifting from an overreliance on the deployment of white supremacy as identity (i.e. the 'white supremacist' or 'white nationalist') to deal with the structural and material embeddedness of white racial hegemony in the world; and 3) understanding white supremacy as constituent of patriarchy, heteronormativity, settler colonialism, mass incarceration, political violence, xenophobia, and western imperial violence in and between societies structured in racial dominance.

Further readings

DuBois, William E.B. 1946. *The World and Africa: An Inquiry into the Part Which Africa Has Played in World History*. New York: Viking Press.

Lee, Butch and Rover, Red. 1998. *Night-Visions: Illuminating War and Class on the Neo-Colonial Terrain*. Oakland: AK Press.

Mills, Charles. 1994. 'Revisionist Ontologies: Theorizing White Supremacy'. *Social and Economic Studies* 43 (3): 105–134.

Wynter, Sylvia. 2003. 'Unsettling the Coloniality of Being/Power/Truth/Freedom: Towards the Human, After Man, Its Overrepresentation – An Argument.' *CR: The New Centennial Review* 3 (3): 257–337. https://doi.org/10.1353/ncr.2004.0015

INDEX

For Product Safety Concerns and Information please contact our EU representative GPSR@taylorandfrancis.com
Taylor & Francis Verlag GmbH, Kaufingerstraße 24, 80331 München, Germany

www.ingramcontent.com/pod-product-compliance
Lightning Source LLC
LaVergne TN
LVHW080808011026
849160LV00015B/1573